The Search for Life in the Universe

DONALD GOLDSMITH

Interstellar Media
Berkeley, California

TOBIAS OWEN

Earth and Space Sciences Department
State University of New York
Stony Brook, New York

The Benjamin/Cummings Publishing Company, Inc.

Menlo Park, California / Reading, Massachusetts
London / Amsterdam / Don Mills, Ontario / Sydney

To Rachel and Jonathan and David

ABCDEFGHIJK-MA-8987654321

The Benjamin/Cummings Publishing Company, Inc.
2727 Sand Hill Road
Menlo Park, CA 94025

Library of Congress Cataloging in Publication Data

Goldsmith, Donald.
  The search for life in the universe.

Bibliography: p.
Includes index.
1. Life on other planets. 2. Astronomy.
I. Owen, Tobias C., joint author. II. Title.
QB54.G58   574.999   79-18653
ISBN 0-8053-3325-8

Credits

Cover photo © Shorty Wilcox 1978.

Halton Arp: Fig. 3-16.
Elso Barghoorn: Fig. 10-1.
Breitkopf & Haertel, Wiesbaden: p. 322, Excerpt from J. S. Bach, Brandenburg Concerto No. 2, PB 4302.
Burndy Library, Norwalk, CT: photograph, p. 228.
Stillman Drake: Fig. 1-3 (left).
David Dressler and John Wolfson: Fig. 8-6.
Editions Hermann: Fig. 1-2.
Sidney Fox: Fig. 9-11.
Robert M. Gottschalk: Fig. 1-3 (right). © 1946.
Hale Observatories: Figs. 2-5, 2-8, 3-3, 3-8, 3-10, 3-14, 3-15, 4-1, 5-4, 6-4, 11-6, 12-5, 16-4, 18-1, photograph, p. 14.
High Energy Astrophysics Division of the Harvard/Smithsonian Center for Astrophysics: Figs. 3-3, 6-8, 7-7.
Houghton Mifflin Company: p. 145, from *The Book of Nightmares*, "Lastness," by Galway Kinnell, copyright © 1971 by Galway Kinnell. Reprinted by permission of Houghton Mifflin Company.
Instituto Nacional de Antropologia e Historia de Mexico: Fig. 21-5.
International Planetary Patrol photographs furnished by Lowell Observatory: Figs. 13-1, 14-1.
Kitt Peak National Observatory: Figs. 3-1, 3-8, 17-1 (upper).
Philip Klass: Figs. 21-1, 21-2.
Lick Observatory: Figs. 2-5, 3-7, 3-12, 4-4, 4-10, 6-7, 6-9, 7-6, 12-10; photograph, facing p. 1.
Little, Brown & Company: p. 323, from *Poems*, "Argonauts," by George Seferis, English translation copyright © 1960 by Rex Warner.
Reprinted by permission of Little, Brown, & Company.
Lowell Observatory: photograph p. 2.
Lund Observatory: Fig. 4-2.
Master and Fellows, Magdalene College, Cambridge: Fig. 1-3 (center).
Carter Mehl: Fig. 4-3.
Meteor Crater Enterprises, Inc.: Fig. 12-6.
National Aeronautics and Space Administration: Figs. 7-1, 12-9, 12-11, 12-12, 13-2, 13-4, 14-2, 14-3, 14-4, 14-7, 14-8, 14-9, 14-10, 14-11, 14-12, 14-13, 15-1, 15-5, 15-6, 16-1, 16-3, 16-4, 16-5, 16-6, 16-7, 16-8, 19-1, 19-3, 20-12; photographs, pp. 144, 226, 268.
National Astronomy and Ionosphere Center: Figs. 20-10, 20-11.
Allison Palmer: Fig. 10-4.
Allen Parker: Fig. 12-7.
Cyril Ponnamperuma: Fig. 12-8.
W. H. Robertson: Fig. 17-1 (lower).
Ben Rose: photograph, p. 228.
Arnold Rots: Fig. 4-6.
Martin Ryle and P. J. Hargrave: Fig. 13-3.
J. William Schopf: Fig. 9-4.
Veronica Schwartz: Fig. 21-3.
Scientific American: Fig. 20-14, from "The Search for Extraterrestrial Intelligence," by Carl Sagan and Frank Drake, copyright © 1975 by Scientific American. All rights reserved.
Frederick Seward: Figs. 3-3, 6-8, 7-7.
Stanford Linear Accelerator Center: Fig. 19-2.
Woodruff Sullivan III: Fig. 20-7.
Paul Trent: Fig. 21-4.
University of Chicago Press: Figs. 2-9, 5-5, 10-5.
United States Air Force: Figs. 21-1, 21-2.

Contents

Foreword

Human beings evolved a few million years ago as one species among millions of others. Our ancestors were not markedly faster or stronger or better camouflaged than their competitors: they were only smarter. This intelligence has led to the invention of tools—wave after wave of extraordinarily clever discoveries and inventions—which have given human beings speed and strength and other powers undreamt of by our ancestors, and has led to a very real, although perhaps brief, dominance of the planet Earth by this single species. As we became more intelligent and more contemplative we wondered about origins and destinies, about the mystery of our own beginnings and about whether, in some distant lands, there are other creatures more or less like us.

In the breathless pace of human discovery the planet Earth has now been all explored. Each culture has found that many other exotic societies also existed. But as our technology advanced, developments in transportation and communications whittled down human cultural diversity. Today we live on a planet which has almost no unexplored land areas and in which all human societies are in breakneck progression towards forming a single global culture. It is natural that our search for other beings and other cultures has transferred itself from the Earth to the stars.

Some set of events has occurred which led to the development of a technical civilization on the planet Earth. But what is the generality of these events? Are there other planetary systems suitable for life; with the early chemical events leading to the origin of life; an environment with a suitable balance between constancy and variability to permit further biological evolution; the emergence of intelligence and technology; a long lifetime for the technical civilization; and the wish to communicate with other beings? All of that has happened here. And if the Copernican and Darwinian traditions are relevant, similar events may well have occurred on countless other worlds through the Milky Way Galaxy.

The question of life and intelligence elsewhere excites our mythic instincts, our sense of wonder, our curiosity about nature, and our deep quest—exemplified in the myths and legends of human culture—for our own origins. To pursue the subject requires knowledge in cosmology, astrophysics, planetary science, organic chemistry, evolutionary biology, radiophysics, psychology, sociology, and politics. One of the virtues of the subject is that it provides an integrated framework for inquiring into, more or less, everything.

For all of human history until the last few decades the search for life elsewhere was exclusively a speculative endeavor, an exercise in fiction or theology or romantic speculation. But lately, in the most recent instant of cosmic time, tool-using humans have for the first time invented the technology to pursue this subject seriously. We have built the first interplanetary space vehicles, which are now performing a preliminary reconnaissance of the solar system in which we live. We have set down on the planet Mars and performed the first tantalizing searches for microbial life there, sent our little probes past Mercury and Jupiter and Saturn, and into the broiling and corrosive atmosphere of Venus. We have discovered organic molecules in meteorites and comets, in the atmospheres of the outer solar system, and in the vast dark of the space between the stars. We have attached messages telling something about our planet and ourselves to four spacecraft, Pioneers 10 and 11 and Voyagers 1 and 2, which are at this moment on trajectories which will take them out of the solar system, the first human emissaries to the realm of the stars and whatever beings may live there. We have constructed giant radio telescopes which, in a halting and provisional way, have begun scanning the skies to see if there are intelligent creatures on planets of other stars beaming radio messages to likely abodes of life—among which, we hope, they will count Earth as one.

Our early efforts to find life elsewhere have not been successful, but we have barely begun the search. We are at the very earliest phase of this grand exploration, and major advances are likely to be made in the relatively near future. The present book, a happy collaboration between two distinguished astronomers with a flair for popularization, is an excellent modern introduction to the entire field of exobiology, with an understandable emphasis on the astronomical factors. Reading this book illustrates one of the delights of the search for life elsewhere: to pursue the subject seriously we are required to know a great many things which are themselves of enormous intrinsic interest.

If, after a long, systematic search, with spacecraft, of the planets and moons of our solar system and, with large radio telescopes, of the distant stars, we fail to find a sign of extraterrestrial life, then we will have calibrated something of the rarity and preciousness of what we take so much for granted on our exquisite home planet. And if we succeed in finding life elsewhere—even very simple life—we will have turned a corner in human history and we will never be the same again: our science, technology, philosophy, and view of ourselves will have made a stunning advance. The search for extraterrestrial life is inexpensive by the standards of modern technological societies. It brings with it many subsidiary benefits. If we have the wisdom to keep looking, it is hard to imagine another enterprise which holds so much promise, whether it succeeds or fails, for the future of the human species.

Carl Sagan
David Duncan Professor of Astronomy
  and Space Sciences
Director, Laboratory for Planetary
  Studies
Cornell University

Preface

In this book we ask how and where we might hope to find other beings in the universe similar to ourselves. Does our galaxy contain millions of civilizations far more advanced than ours? Or does it have at best only a few planets with relatively primitive forms of life?

Our book provides a survey of these problems for the educated layperson, and a text suitable for nonscience majors in the first or second year of a college curriculum. We have adopted a nonmathematical approach in describing the search for life in order to make the book accessible to a wider audience. Standard courses in biology, geology, and astronomy can benefit from the perspectives provided here, which include the presentation of fundamental concepts in a new context. But the main function of our book as a text is its use in a one-quarter or one-semester course on astronomy emphasizing the search for extraterrestrial life.

Since we must often attempt to interpret ideas, observations, and experiments from the frontiers of several sciences in this book, the reader must not expect to find established dogma. Instead, we hope to convey a sense of the excitement of trying to reach beyond what we know, while still attempting to stay within the bounds of scientific thought. The tension between rigorous proof and free-ranging speculation provides one of the most enjoyable aspects of scientific research; we hope that some of this pleasure can be found by considering the various riddles that permeate the search for life in the universe.

To answer the questions of the origin of life and its cosmic distribution, we must summarize our knowledge of the physical universe: of space and time, the origin of matter, and the environments and chemical processes that determine the prevalence of life. To answer the question, "What is life?" we have only a single example, life on Earth. We must imagine the variations on our life that might occur within the sun's family of planets and satellites—where we can test our predictions with detailed investigation—and among the myriad stars of the Milky Way galaxy and beyond.

A single theme underlies this effort: the attempt to know ourselves, who we are, where we came from, our future as potential members of a galactic community. As babies, we humans each felt unique, the center of the world; as we grew, we saw other children like ourselves, and acquired a new sense of identity. As adults, we must cope with the conflict between our desire to feel unique as persons and our commonality with other humans, as well as the totality of life on Earth.

In a manner analogous to the growth of infants, human beings have now moved from an ancient belief in the Earth as the center of the universe to a proper appreciation of our relatively insignificant place in the cosmos. We now turn to the question of whether human intelligence and self-consciousness are unique, or nearly so, or whether the events that developed intelligent beings from chemical interactions on Earth may have occurred over and over again throughout the universe.

Despite the "averageness" of our sun among stars, the human sense of uniqueness dies slowly. The lingering belief that we must be special appears in some of the UFO reports that we read, and in the phenomenal interest in the idea that we humans are descended from, or educated by, ancient astronauts, superhuman visitors from the great beyond.

To progress beyond arguments based on belief and desire, we must use the methods of science and rely on experiments to test our hypotheses. But with only a single example of life to examine, our generalizations cannot be unquestionable. Instead, we must stretch our knowledge and our theories as far as we can, to see what insights we can reach: certain possibilities appear more likely than others. The ultimate experiment would be an effort (and a successful one!) to contact extraterrestrial intelligent beings. And if such contact should render this book obsolete, no one would be more delighted than its authors.

No book on this subject can fail to be indebted to the pioneering work *Intelligent Life in the Universe,* by Josef Shklovskii and Carl Sagan. We are especially aware of this influence since we have admired the book ever since its appearance in 1966, and have used it repeatedly in our courses. The major discoveries since then in astronomy, biology, and geology have led us to write the present book, but the basic perspective of Shklovskii and Sagan's remarkable work remains intact.

We are grateful to Jon Arons, Elso Barghoorn, William Baum, Klaus Biemann, John Billingham, Victor Blanco, Geoffrey Briggs, Elof Carlson, Karl Kamper, James Lawless, Mikhail Marov, Allison Palmer, Michael Papagiannis, William Robertson, Tom Scattergood, J. William Schopf, Frank Shu, Michael Soulé, Jill Tarter, William Ward, James Warwick, Richard Young, Ben Zuckerman, and especially to Frank Drake and Lynn Margulis for their comments on various parts of the manuscript and assistance with its text and illustrations. Much of the planetary research described in this book was supported by NASA. Henry Marien helped greatly with the proofreading, and Patti Rosen provided essential research assistance. But any mistakes are our own; like Shakespeare's Cassius, we must admit that the faults lie "not in our stars but in ourselves."

Donald Goldsmith
Tobias Owen

PART ONE

Why Do We Search?

To see a World in a Grain of Sand
And a Heaven in a Wild Flower,
Hold Infinity in the palm of your hand
And Eternity in an hour.
                    —WILLIAM BLAKE

    The history of human awareness of the universe has brought a steadily growing desire to find the roots of our existence and to understand our relationship to the cosmos in which we live. We now stand poised on the threshold of determining how life arose on this planet, and of applying this information to the quest for life on other planets circling other stars. But it is worth pausing as we do so to ask ourselves: Why do we search? How has the search for our own origins and for evidence of our cosmic kin proceeded in the past? And what does the search for extraterrestrial life tell us about our attitude toward the universe around us?

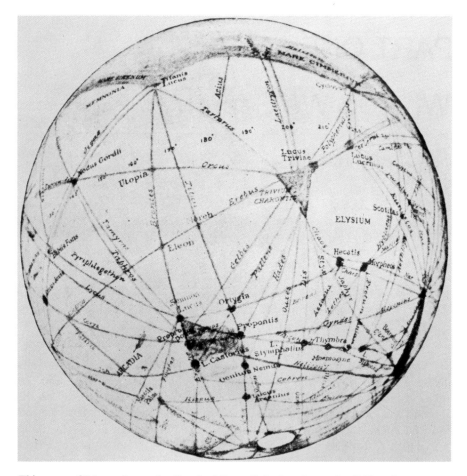

This map of Mars, drawn by Percival Lowell during the early 1900s, shows some of the "canals" which Lowell imagined to cover the red planet.

CHAPTER 1

The Search
from the Human Perspective

The Search for Life's Origins

Just over a hundred years ago, in May, 1876, Her Majesty's Ship *Challenger* returned to port at Sheerness on the Thames after more than three years away from England. During the ship's voyage around the world, scientists aboard the vessel had systematically dragged the ocean bottoms for the first time. Day after day, the ship's crew brought up samples of water and mud from the abyssal deeps, the bottom layers of the oceans of the world, which contained marvelous new sea creatures previously unknown to humanity. For the scientists on board the *Challenger,* the special fascination of the expedition had been the hope that they would find living fossils, early forms of life on Earth, happily ensconced in the great depths where conditions had barely changed since the time that life began. But there was still more to the scientists' expectations. When the first transatlantic telegraph cable had been laid ten years before, the ship's crew had discovered at the bottom of the ocean a gelatinous ooze, which, according to several leading scientists, was probably the primitive protoplasm from which all life had descended. A careful study of this *Urschleim* (original slime) would surely unlock the secret of how life began on Earth.

Alas! No living fossils were found by the *Challenger,* and the mysterious ooze turned out to be totally inanimate. Although the ooze appeared to undergo chemical changes reminiscent of life processes, these lifelike characteristics could be reproduced nicely by adding a strong solution of alcohol to ordinary sea water. Chemistry, not biology, rules the ocean floors.

A century after the *Challenger* expedition, we know much more about the Earth and its oceans, and have found that the history of primitive events

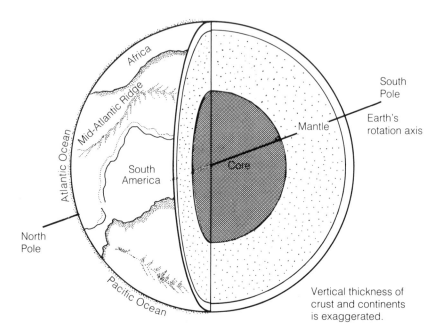

Fig. 1-1   The Earth's continental plates, which form the Earth's crust, slowly grind against one another, sometimes overlapping upon collision. These plate-tectonic encounters eventually carry what was once the crust down into the Earth's mantle, which fills about seven-eighths of the Earth's total volume, surrounding the iron-rich core.

on this planet has vanished forever. The geological records of the first billion years on Earth have been erased by erosion and the motions of the giant plates that form the Earth's crust. As the plates have moved, slowly but inexorably, they have dragged eroded material that once formed the Earth's surface down below the present crust of our planet (Figure 1-1). Thus, the most direct means of uncovering the earliest terrestrial history has been destroyed through 4.6 billion years of erosion and plate-tectonic activity. We do find around us an amazing variety of living organisms, and we can study the fossil record that extends over just 3 billion years. But we cannot reach back through this record to the point at which life differentiated itself from inanimate matter.

Despite our lack of definite information, no doubt exists concerning the *interest* of human beings in the origin of life. Every culture has creation myths, and even our own "sophisticated" civilization cares deeply about its origins. We find this pervasive interest at the root of diverse religious beliefs, and we find also that humanity has always been fascinated by the possibility that life may exist elsewhere in our solar system or somewhere farther out in space, among the myriad stars in the sky. The idea of visitors coming to Earth from somewhere "out there" also has an ancient history, and appears among the earliest writings that archaeologists have deciphered.

All of our speculation about other forms of life, and about possible visits from other civilizations, must confront the two key questions that concern the origin of life on Earth: How did life begin here? And how likely is such a process, given a certain set of conditions and the availability of certain basic materials? Only during the last few years have we been in a position to provide some scientific answers to both of these questions, though we still cannot give anything like a complete unraveling of the mysteries they contain.

We can imagine two extreme approaches to providing the answers. One method constructs theories of the processes that began life and then attempts to duplicate these processes by laboratory experiments. The second method seeks to find examples in nature that we can study, in the hope that these examples will reveal the essential clues to explain how life started. Chapter 9 presents the first method of attack, which has been modestly successful. We have already seen that the second method cannot be pursued very far on Earth, for the early record of the planet's history no longer exists. But what about the other planets? With the coming of the space age, we have suddenly achieved an enormous broadening of our scientific horizons. We can now investigate other planets, and can carry out experiments within their atmospheres and on their surfaces that should start to answer some of the questions that have fascinated human brains since intelligence first developed on the Earth.

The Importance of Mars

Different cultures set different estimates of the number and kind of advanced civilizations that may exist in the heavens. Even within what we call "Western civilization," the prevailing opinion has varied widely with time. During the seventeenth century, as experimental science began to catch on as an interesting way to enjoy nature, the idea that *all* the planets in our solar system are inhabited became relatively widespread. Christian Huygens, famous for his achievements in the field of optics, wrote an entire book on the subject of life on other worlds, in which he speculated on the characteristics that the inhabitants of the different planets must have in order to survive comfortably under the extremes of gravity and atmospheric composition that Huygens imagined. Half a century later, the great satirist Voltaire imagined a giant inhabitant of the planet Saturn visiting Earth and eating mountains for breakfast. During the next two centuries, belief in the possibility of life outside the Earth had its ups and downs, as new discoveries concerning the nature of life on Earth, and the conditions on our planetary neighbors, emerged from new advances in scientific inquiry.

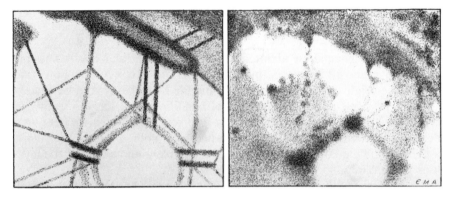

Fig. 1-2 When Giovanni Schiaparelli drew maps of Mars, he saw numerous lines or "canals" (left panel). Eugenio Antoniadi drew the same region of Mars, but he saw no such canals (right panel). We now know that these canals were only optical illusions, despite the claim that they showed the technological handiwork of intelligent Martians.

All these imaginings, different as they were, shared one common feature: In them, the planet Mars seemed an especially attractive haven for extraterrestrial life. The reasons for this fixation on Mars are easy to find. The planet's reddish color and its peculiar apparent motions in the sky had won it both special attention and fear from those who watched it ever since the dawn of history. Tycho Brahe's careful observations of Mars had led Johannes Kepler to formulate his three laws of planetary motion in 1609. During Kepler's lifetime, the invention of the telescope allowed scientists to discover that Mars shows permanent markings on its reddish disk. Thus, it is the planet's firm surface we see, not changing cloud bands like Jupiter's or a featureless cloud cover like Venus's. Observations of Mars made during the next two hundred years revealed the polar caps, the Martian clouds, and seasonal changes in the dimensions of the caps and in the contrast of the light and dark markings on the planet.

In the last decades of the nineteenth century, a new dimension appeared in the Mars puzzle. Several European astronomers, among them Giovanni Schiaparelli of Milan, discovered faint, straight markings on the surface of Mars, the famous Martian "canals."[1] We know now that these canals simply do not exist; they are an optical illusion that makes the eye see lines when only dots are present. The "canals" caused a series of bitter controversies, with such experienced observers as Edward Barnard in the United States and Eugenio Antoniadi in France firmly convinced that the canals did not exist, while others claimed not only that they were visible to the eye, but also that they could be photographed (Figure 1-2).

---

[1]Schiaparelli used the Italian word *canali,* which means "channels" as well as "canals."

The high point of the belief in the canals of Mars was reached by Percival Lowell, who built his own observatory in Flagstaff, Arizona, where he carefully studied Mars with a 24-inch telescope and drew maps of a fine network of canals that completely covered the planet except for the polar frost caps. This network appeared to undergo seasonal changes that were synchronized with changes in the large, dark areas on Mars: The canals would always grow darker, and would sometimes become double, as the Martian summer began. Lowell suggested that intelligent Martians had constructed the canals to bring water from the melting polar caps to irrigate their fields (the dark areas), which otherwise would remain incapable of producing crops. The seasonal increase in contrast between the light and dark areas, Lowell said, was the straightforward result of plant growth in the dark areas during the summer season. He thought that the apparent darkening (and doubling) of the canals occurred as vegetation along these watercourses flourished with the onset of the temperate season, thereby increasing the visibility of the canals to a remote observer.

Lowell's arguments made up with passion what they lacked in proof. During the first few decades of this century, astronomers developed the techniques of modern astrophysics, and their applications to studies of Mars struck hard at Lowell's hypotheses. Radiometric observations revealed that the average temperature on Mars lies well below the freezing point of water, a full hundred degrees below during the Martian night. Astronomers could detect no water vapor or oxygen in the Martian atmosphere, and the conclusion grew stronger that this atmosphere must be extremely thin.

Despite the scientific evidence that Mars should be hostile to life, and despite the negative results from a search for radio signals from Mars that was made in 1924, the idea that an advanced civilization might exist on the planet became so well entrenched in the public mind that on October 31, 1938, a serious panic erupted when Orson Welles presented a radio dramatization of H. G. Wells's novel *War of the Worlds*. Hundreds of thousands of listeners temporarily abandoned the safety of their homes and rushed outdoors to meet, or to flee from, the wave of Martian invaders who, Welles reported, were overrunning New Jersey.

Of course, the explanation that the broadcast was merely a play soon satisfied everyone, and within a few months it was hard to find anyone who admitted to having been fooled by Welles's narration. Were the performance to be repeated today (of course, we would now require facsimiles of Martians for television), we can be sure that millions of viewers would be easily taken in, not because people are always fooled so easily, but because we retain a sizable urge to believe in the existence of other civilizations. The Welles broadcast also reminds us of our basic ambivalence toward extraterrestrial visitors. We expect them to be far wiser, stronger, and more sophisticated than ourselves, but we can't decide whether they will be gentle and loving or belligerent and tyrannical. Humans have always feared and loved their gods, and it should not surprise us that these emotions arise when we speculate about advanced forms of extraterrestrial life.

Sixty years after Lowell, and 25 years after Orson Welles's broadcast, humanity sent the first terrestrial spacecraft to Mars. The story of Mariner 4, which took the first close-up pictures of the red planet, of the succeeding Mariner missions, and of the Viking landings on Mars, appears in Chapter 14. With these and with other spacecraft, a few dozen years have brought humanity from distant spectatorship into the era of planetary exploration, and one major focus of this exploration has been the search for life on Mars.

The plain fact remains, however, that today we have no convincing evidence for the existence of any kind of life anywhere in the universe except on Earth. Thus, for the time being at least, we must continue our exploration of our own solar system and our search out among the stars for the secret of our origin and some indication of other life.

The Scientific View of the Universe

This book describes the universe from the moment of its birth to the present day, always looking for clues to these two mysteries. Because we still have no definite answers, we must attempt to understand the facts of life as we know it on Earth and to rely on informed speculation to make the final leaps in judging the probability of finding other kinds of life elsewhere in space. We shall see how humanity has increased not only our knowledge of the universe, but also our ability to deal with its strangeness. This process has not been easy. Even today, many people find the universe outside the Earth so strange that either they never think about it, or else they believe that anything goes, that no sort of life should be more improbable than another, or that laws of nature as yet unknown to us appear in everything from extrasensory perception to the Bermuda Triangle.

In contrast to these views, the scientific attitude toward the world holds that we must proceed carefully from what we understand—through multiple observations and experiments—toward our speculation about things that we don't understand. Each separate piece of evidence about the universe fits to a greater or lesser degree into the framework of understanding that scientists have constructed. If something new and startling seems to contradict this series of mental pictures, scientists remain reluctant to change their framework *until* they have become convinced that they can exclude as reasonable any other explanation that would *not* require such a change. Therefore, most scientists will not, for example, consider UFO reports to be evidence of extraterrestrial spacecraft until they have eliminated, to their satisfaction, explanations of the reports that assign human error, psychological reactions, natural phenomena, or fraud as the causes of the UFO reports.

The scientific way of looking at the universe does not, of course, form the only way to see the world. In fact, only a small fraction of humans try to maintain a scientific outlook constantly, because to do so often violates our human intuition, with which we maintain a system of beliefs formed long before the scientific outlook entered our view of life. What makes science look good as a way to see things is that it works: The model of the physical universe that scientists use can successfully explain observations and make accurate predictions of future events. Furthermore, it permits changes in the framework of our understanding to occur as new discoveries are made. As these changes arise, they provoke great debate among scientists, but the scientists agree on the *principles* by which they alter their framework of knowledge, through written and verbal arguments (Figure 1-3). The basic subject matter of this book involves many areas of knowledge where profound changes are taking place, so the reader will continually encounter passages in which we quote more than one informed opinion. This is the most exciting part of science—trying to reach past the knowledge we have to new knowledge that will change our perspectives about the universe and about ourselves.

Fig. 1-3 We owe our understanding of the physics of motion mainly to three great scientists: Galileo Galilei (left), Isaac Newton (center), and Albert Einstein (right). Newton built on the framework of ideas that Galileo had made, and Einstein refined Newton's laws of motion, showing how they would apply at speeds close to the speed of light.

Speculation about Extraterrestrial Life

Our discussion of the attitudes of scientists and nonscientists has a direct bearing on the search for life outside the Earth. When scientists consider the possibility that other civilizations exist and that we may be able to communicate with them, they rely on the knowledge that we have already gathered to make estimates of the difficulty of sending or receiving messages, or of traveling from one star system to another. Most nonscientists have a much greater willingness than scientists do to believe that other civilizations surely "know things we don't," and thus can perform any feat of engineering, violate any "law" of science known to humanity, appear or disappear as they choose, and generally raise havoc with the universe. If we think about this point of view, we can see that it embraces a continuing human tradition that sees the universe as a mysterious place, full of powerful beings who have knowledge denied to us. Does this tradition arise from extraterrestrial visits in bygone eras? The only good evidence for such visits (see Chapter 21) rests in the stories of godlike creatures from heaven; for some people, this evidence suffices. For scientists, the evidence clearly falls below the minimum persuasive level. We can keep on looking for more evidence, but let us consider, for a moment, where the tradition has taken us.

Once upon a time, humans generally believed that an ever-watching cosmic force guided all of the events on Earth. During the past few thousand years, some of the human population has lost this view, believing instead that we have responsibility for what we do, even if a deity exists. This change in viewpoint, which scientists in particular have often led, has tended to emphasize the importance of humanity, since a greater sense of our own power over our lives has naturally produced a greater awareness of our ability to affect the world. Indeed, once humans saw the possibilities even dimly, they set out with vigor at reshaping our home planet.

But the price we have paid is loneliness, the feeling that we are alone— alone in our self-awareness, no longer a part of the wonderful cosmic web that once seemed to support us effortlessly as part of nature. Although this feeling of separation may not be grounded in reality (for we obviously remain part of the universe, whatever we do), the *feeling* is real and pervasive. By attempting to understand the world through categories and logic, we have separated our mental outlook from the totality of the universe, and must pass through the pain of separation before we can once again see how we fit the universal pattern.

This book aims at helping this process along, but we should first notice where human consciousness has wandered since the time that humans began to look carefully at how the world works. As our sense of separateness from the universe grew, human beings felt real fear of the idea that once the Earth had no people upon it. The best-known religious texts devote only a tiny fraction of their story to the prehuman Earth, which in fact covers more than 99.9 percent of the Earth's history.

During the last century, Charles Darwin proposed a theory of the evolution of species to explain how a planet without human beings came to acquire them through the transformation of species types. The widespread opposition to Darwin's theory demonstrated, among other things, how much most people cherished the idea of an Earth that had always had humans living upon it. Darwin's theory received general confirmation as scientists uncovered the fossil history of life on Earth, extending a billion years and more into the past.

Today, as we search for life outside the Earth, astronomers' efforts in this search are sometimes considered foolish: How can we study what we know nothing about? The answer to this question lies in the fact that we do know a great deal about the general principles of life in the universe, assuming that life on Earth provides a reasonably typical example of life in general. If evolution made us the way we are, then roughly similar conditions could have led to the development of roughly similar forms of life elsewhere. Thus, our study of the evolution of the universe extends our perspective from the many, varied forms of life on Earth and in its fossil record, to the still more varied possibilities for life in other places, perhaps now, perhaps in the past, perhaps only in the future.

To study how life might have evolved far from Earth, we must study how the universe itself has reached its present configuration. Then we can find the likeliest sites for the origin and development of life, and can consider the probability that our own galaxy may hold millions of other civilizations, perhaps already in mutual contact. The astronomical goal of our search will be the comparison of the conditions under which life on Earth has arisen with the conditions that prevail in different places throughout the galaxy and beyond.

Summary

Ever since human beings realized that our own evolution proceeds through the same processes as other events on Earth, we have looked for past and present evidence of other kinds of life on our own planet, and have speculated about the possibility of finding living creatures on other planets. Now we stand at the threshold of the era in which we shall be able to enter into communication with other civilizations, should they be relatively numerous in our galaxy. In this effort, scientists will be guided by the approach that has proven so useful in the past, that of refining the framework of knowledge tested in experiment after experiment. From such tests, we have learned much about the conditions under which life has developed on Earth, and in which it might be able to arise in other environments. As we follow scientific speculation to estimate the likelihood of other forms of life and other civilizations, we can grow in knowledge about ourselves, better appreciating how our own existence fits within the universe.

Questions

1. Why has the geological record of the Earth's first billion years completely disappeared?

2. How can you explain the particular fascination that Mars has exerted on the human imagination over the past few hundred years? Do astrological theories concerning the red planet provide any clue?

3. What are the "canals" of Mars? Do they really exist to carry water from the Martian polar caps to the warmer equatorial regions?

4. Compare the scientific approach to estimating the probability of life elsewhere in the universe with various other approaches, such as theological, spiritual, and astrological arguments. Do you feel that the scientific attitude is basically different?

5. Do you feel a sense of loneliness in contemplating the vast spaces that separate planets and stars from one another? How do you imagine that your great-grandparents felt when they thought about the planets and the stars?

6. Do you believe that human beings have evolved slowly from other forms of living creatures, or that humans have their origin in the relatively recent past—say, a few hundred thousand years ago? What arguments can be made for each point of view?

Further Reading

Berendzen, Richard, ed. *Life beyond earth and the mind of man.* Washington, D.C.: U.S. Government Printing Office.

Hoyle, Fred. 1977. *Ten faces of the universe.* San Francisco: W. H. Freeman and Company.

Jevons, W. Stanley. 1958. *The principles of science.* New York: Dover Books.

Krupp, Edwin. 1978. *In search of ancient astronomies.* New York: Doubleday & Company.

Ley, Willy. 1963. *Watchers of the .kies.* New York: Viking Press.

Lowell, Percival. 1908. *Mars as the abode of life.* New York: Macmillan.

Neyman, Jerzy, 1974. *The heritage of Copernicus.* Cambridge, Mass.: The M.I.T. Press.

Sagan, Carl. 1975. *The cosmic connection: An extraterrestrial perspective.* New York: Dell Publishing Company.

PART TWO

The Universe at Large

The stars are threshed
And the souls are threshed from the husks.
                    —WILLIAM BLAKE

In order to organize our thoughts and conclusions about the possibilities of life throughout the universe, we must learn how the universe itself is organized, starting from the immense clusters of galaxies that extend through billions of light years of space around us. From our study, we can learn how stars are born, live and die; how they cluster into groups of various sizes, such as our own Milky Way galaxy; and how their distribution and their histories may determine the likelihood of life in the spaces between them. The story of the evolution of the universe embraces the crucial chemical history of matter, which apparently began as protons, electrons, helium nuclei, and a dash of deuterons. Only later a tiny fraction became the carbon, nitrogen, and oxygen nuclei that may be essential for life anywhere. By following this history, we can understand what happens not only to the stars themselves, but also to the raw material for life which they contain and alter as they grow older.

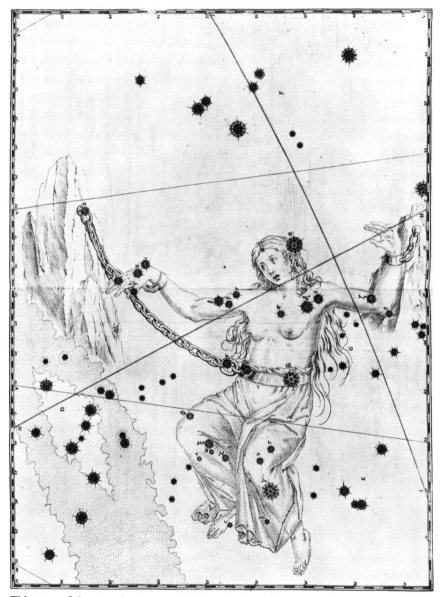

This map of the stars in the constellation Andromeda appeared in the first published star atlas, Johann Bayer's *Uranometria* (1603). The famous Andromeda galaxy, not shown in Bayer's map, would form part of the chain below Andromeda's right elbow.

CHAPTER 2

Space, Time, and
the History of the Universe

If you look at the night skies on a clear evening, you can see perhaps a thousand stars in the Milky Way galaxy. Each of these stars, basically similar to our sun, has been shining for millions or billions of years, and the light that you see has taken anywhere from eight to two thousand years to travel through interstellar space into your eyes.

Although we have begun to explore our own planetary system in person and with spacecraft, our knowledge of the universe outside the solar system rests on our study of the light rays, radio waves, X-rays, and infrared and ultraviolet radiation from celestial objects. Remarkably enough, we have nonetheless reached some sweeping conclusions about the universe simply from studying this radiation in detail.

What we call the universe—all of space and all the matter that exists— has been expanding for the last 15 or 20 billion years, ever since the universe began in a fiery "big bang." The universe today consists of clumps of matter called stars, which cluster by the billions into *galaxies,* which themselves gather in galaxy clusters. Our own star, the sun, has a family of planets in orbit, small aggregates of gas and rock which, unlike the stars, shine not by thermonuclear fusion reactions but by reflected light. In addition to the stars in galaxies, and to the planets that we believe orbit around many stars similar to our sun, we find diffuse matter, gas and dust particles, spread through the space within galaxies and even among the galaxies in a galaxy cluster.

The Distances to Stars

The record of human awareness of the heavens testifies to a steadily increasing perception of the tremendous distances, vast beyond human intuition, that characterize the universe. Early philosophers who thought about the distances to the stars concluded that the stars must be incredibly far away. They based this attitude upon a feeling that the Earth, made of changeable matter, must be isolated from the immutable, eternal stars. Although the philosophers were wrong to consider the stars eternal, they were right to imagine them at great distances from us.

Accurate measurements of the distances to stars rely on the *parallax effect,* the apparent displacement of an object's position, relative to more distant objects, as its observer moves. As you drive a car down the highway, you will see nearer objects, such as houses and telephone poles, shift their positions relative to more distant objects, such as hills and clouds. In a similar way, once we know that the Earth orbits around the sun, we can expect to see the closer stars appear to shift their positions with respect to the more distant stars during the course of a year (Figure 2-1). This parallax shift is difficult to measure, because *all* of the stars have such great distances from the sun that the Earth's orbit, 300 million kilometers in diameter, spans only a tiny fraction of the gap between the sun and the next closest star. Therefore, the apparent changes in the positions of the closer stars against the background of farther stars amounts to a tiny angular shift, one that early astronomers could not measure at all.

Before the invention of the telescope in 1609, astronomers could pinpoint the locations of stars on the sky with an accuracy of about 1 minute of arc: one sixtieth of a degree, about equal to the angular size of a dime seen from a distance of 60 meters. With this limited ability to measure angles, we could discover and measure the parallax shift of a nearby star only if the back-and-forth change in the star's position on the sky exceeded

Fig. 2-1   The Earth's motion around the sun makes the nearer stars appear to shift their positions against the background of much more distant stars. The amount of this parallax shift *decreases* as the distance of a star *increases,* so astronomers have progressively more difficulty in trying to measure the parallax shifts of more distant stars. If a star has a distance from us of 1 parsec, it will change its position by 1 second of arc (in either direction from its average position) during the course of a year.

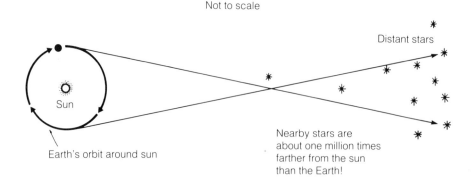

Not to scale

Sun

Earth's orbit around sun

Distant stars

Nearby stars are about one million times farther from the sun than the Earth!

1 minute of arc. Since *greater* distances imply *smaller* angles of shift (Figure 2-1), pretelescopic astronomy could detect parallax shifts only if stars were close enough to have parallax shifts greater than a minute of arc. By trigonometry, such as that shown in Figure 2-1, we can determine that a star's distance from us must be less than 3500 times the distance from the Earth to the sun, if we are to measure a parallax shift greater than 1 minute of arc.

Unfortunately for early attempts to detect the parallax shift, the closest star to the sun, Alpha Centauri, has a distance from us that is 260,000 times the Earth-sun distance, more than 60 times the distance that would produce a parallax shift as large as 1 minute of arc. Thus, more accurate telescopes, as well as the measuring machines to go with them, had to be built during the first half of the nineteenth century before even the closest stars' parallax shifts could be determined.

The huge distances to stars demand new units of measurement, just to avoid the use of huge numbers. The Earth orbits the sun at a distance of 149 million kilometers, already a fantastic distance in human terms. When we multiply this figure by 260,000 to obtain the distance to Alpha Centauri, we find the result equals 40 trillion kilometers, or 4×10^{13} kilometers. Clearly, we need a new measure of distance to avoid the constant repetition of "trillions" and "quadrillions" in discussing stellar distances in terms of our Earth-based kilometers.

Because of the importance of the parallax effect in measuring the distances to nearby stars, astronomers invented a unit of distance called the *parsec.* One parsec equals the distance from us that a star would have if we observed its annual back-and-forth parallax shift to be 1 second of arc. (*Par*allax plus *sec*ond of arc gives parsec.) One parsec equals 206,265 times the distance from the Earth to the sun, or about 30 trillion kilometers. Even light, which travels at 300,000 kilometers per *second,* takes 3¼ years (about 100 million seconds) to cover 1 parsec of distance. Thus, 1 parsec equals 3¼ *light years,* for the "light year" measures the distance that light can travel in one year. Modern astronomers measure distances almost exclusively in parsecs, kiloparsecs (thousands of parsecs), and megaparsecs (millions of parsecs), and only rarely in light years.

From the fact that kiloparsecs (kpc) and megaparsecs (Mpc) are often used in astronomy, we can quickly conclude that even the huge number of kilometers spanned by a parsec barely covers some significant astronomical distance. Alpha Centauri, the star nearest the sun, is 1.3 parsecs away, while Sirius, the brightest star in the sky, has a distance of 2.5 parsecs. Because the size of the parallax effect *decreases* as the distance to a star increases (Figure 2-1), we must measure smaller and smaller parallax shifts as we observe more and more distant stars. The bright star Vega has a distance of 8 parsecs, and its annual parallax shift equals only one eighth of a second of arc. Today we cannot determine stars' parallax shifts if they are smaller than about one thirtieth of a second of arc, because our atmosphere blurs the stars' images on the photographic plates that we measure. As a result, we must use methods other than the parallax shift to determine the distances to stars that are farther away than 30 parsecs (100 light years). These other methods, however, lean heavily on the measured distances to the brightest of the thirty thousand or so stars that lie within 30 parsecs of the sun.

The distances of stars that lie too far away to have a measurable parallax shift can be found by an indirect approach. This method relies on an important physical fact: *The observed brightness of a source of photons will appear to decrease as 1 over the square of its distance from us.* For example, a car's headlights will appear one fourth as bright at a distance of 200 meters as they would if the car were 100 meters away from us.

The trick to measuring stellar distances is to *compare stars' apparent brightnesses.* Then, *if* we know that two stars have the same intrinsic, or true brightness—by analogy, if we know that two cars have headlights of the same type and the same conditions—we can conclude that if one star appears, say, a hundred times fainter than another, the fainter star must be 10 times farther away from us than the star with the greater apparent brightness. Our difficulties in using this technique arise from the fact that stars do not all have the same true brightness, so that we are sometimes looking at the stellar analogues of a bicycle headlamp and a semitrailer on high beams. Through years of patient study, however, astronomers have overcome this problem in comparing stars by learning how to recognize those stars that do have the same true or absolute brightness, no matter what the stars' distances from us may be.

The Spectra of Stars

The key to this recognition lies in the stars' spectra of photon energy; that is, in the distribution of the stars' *photons* with various energies (Figure 2-2).

Light waves consist of photons, which we may think of as bundles of energy with no mass at all. Each photon, traveling through space at a speed of 300,000 kilometers per second, carries a definite amount of energy. The energy of an X-ray or a gamma-ray photon can be trillions of times greater than the energy of a radio-wave photon (Figure 2-3), but all photons have a basically similar nature. We may imagine photons to be vibrating particles that oscillate back and forth as they move through space at the speed of light (Figure 2-3). The *frequency* of their vibration (number of wiggles per second) varies in direct proportion to the photons' individual energies. In contrast, the *wavelength* of vibration (distance between two successive wiggles) varies as 1 over the frequency (or 1 over the energy): More rapid wiggling means less distance between wiggles (Figure 2-3). We can specify

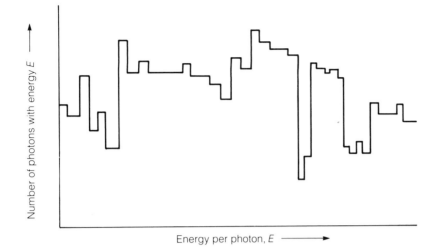

Fig. 2-2 The spectrum of light from a star can be illustrated by a graph that shows the number of photons with *each* particular energy as a function of the energy per photon.

a given photon's characteristics by naming either its energy, its frequency, or its wavelength, since if one of these three quantities is known, we can always compute the other two. Our eyes recognize the energy of photons as color: Those visible-light photons with larger energies and frequencies

Fig. 2-3 We can imagine a photon to be a bundle of energy, traveling through space at the speed of light (300,000 kilometers per second) and vibrating as it travels. The photon's energy of motion varies in *direct* proportion to the frequency of its vibration, while its wavelength varies in *inverse* proportion to its frequency and its energy. Different types of photons (gamma-ray, X-ray, ultraviolet, visible-light, infrared, and radio) differ in their energies (or equivalently, in their frequencies and wavelengths).

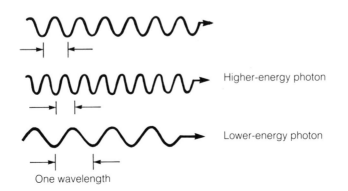

All photons travel at the speed of light.

Higher-energy photon

Lower-energy photon

One wavelength

Before   After

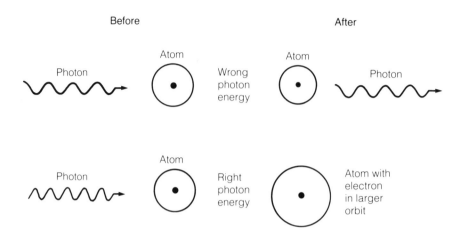

Fig. 2-4   When photons encounter atoms, the photons will not interact with the atom at all, *unless* a particular photon has just the right energy of motion to make one of an atom's electrons jump into a larger orbit. Such an excitation of the atom causes the photon to disappear.

appear bluer, while those with lower energies and frequencies (hence larger wavelengths) appear redder.

An interesting effect occurs when photons pass among atoms. For example, among the atoms in the outer layers of a star, only certain photons with particular energies will interact with the atoms (Figure 2-4). Photons that do not have the particular energies to make an atom's electrons move into a larger orbit will pass by the atoms, unaffected by the encounter. On the other hand, photons that *do* have the correct energies will excite the electrons into new, larger orbits, and in doing so will disappear (Figure 2-4). Thus, atoms act as filters upon beams of light that pass among them, since they remove photons of certain definite energies and leave the remainder unchanged. Each kind of atom, each atomic *element,* acts as a different kind of filter in this process.

Decades of observations have taught astronomers how to recognize the presence of a particular kind of atom (or a particular kind of molecule, made from a combination of atoms) from the signature left by the filtering process. By spreading a beam of light into its various colors, each of which corresponds to a particular photon energy, astronomers can tell whether the photons come from a star whose mixture of elements resembles the mixture in, for example, our sun. With enough care, astronomers can deduce from the number of photons with each energy—the *spectrum* of the star—not only the abundances of the different elements in the star, but also the star's surface temperature and the density of matter in its surface layers. All this information comes to us from the way that the star's outer layers filter the light that emerges from the star's interior.

By their studies of stellar spectra, astronomers have learned how to determine which stars resemble one another in element abundances, surface temperatures, and surface densities. The next step, backed by many observations and a long chain of reasoning (see Chapter 5), is to assume that any two stars that are nearly identical in the three observable quantities—surface temperatures, densities in the surface layers, and relative abundances of the various elements—can be assumed to be nearly identical in all their characteristics. In particular, we can assume that such stars have almost the same intrinsic or absolute brightness. Then, if we feel reasonably sure that two stars do have the same absolute brightness, we can look at the stars' relative *apparent* brightnesses to find how the distance to one star compares with the distance to the other.

As an example, consider the two stars Altair and Alderamin, which are the brightest stars in the constellations Aquila and Cepheus, respectively.[1] Examination of the spectra of light from these stars shows that they must be almost identical. Altair, the tenth brightest star in the sky (in *apparent* brightness), seems to be about nine times brighter than Alderamin. We conclude from this that Alderamin must be three times farther from us than Altair is, because the apparent brightness decreases as the square of the distance. Now Altair's distance from us, 5 parsecs, can be easily determined by measuring its apparent parallax shift (see p. 18). The distance to Alderamin must be about three times this, or 15 parsecs. We can verify this fact by measuring the parallax shift for Alderamin (0.067 seconds of arc), which is one third of the parallax shift for Altair (0.2 seconds of arc). If we were to look at a star three times farther away than Alderamin, its parallax shift would be too small to be measured directly. Then we should rely on the comparison of the apparent brightnesses of two stars that have essentially identical spectra as our means of finding the distances to the farther stars.

The technique that works for individual stars will also work for entire galaxies, with suitable modification. If we observe two galaxies that appear to have the same shape and the same general mixture of stellar types, then we may assume that the two galaxies are virtually identical (Figure 2-5). This assumption may be less valid than the assumption based on the spectra of individual stars, but astronomers are often forced to rely on it just the same. If we could somehow determine the distance to a relatively nearby galaxy, we could estimate the distance to a similar-looking galaxy that lies much farther away, once again by comparing apparent bright-

[1]Following a tradition two thousand years old, astronomers like to call the brightest star in a constellation Alpha, the second brightest Beta, and so on through the Greek alphabet. These designations, which always refer to *apparent* brightness, make Altair the same as "Alpha Aquilae" and Alderamin into "Alpha Cephei." Notice that the Latin names of constellations (Aquila and Cepheus) appear in the genitive case. Altair and Alderamin are Arabic names more than a thousand years old.

Fig. 2-5 If we observe that two galaxies are just about identical in their structure, we can tentatively assume that the galaxies are also identical in their intrinsic light output.

nesses. If one galaxy appears to be, say, a hundred times brighter than another, then the fainter galaxy must be 10 times farther away, *if* the two galaxies are indeed identical.

Our comparison of apparent brightnesses to find galaxies' distances will do us little good unless we know the distance to at least one of the galaxies (presumably the nearer one). Since the galaxies outside our own Milky Way have distances of anywhere from fifty thousand up to several billion parsecs, estimating their distances has proven to be difficult indeed. Parallax methods obviously will not work, for the galaxies are thousands of times farther away than the farthest stars whose parallax shifts can be measured.

Variable Stars as Distance Indicators

The key that unlocked the mystery of the distances to other galaxies and thus established the rest of the cosmic distance scale came from the discovery of *variable stars* in the nearest galaxies. A variable star changes its brightness, often in a regular cycle of variation, over time periods familiar to us: hours, days, weeks, or months. A certain class of variable stars, the *Cepheid variables,* showed particularly regular changes in their brightnesses, changes that repeated on a cycle of anywhere from two days to two months depending on the star in question.

In the early years of this century, Henrietta Swan Leavitt discovered that Cepheid variable stars show a definite connection between their absolute brightnesses and periods of light variation. Cepheid variables with *larger* absolute brightnesses take *longer* to complete each cycle of light variation (Figure 2-6). Once this fact had been established, the Cepheid variables could be used as distance markers in galaxies where we could recognize Cepheids by their characteristic light variation.

Since the closest Cepheid variable in our own galaxy has a distance greater than 200 parsecs from us, we cannot use the parallax effect to measure any Cepheid's distance. As a result, astronomers had to wait a long time to estimate the distances to Cepheid variables, and thus to discover their true brightnesses. Such distance determinations rely on careful observations of the stars' motions in space, observations made over the course of many years, and can give the stars' distances only approximately. But for the *comparison* of distances, we need know only that longer-period Cepheids have greater true brightnesses, and that two Cepheids with the same period of light variation have the same true brightness.

In 1923, an American astronomer named Edwin Hubble discovered, from repeated observations of the Andromeda galaxy, that this nearby agglomeration of stars contains several Cepheid variables. Hubble was quick

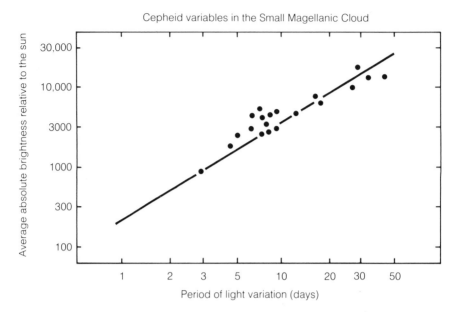

Fig. 2-6 This graph shows the correlation between the average absolute brightnesses and the length of time for each cyclical change in brightness for the Cepheid
variables in the nearby galaxy called the Small Magellanic Cloud, a satellite of our
own Milky Way.

to compare the apparent brightnesses of these Cepheids with those of
Cepheids in our own galaxy known to have the same periods of light variation. Hubble thus demonstrated that the Cepheid variables in the Andromeda galaxy, and so the galaxy itself, must have a distance from us of
300,000 parsecs, equal to 10 times the diameter of the entire Milky Way
galaxy. Since 1923, more refined analyses of Cepheid variable stars have
revised this distance upwards to 600,000 parsecs.

The Expanding Universe

The crowning discovery of Hubble's career, which also marked the beginning of modern cosmology, came just over 50 years ago, in 1929, the year
that, coincidentally, the American stock market collapsed. As the 1920s
passed by, filled with the impressive feats of Charles Lindbergh, Babe
Ruth, and Calvin Coolidge, Edwin Hubble went on measuring the distances
to other galaxies. Night after night, Hubble and his colleague Milton Humason photographed distant galaxies to determine the brightness variation
of possible Cepheid variable stars within them. When they had a reliable
distance estimate for a nearby galaxy, they compared the distance they had
determined with the galaxy's overall *motion* towards us or away from us.

This motion could be found from the *Doppler effect,* which changes the spectrum of a source of photons if the source moves toward or away from an observer (Figure 2-7). Photons that arrive from a source moving away from us will all have lower energies (hence, lower frequencies and larger wavelengths) than the photons from the same source if it is stationary with respect to us. This Doppler effect also occurs in sound waves, when it appears as the decrease in the pitch, or frequency, of an ambulance siren speeding away from us. Whether we move away from the source or the source moves away from us does not matter; nor does the amount of distance between us and the source play any role in the Doppler effect. The velocity of recession between the source and observer makes the photon energies decrease in comparison to a stationary standard, and the amount of this decrease, the *redshift* to lower photon energies and thus to the red end of the visible-light spectrum, grows larger as the relative velocity of recession increases.

Similarly, if we move toward a photon source or the source moves toward us, then we shall observe photons with larger energies (thus, larger frequencies and smaller wavelengths) than the photons from the same source when it is motionless with respect to us. If the velocity of the source with respect to us, denoted by *v,* is much less than the speed of

Fig. 2-7   The Doppler effect increases the energies of photons from a source moving towards us, and decreases the energies of photons from a source moving away from us, in comparison with the energies we see for the light from a stationary source.

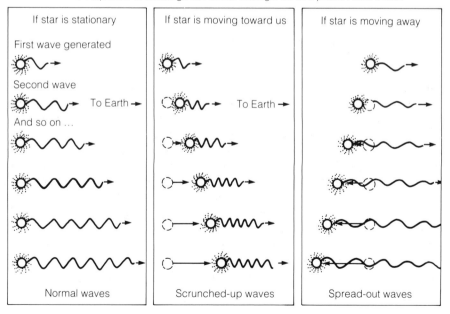

A time sequence of drawings shows the emergence of a photon from a star.

If star is stationary	If star is moving toward us	If star is moving away
First wave generated		
Second wave	To Earth →	
And so on ...	To Earth →	
Normal waves	Scrunched-up waves	Spread-out waves

light (c = 300,000 kilometers per second), then the shift in the photon energies varies in direct proportion to v divided by c. If the source approaches us or we approach the source, then we see a *blueshift,* to higher photon energies and therefore to the blue end of the visible-light spectrum. If the source recedes from us or we recede from the source, we observe a *redshift* to lower photon energies.

Galaxies contain billions of stars, most of them basically like the sun (though perhaps a few thousand times fainter or brighter in absolute brightness). The stars' surface temperatures may be a few times greater or less than the sun's, but a general similarity in size, energy output, and composition characterizes most of the stars in a galaxy. Thus, the spectrum of the light from an entire galaxy, which shows the combined output from billions of stars, can be analyzed in the same way as the light from a single star. Indeed, these spectra reveal the same features that are prominent in most stellar spectra; for example, the emission of light at a particular photon energy that comes from hot oxygen atoms, or the absorption of photons with a particular energy by calcium ions (calcium atoms with an electron missing).

In the spectra of the light from galaxies, the energies at which, for example, calcium ions absorb photons do not appear at exactly the energy we would measure in an Earthbound laboratory. The general features of the spectrum seem the same from galaxy to galaxy, except that they occur at slightly different frequencies. We may therefore conclude that the motion of the entire galaxy, relative to ourselves, has produced a change in the observed photon energies, the result of the Doppler effect. If the calcium absorption occurs at a lower energy than we would expect, we have a redshift, so the galaxy must be receding from us (or we from it); if the absorption appears at a higher energy than usual, we have a blueshift, so the galaxy must be approaching us (Figure 2-8).

Once Hubble had discovered Cepheid variables in other galaxies and had used them to find the galaxies' distances from our own, he proceeded to use the measurements of the Doppler effect in the light from nearby galaxies that had been made by Vesto Slipher. When Hubble combined Slipher's observations of the galaxies' motions along the line of sight (toward us or away from us) with his own estimates of the galaxies' distances from us, he found a trend: Galaxies are generally receding from us, and the more distant galaxies are receding at proportionately greater velocities (Figure 2-8). Today we call this relationship *Hubble's Law.*

Since Hubble was a cautious and a thorough researcher, his work convinced astronomers that galaxies' speeds of recession vary in direct proportion to the galaxies' distances from us. During the two years after his initial report in 1929, Hubble and Humason extended their distance estimates to far more distant galaxies, determining the velocity-distance relationship shown in Figure 2-9. With this work they won over the last doubt-

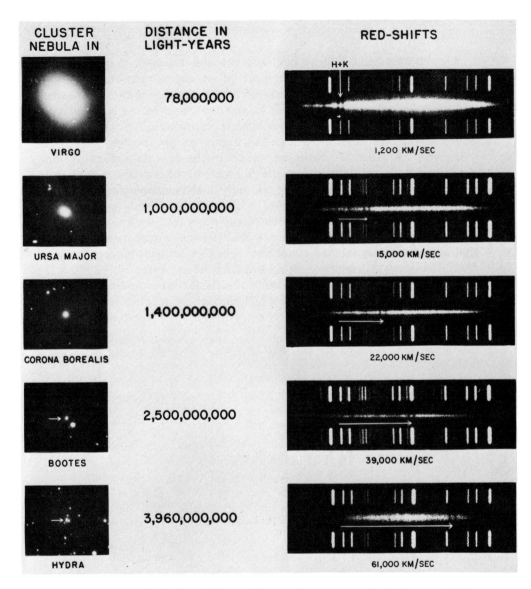

CLUSTER NEBULA IN	DISTANCE IN LIGHT-YEARS	RED-SHIFTS
VIRGO	78,000,000	H+K ... 1,200 KM/SEC
URSA MAJOR	1,000,000,000	15,000 KM/SEC
CORONA BOREALIS	1,400,000,000	22,000 KM/SEC
BOOTES	2,500,000,000	39,000 KM/SEC
HYDRA	3,960,000,000	61,000 KM/SEC

Fig. 2-8 The galaxies in five different clusters show progressively larger redshifts for more distant clusters. For each of these galaxies (mistakenly called "nebulae"), the redshift can be measured from the change in frequency of the two absorption lines in the spectrum that are produced by calcium ions, labeled H and K.

ers: Except for the galaxies in our own group, all of the galaxies that we observe are in relative motion away from us, with recession velocities that increase in proportion to the galaxies' distances from us.

What does this mean? Why should all galaxies be receding from us? The answers to these questions came quickly, as a mind-stretching bombshell: The entire universe must be expanding!

Now it may seem bold to conclude that the universe, everything that exists, must be expanding, simply because we see galaxies moving away from *us*. But if we believe that our view of the universe is a representative one, then *any* observer in *any* other galaxy must see what we see: Galaxies are moving away from that observer, with recession velocities that increase with distance from that observer. And if galaxies are moving away from every observer, then the entire universe must be in a state of expansion.

We can never tell what the undetected parts of the universe may be like, so we cannot be entirely sure that our view of the universe does represent the average. Astronomers, however, prefer this hypothesis to any other, since any other would require some assumption that we are special, not representative of the universe as a whole. Hence, astronomers conclude that all galaxies are moving away from all other galaxies (except perhaps from their nearest neighbors), everywhere in the universe.

But how can this be true? Surely if galaxies are receding from one another here, somewhere they must be approaching one another? No, they

Fig. 2-9   This graph, produced by Hubble and Humason in 1931, shows a straight-line relationship between the distances of galaxies and their speeds of recession from us.

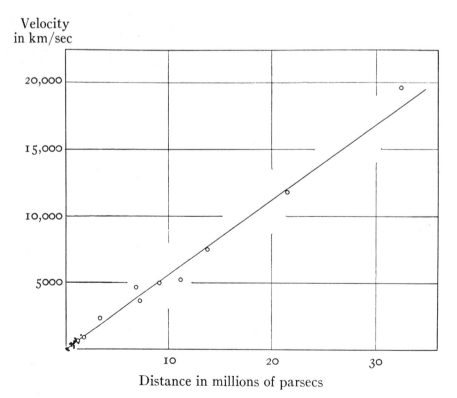

Distance in millions of parsecs

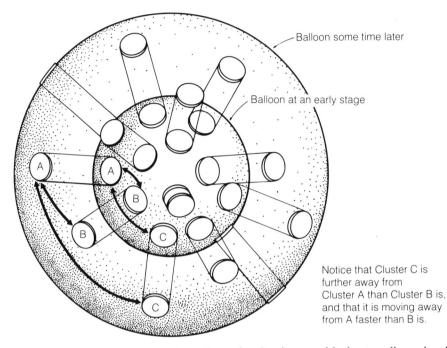

Balloon some time later

Balloon at an early stage

A
A
B
B
C
C

Notice that Cluster C is
further away from
Cluster A than Cluster B is,
and that it is moving away
from A faster than B is.

Fig. 2-10   We can model the three-dimensional universe with the two-dimensional surface of an expanding balloon. To use this model, we must imagine that light travels only on the surface and that nothing inside or outside the balloon exists. As we blow up the balloon, every dot moves away from every other dot; the relative recession velocity between any two dots will vary in proportion to the distance between the two.

need not be. For the entire universe to be expanding, all of the galaxies (more precisely, all of the galaxy clusters) must be moving away from one another. Consider, as an imperfect model, the surface of an expanding balloon, with dots to represent the clusters of galaxies (Figure 2-10). As the balloon expands, every dot moves away from every other dot. If we measure distances around the balloon's surface, then the mutual speeds of recession vary in direct proportion to the distances between the dots. The balloon's surface has no center of expansion, and each dot can see all the other dots receding in accordance with a two-dimensional version of Hubble's Law: Greater distances imply greater recession velocities.[2]

---

[2]Notice that the balloon does have a center, but that the center does not lie on the balloon's surface. We are using the *surface* to model the universe, and can do so with ease because we have an extra, third, dimension in which to visualize the two-dimensional surface of the balloon. If space in the universe is curved (see page 38), then space could be something like the balloon's surface—with a "center" only in the fourth dimension (which exists only as a mathematical construct, unlike the three dimensions of real space). *No* center of the universe would exist within those three dimensions.

The Early Moments of the Universe

Since matter in the universe is moving farther apart, we can conclude that the matter used to be closer together. In other words, the average density of matter was larger in the past; as we travel farther and farther into the past, we reach greater and greater densities. (An alternative, *steady-state* model of the universe, which holds that new matter appears as the universe expands so that the density remains constant at all times, has been ruled out by the observations described on page 36). If we pursue this line of thought, we reach a moment in past time when the density appears to become infinitely large (Figure 2-11). This moment, called the *initial singularity*, or the *big bang*, marks the start of the expansion, and, for all astronomical purposes, the beginning of the universe. We simply cannot tell with any accuracy what the universe was up to before the big bang. It may have been contracting toward this enormous density, or it may not have existed at all.

Let us consider the universe as it was during the first few minutes after the big bang. From the graph in Figure 2-11 we can see that the densities of matter were enormous. These fantastic densities created physical con-

Fig. 2-11 The average density of matter in the universe has been decreasing ever since the big bang because all parts of the universe have been moving away from all other parts.

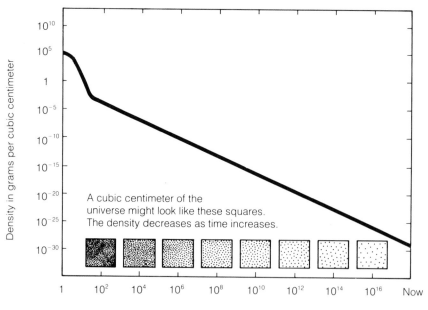

A cubic centimeter of the universe might look like these squares. The density decreases as time increases.

Density in grams per cubic centimeter

Seconds of time since the beginning

ditions entirely different from those we encounter now. No galaxies, stars, planets, or other individual objects existed; instead, a featureless, seething broth of particles roiled and boiled as countless collisions created and destroyed all sorts of particles and antiparticles.

As the universe expanded, the density of matter kept decreasing so that particles collided less frequently and with less energy of motion. After the first half hour A.B.B. (after the big bang), the basic mixture of particles had emerged. The universe then contained, as it does now, mostly protons, neutrons, electrons, and helium nuclei, plus photons, neutrinos, and antineutrinos in great numbers. The last three types of particles have no mass, though they can carry energy of motion through space as they move at the speed of light.

Matter and Antimatter

Each type of elementary particle has a corresponding antiparticle, with the same mass but the opposite sign of electric charge. When a particle collides with its antiparticle, the collision turns all of the particles' energy of mass into energy of motion. Thus, for example, a proton and antiproton each have the same mass m, and the same energy of mass given by $m \times c^2$, where c is the velocity of light. A proton-antiproton annihilation will turn the total energy of mass, $2m \times c^2$, into the energy of motion of photons, neutrinos, and antineutrinos that emerge from the annihilation.

The predominance of matter over antimatter within our galaxy reflects the balance that remained once most of the antimatter had annihilated along with the corresponding particles of matter. We cannot tell whether other galaxies consist primarily of matter or of antimatter, because the photons from either sort of galaxy are identical. Photons and antiphotons are entirely the same, and we can call them all photons: They carry no information about whether they come from matter or from antimatter stars. But if a visitor from an antimatter planet reached the Earth, this encounter could prove highly destructive: A man made of matter and a woman made of antimatter could, upon contact, annihilate with the release of the same amount of energy produced by the most powerful hydrogen bombs!

Within the early universe, the mutual annihilations of particles and their antiparticles left behind—at least as far as we can tell—mostly protons, neutrons, helium nuclei, and electrons, along with the massless particles, photons, neutrinos, and antineutrinos. However, the neutrons that were not part of helium nuclei (each helium nucleus contains two protons and two neutrons) soon *decayed* or separated into other particles (Figure 2-12). These neutron decays produced more protons, electrons, and antineutrinos. Only a few *days* saw the disappearance of the neutrons not inside helium nuclei. Once the neutrons had vanished, about 75 percent of the

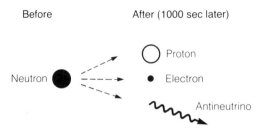

Before   After (1000 sec later)

Neutron   ⟶   ○ Proton
         ● ⤍ ⤍ ⤍ ⤍ ➤   ● Electron
                              ⟿ Antineutrino

Fig. 2-12   Any neutron that is not part of an atomic nucleus will decay after about 15 minutes into a proton, an electron, and an antineutrino.

mass in the universe was in the form of protons and 25 percent in helium nuclei. The electrons had only about $^1/_{20}$ of a percent of the total mass, because each electron has $^1/_{1836}$ of a proton's mass.

The Production of Heavy Elements

The great violence of the early moments of the universe fused many of the elementary particles into helium nuclei, each with two protons and two neutrons. What about nuclei larger than helium, which astronomers call *heavy* nuclei? These nuclei, such as carbon, nitrogen, oxygen, aluminum, silicon, and iron, form the centers of the most common atoms on Earth. Did the early moments after the big bang make these nuclei along with the hydrogen and helium?

The answer turns out to be negative. Violent as the clash of particles in the early universe was, the time to form nuclei passed so quickly, because the density of particles decreased so rapidly, that an insignificant number of heavy nuclei appeared. Calculations of the behavior of the early universe show that the only nuclei produced in significant amounts were protons, deuterons (one proton plus one neutron), helium-3 nuclei (two protons plus one neutron), and helium-4 nuclei (ordinary helium, with two protons and two neutrons in each nucleus). All heavier nuclei amounted to *less than* $^1/_{5000}$ *of 1 percent* of the number of protons (see Table 2-1).

To explain where the heavy nuclei came from, we must look to the stars. As we shall discuss in Chapter 6, stars turn hydrogen and helium nuclei into heavier and heavier nuclei as they age. Some stars explode at the end of their lifetimes, spewing forth heavy nuclei throughout the galaxy that contains them. Stars born later on in that galaxy, and the planets that form with them, can include some of this star-cycled matter. In this way, the sun managed to form with 1 percent of its mass in the form of nuclei heavier than helium; thus, too, Earth obtained its own supply of heavy elements. In short, stars "died that we may live," for our own bodies and

TABLE 2-1

FORMATION OF NUCLEI OF DIFFERENT ELEMENTS DURING THE FIRST FEW MINUTES A.B.B.

Element	Atomic Number (number of protons)	Number of Nuclei (per 10^{12} protons) Produced during the First Few Minutes A.B.B.[a]	Present Number of Atoms (per 10^{12} hydrogen atoms) Found in the Solar System and in Stars Like the Sun
Hydrogen	1	1,000,000,000,000	1,000,000,000,000
Helium	2	80,000,000,000	80,000,000,000
Carbon	6	1,600,000	370,000,000
Nitrogen	7	400,000	115,000,000
Oxygen	8	40,000	670,000,000
Neon	10	180	110,000,000
Sodium and all heavier elements	11 or more	2500	140,000,000

[a]Different models of the early universe produce different numbers of the nuclei heavier than hydrogen as a result of the first few minutes A.B.B. The third column in this table lists the *maximum* number of nuclei that could have been produced under the most favorable conditions thought to be reasonable by cosmologists. The elements heavier than helium are made in only tiny abundances during the first few minutes.

the environment in which we survive consist largely of heavy elements, all of which came not from the first few minutes after the big bang but rather from the exploded wisps of countless long-vanished stars.

The Cosmic Background of Photons

By far the most abundant particles in the universe, ever since the first few moments after the big bang, are photons. For every proton in the universe, there are *100 million* photons; but we care more about matter, hence a book such as this deals with these photons only briefly.

The enormous number of photons arose during the first half hour after the big bang, as the annihilation of particles and antiparticles created huge numbers of photons, most of them made from the annihilation of electrons and antielectrons. As the universe expanded, the photons created during its first half hour continued to travel through space—all of it—at the speed of light. If we use our balloon analogy for the universe, we can think of the photons as appearing everywhere on the balloon's surface at the time when the balloon was relatively small. The photons that reach *any* observer at *any* time must have been emitted by a source (some other part of the universe) that is moving away from that observer, because the entire universe is expanding.

As the universe expands, any observer will notice that the photons that arrive tend to have less and less energy per photon. All of the original photons appeared at about the same time in universal history, during the first half hour A.B.B. Any photon that reaches an observer has been traveling at the speed of light ever since then. Because of the tremendous expansion of the universe during the past 15 or 20 billion years since the big bang, the photons created during the first half hour have been tremendously redshifted by the expansion itself. As a result, a photon that we now detect from this long-vanished epoch has far less energy than it did at the time of its creation.

A Russian-born astrophysicist named George Gamow was the first to realize what must have happened to the high-energy photons made soon after the big bang. Gamow predicted that the entire universe should be filled with a sea of relatively low-energy photons, the redshifted remnants of the first half hour A.B.B. Ten years after Gamow made his prediction, Arno Penzias and Robert Wilson used an antenna at the Bell Telephone Laboratories to discover the sea of photon radiation that has surrounded the Earth throughout its history. The discovery of this *cosmic background radiation,* and its later confirmation in greater detail, provides us with the most straightforward evidence that the big bang theory of the universe has solid roots in reality.

The cosmic background of microwave photons provides a good way to remind ourselves that we are *inside* an ever-expanding universe with *no* center of expansion. From where do the photons arrive? From all directions, and in the same amounts from all directions, thus demonstrating that the early universe must have had the same density of matter, which produced the same number of photons, in different places at the same time. That is, the early universe must have been homogeneous, not highly clumped. But what produced these photons? The early universe, which no longer exists. And why are the photons so greatly redshifted? Because the now-vanished early universe has an enormous velocity of recession in *every* direction. If we imagine ourselves on the surface of an expanding balloon, we receive photons around the balloon's surface from its early years coming from all directions, but always from more and more distant regions, thus with greater and greater redshifts. The farther away any two objects are on the balloon's surface, the greater will be the relative velocity of recession of the two objects (Figure 2-10).

Is the Universe Finite or Infinite?

We cannot now determine whether the universe—all of space and everything in it—spans a finite or an infinite extent, and includes a finite or an infinite amount of matter. This problem has an intimate connection with the problem of whether or not the universe will expand forever, and both may be resolved within the next generation or so of astronomical effort.

What does it mean to say that the universe could be finite? Could the universe have a boundary? Certainly not: The universe would have to include the boundary and anything on the other side. In a finite universe, space *curves* so that all seemingly straight lines bend back to join themselves after a long but finite distance. (This kind of curvature has the technical name *positive curvature*.) A straight-line rocket journey in a finite universe would eventually bring us back to our starting point.

We may use the Earth's surface as a suggestive, though unsatisfactory, model of a finite universe. Today we all know that we live on a spheroid, and that travel over the Earth carries us over the *curved* surface of this spheroid. If we travel far enough, we will end up back where we started. Now the Earth's surface has only *two* dimensions, north-south and east-west. Travel in the *third* dimension, up or down, takes us completely off the surface, and this fact makes it hard to compare the Earth's surface with the entire universe. If space in the universe is curved, then not simply the *two* dimensions of a spherical surface have curvature, but rather all *three* dimensions of space do. We have no way to back away from a part of space in the universe, as we can for an isolated sphere like the Earth, to get a good look at it, either in a physical or in an imaginary context.

Hence, we cannot hope to acquire an intuitive understanding of *how* three-dimensional space could be curved. But we do know that the curvature might exist, and that such curvature would be detectable and measurable.

Suppose that we ourselves had only two dimensions and spent our lives sliding around the surface of a smooth sphere (Figure 2-13). When the subject of curved space came up, we would say, "I don't see how space could possibly curve back on itself," since we would be unable to conceive of all two-dimensional space (that which was accessible to us) as a quantity with curvature. We might, however, discover this anyway: If we slid all the way around the sphere, we would return to our starting point, and we would know that something odd had occurred, even though we couldn't understand how we had returned. Thus, despite the absence of comprehension, the *fact* of curvature would eventually leap out at us.

If you have difficulty in imagining a finite universe, consider the alternative possibility, an infinite universe. Just imagine space extending forever and ever and ever, totally without limit, including an infinitely large number of stars, galaxies, planets, atoms, and particles. Consider what an infinite universe would contain. With an infinite set of situations, *everything not forbidden would occur somewhere, not once but an infinite number of times!* Other forms of life would duplicate our own, as well as every other conceivable possibility, over and over again in all variations, an infinite number of times for each individual possibility. This book would exist in every conceivable version, in every possible language (human or other), in

Fig. 2-13   If we were flat creatures sliding around the surface of a smooth sphere, we might not be able to imagine what it meant to say that our universe would be "curved." If we found that we had returned to our starting point after sliding in a "straight line," however, we would conclude that our universe, the two-dimensional surface of a sphere, must include only a finite amount of area.

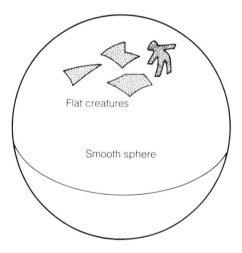

Flat creatures

Smooth sphere

every conceivable style of printing, photography, or paper. Sometimes the reader would have written it, sometimes not. Sometimes one word would be different from this text, sometimes two, sometimes more, and each possibility would be reality not just in one place, or in a few places, but in an infinite number of places.[3]

We have presented these facts about life in an infinite universe to underscore the enormous difficulties that an infinite universe also presents for our imagination. We reside *inside* the universe, but to picture the universe inevitably requires us to step outside it, physically or mentally, for an accurate look. This we can never do.

If we search for other forms of life, then we can rest assured that in an infinite universe, not only life but an infinite variety of life exists, of every possible function or appearance. A finite universe may seem less interesting, but since we know that the universe contains at least 100 billion galaxies, each with billions of stars, we still face an enormous array of possibilities in a finite universe.[4]

On Earth, 3 million years of human development have made us aware of our situation within the universe. We now stand ready to discover whether or not the history of the universe, 7,000 times the history of hominids on Earth, points to its eventual collapse or to endless expansion. Someday soon, using large telescopes in orbit above the Earth's blurry atmosphere, we may find out whether the universe is finite or infinite and whether or not it will expand forever by studying the distances and the apparent brightnesses of the farthest galaxies that we can see, galaxies too faint to be observed from inside our protecting sea of air.

The Future of the Universal Expansion

Once people learned that the entire universe is expanding, two questions immediately arose: How does this expansion affect us? And will the universe continue to expand forever?

We have seen the answer to the first question, as we have probed the history of the universe, whose first few minutes yielded the basic mixture

---

[3]The reader may think at first glance that this is a joke, but it most definitely is not. No matter how improbable an event may be in an infinite universe, the fact of infinity implies that it must occur not once, or a thousand times, or any finite number of times, but an infinite number of times. However, we can hope to observe, even in principle, only a finite region of the universe, and thus can see only an infinitesimal fraction of this infinite variety.

[4]Because only a finite amount of time has passed since the big bang, we can interact with only a finite volume of the universe, no matter how large the entire universe may be. At any time after the big bang, we can know about only those regions of the universe whose distance from us does not exceed the speed of light times the age of the universe.

of particles that we find today. Later years brought the formation of galaxies and of stars, some of which have exploded to seed heavy nuclei, including those essential to life, throughout galaxies such as our own. Although these heavy nuclei amount to only 1 percent, at most, of the number of hydrogen and helium nuclei, they are absolutely essential to life. Hence, we may fairly say that for life to appear, we need both the early universe, which made the hydrogen, helium, and electrons, *and* we need later exploding stars. Incidentally, *we* are not expanding at the same rate as the universe: The mathematical calculations that deal with the universal expansion show that only on the largest distance scales, those between clusters of galaxies, does the universe expand according to Hubble's Law. More local objects, such as people, the Earth, the solar system, and even the Milky Way galaxy, are not expanding in any significant way as the galaxy clusters recede from one another.

Meanwhile, we can hope to determine whether the universe is finite or infinite if we can discover whether or not the universe will expand forever. Cosmological calculations based on the theory of relativity point to a key result: If the universe is infinite, it will expand forever. On the other hand, if the universe is finite, it will eventually cease its expansion and begin to contract.[5] The riddle of the universal expansion wraps around the *size* of the universe. Finite models of the universe reverse their expansion, but an infinite universe grows ever larger.

The determination of which future—eternal expansion or cosmic recycling—lies in store for the universe presents astronomers with one of the most significant problems with which they grapple. The attack on this puzzle rests on the fact that *if* the universe ever starts to contract, it will do so because of gravity. Gravitational forces, as far as we know, provide the only chance to overcome the tendency to expand which the universe acquired at the moment of the big bang, and which has dominated the universe ever since. Every piece of matter in the universe attracts every other through gravitational forces, and these mutual attractions constantly try to overcome the universe's expansionist tendencies. As a result, the universe is no longer expanding as rapidly as it used to: The time for distances between galaxy clusters to double has steadily increased. But will the distances ever stop increasing and start decreasing?

---

[5]The theories we are describing, which provide this match between finite-with-contraction and infinite-with-no-contraction possibilities, are those that assume a *zero cosmological constant*. A minority of cosmologists believe that theories with a nonzero cosmological constant are plausible; then not all the finite universe possibilities must contract eventually. *Cosmological constant* is a term first suggested by Einstein in the equations he wrote to describe the behavior of the universe. Einstein's equations implied that the universe must be expanding; since he did not then know that the universe in fact *is* expanding, he introduced the cosmological constant, which amounts to a new kind of force, to keep the universe at a constant size. Later, when he learned of the universal expansion, Einstein called this term a ''great blunder.''

The possible reversal of the universe's expansion depends on the strength of gravitational force, compared with the initial acceleration of all parts of the universe (including space!) that emerged from the big bang. To find out whether or not the universal expansion will ever cease, we have two lines of attack. The first and more straightforward one consists of determining the average density of matter in the universe. If this density now exceeds a certain critical value (close to 10^{29} gram per cubic centimeter), then the expansion will eventually stop, to be replaced by a universal contraction. Conversely, if the average density is now less than the critical value, then the universe will expand forever.[6]

The second approach in determining the future turns to the *past,* by looking as far out into space as we can. Since light travels at a finite speed (300,000 kilometers per second), our observations of more and more distant objects take us farther and farther back in time to the moment when the photons were emitted to carry their energy outward. Our look into the past can now reveal how things were billions of years ago if we look at objects that are billions of light years away from us. If we can determine how fast the expansion proceeded in the past, and how fast it goes now, we can extrapolate to find out whether or not the rate of expansion will ever fall to zero. To do this, we must determine the distances to faraway galaxy clusters and their recession velocities with respect to us. These two quantities, measured for different distances (that is, for different times in the past), can reveal whether or not the universe will always expand.

The two methods for resolving the question of the future expansion have so far given somewhat contradictory results. When we attempt to determine the average density of matter in the universe, we can easily locate only those objects which shine (emit photons), either in visible light or in radio waves. If these relatively familiar objects—the stars in galaxies and galaxy clusters—contain most of the matter in the universe, then the universe seems destined for eternal expansion, because the average density of luminous matter falls short of the critical density needed to reverse the expansion by a factor of about 20. On the other hand, the second method, that of looking far back in time, suggests that the universe might have just enough density to reverse its expansion (Figure 2-14). This method cannot be considered as reliable as we would like, because we must determine the distances to faraway galaxy clusters with high precision. The best results now available do indicate that the universe has an average density of matter that is not greater than the critical density but that appears to be at least several times greater than the density we derive from our observations of stars in galaxies.

---

[6]The value for the critical density decreases as the universe expands, but in exactly the same way as the actual density decreases. Therefore, if the actual density of matter exceeds the critical value at any time, it will exceed it for all time. Conversely, if the actual value falls below the critical value at any time, then it will always fall below the critical value.

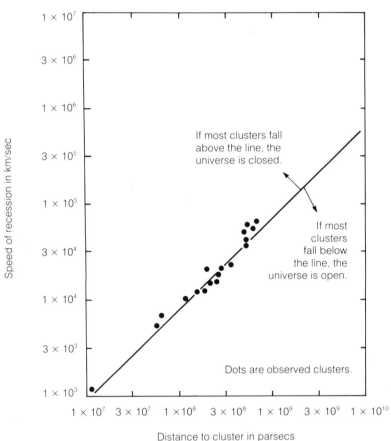

Fig. 2-14 If we plot the redshifts of galaxies, which give their speeds of recession, against their distances, which we can estimate from the galaxies' apparent brightnesses, we can hope to determine whether the trend corresponds to that for eternal expansion or eventual contraction. So far, this approach suggests that the universe will eventually contract, but the data are not accurate enough for us to be certain.

Another indication of a low value for the average density of matter in the universe comes from the measured abundance of deuterium nuclei. Each deuterium nucleus consists of one proton and one neutron, thus representing a rare isotope of hydrogen, the most abundant element. Calculations show that the abundance we can expect for deuterium *now* must depend in a sensitive manner on the average density of the universe, which determines how many deuterium nuclei were made in the early minutes after the big bang. If we rely on these calculations, then the observed abundance of deuterium nuclei implies that the universe's average density of matter falls well below the value that would make the universe eventually contract.

On the other hand, if either our calculations or our measurements concerning the deuterium abundance should prove to be wrong, then the universe could turn out to have a density large enough to pull itself together. In this case, we must conclude that *most* of the matter in the universe does *not* emit photons as well as stars do. Such underluminous matter might exist as diffuse clouds of gas, as stars too faint to be detected, or as black holes in space (see Chapter 7).

Summary

We can conclude from astronomical observations, and from our own logic as applied to these observations, that the entire universe has been expanding for the past 15 to 20 billion years, ever since the initial big bang brought the universe into being. We observe that almost all galaxies are moving away from us, with speeds that increase in proportion to the galaxies' distances from us. If we inhabit an average region of space, then *all* galaxies should find the same proportionality (Hubble's Law) between the distances and recession velocities of the galaxy clusters that surround them. Hence, the entire universe must be expanding, and we can determine the age of the universe, at least approximately, from our observations of this expansion.

The enormously high densities and temperatures in the first minutes of the universe produced all varieties of elementary particles from tremendous numbers of high-energy collisions. After the first half hour, the basic mixture of particles that we see today had been established. Photons emerged in huge quantities from the universe's first half hour, because many photon-producing reactions among particles had occurred. These photons have steadily lost energy, as measured by any observer, as the result of the redshift that arises from the universal expansion; they now form a cosmic background radiation that fills the universe.

To determine whether the universe will ever stop expanding, we must look backward in time by observing more and more distant galaxies whose light has taken longer and longer to reach us. Such observations can tell us how the relationship between galaxies' distances and their velocities of recession has changed with time, and thus how the universal expansion will change in the future. In addition, measurement of the average density of matter in the universe will tell us how well the universe can pull itself together: If the average density exceeds a certain critical amount, the universe will eventually contract. At the present time, both our observations of the average density of matter and our study of distant galaxies have too much uncertainty to resolve the question of whether the universe will ever cease expanding and start contracting. Similarly, these observations fall short of being able to tell us whether the universe is finite or infinite.

Questions

1. Why do we think that the entire universe must be expanding simply because we see galaxy clusters moving away from us?

2. Why does the parallax shift *decrease* as the distance to the stars we observe *increases?*

3. Suppose that two galaxies are almost identical in structure, but we observe that one galaxy appears 400 times fainter than the other. Which galaxy is farther away from us? By how much?

4. Suppose that a certain photon does not have enough energy to knock the electron loose from a hydrogen atom. To make this ionization occur, do we need a photon with a greater or a smaller wavelength? With a smaller or greater frequency?

5. How has the Doppler effect changed the energy of the photons made during the first half hour after the big bang? How will the photons' energy change in the future?

6. Where do the photons in the cosmic background come from? From what directions do they arrive? What does this tell us about the location of the center of the universe?

7. Which parts of the universe consist of antimatter? Why do we have difficulty in resolving this question?

8. What does it mean to say that the universe could be finite? If we model a finite universe with the surface of an expanding balloon, where does new space on the surface come from as the balloon expands? Where does new space in the *universe* come from as the universe expands?

9. Why does the average density of matter in the universe determine whether or not the universe will expand forever? To reverse its expansion and start contracting, does the universe need a greater or a smaller density than stars in the galaxies provide?

10. How can we tell whether or not the universe will expand forever? Which alternative now appears to be more likely?

11. The galaxies in the Coma cluster are moving away from us at 7000 kilometers per second, while the galaxies in the Corona Borealis cluster have recession velocities of 21,000 kilometers per second. Which cluster of galaxies is farther away? How much farther away?

12. Suppose that the brightest galaxy in the Coma cluster has the same true brightness as the brightest galaxy in the Corona Borealis cluster. Which of those two galaxies will appear brighter? How much brighter?

Further Reading

Bonnor, William. 1964. *The mystery of the expanding universe*. New York: Macmillan.

Charon, Jean. 1973. *Cosmology: Theories of the universe*. New York: McGraw-Hill Book Company.

Davies, P. C. W. 1977. *Space and time in the modern universe*. Cambridge: Cambridge University Press.

Ferris, Timothy. 1977. *The red limit*. New York: William Morrow & Co.

Gingerich, Owen, ed. 1977. *Cosmology + 1*. San Francisco: W. H. Freeman and Company.

Goldsmith, Donald, and Levy, Donald. 1974. *From the black hole to the infinite universe*. San Francisco: Holden-Day.

Kaufmann, William. 1974. *Relativity and cosmology*. New York: Harper & Row. 2nd ed.

Weinberg, Steven. 1977. *The first three minutes*. New York: Basic Books.

Science Fiction

Asimov, Isaac. 1972. *The gods themselves*. New York: Fawcett Crest Books.

CHAPTER 3

Galaxies

Galaxies, each composed of many millions or billions of individual stars, form the basic visible units of the universe outside our Milky Way, a giant spiral galaxy that contains about 400 billion stars. Within the few billion parsecs around our Milky Way, the region of the "known universe," billions of individual galaxies, usually members of galaxy clusters, have formed from what was once a featureless and nearly uniform medium. In our search for life in the universe, the *ages* of these galaxies, and of the stars that form them, are of paramount importance, because life needs time to evolve as well as fairly stable conditions during this evolution.

When we study the formation and the later history of galaxies, we must therefore look closely at the various types of galaxies, and of the stars within them, to locate the best hopes for finding other life. Though the closest civilizations to our own are (we hope!) within our own Milky Way galaxy, we must not overlook the fact that our galaxy represents a single grain of sand in the almost uncharted ocean that we call the universe. To obtain a true perspective on the question of life in the universe, we should consider the proportions of different types of galaxies, and of different kinds of stars within a galaxy, to see how representative our own galaxy is, just as we must study our planetary neighbors in detail to decide how representative of planets our Earth may be.

Spiral Galaxies

We inhabit an outer corner of the Milky Way galaxy, a giant spiral system of stars, gas, and dust that resembles the galaxy NGC 6744 (Figure 3-1).

Fig. 3-1 The spiral galaxy NGC 6744 has a distance of about 15 million parsecs from us.

Since our present technological abilities do not allow us to travel outside our galaxy, we cannot yet obtain an overall view of the Milky Way like that of NGC 6744, but we can try to map the structure of our galaxy from the inside, despite the difficulties of locating the forest amid the starry trees around us. During the past 60 years, astronomers have come to recognize the sun's location in the Milky Way: It is close to the galaxy's plane of

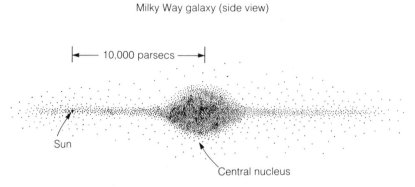

Fig. 3-2   Spiral galaxies such as our own Milky Way consist of a thin disk of stars
and gas, with a small central bulge or nucleus. The sun lies about 10,000 parsecs
from the nucleus of the Milky Way.

symmetry but quite far from the galactic center: We occupy an outer part
of the galaxy, a sort of Milky Way suburb (Figure 3-2).

During the past century, the study of galaxies progressed by observing
the visible light they produce and by analyzing its spectrum. The last few
decades, however, have added new spectral regions, including radio waves,
infrared light, ultraviolet light, and most recently, X-rays and gamma rays.
New observations have brought new discoveries; for example, with X-ray
observations we can see much farther inside the Andromeda galaxy than
with visible light photographs (Figure 3-3). This new knowledge adds to
our understanding of the structure and evolution of galaxies.

Spiral galaxies such as the Milky Way show two striking characteristics in their general structure. First, they resemble flat plates, for their
thicknesses barely equal 1 percent of their diameters (Figure 3-2). Second,
matter in spiral galaxies appears within their disklike structure predominantly within the *spiral arms,* as we see in Figure 3-1. But this dominance
by the spiral arms *looks* more impressive than it actually is. The spiral
arms contain the brightest, youngest stars in a spiral galaxy, stars that
have only recently (that is, a few tens of millions of years ago) begun to
shine. Ironically, these brightest stars are completely useless as centers
for life because they burn themselves out in much less time than we think
it would take for life to develop on any planets that might orbit them. Older
stars in a spiral galaxy, such as our sun, appear both between and within
the arms of a spiral galaxy, giving the disk a fairly even density of stars.

Why, then, do the youngest stars appear only inside the spiral arms of
a spiral galaxy? These galaxies owe the existence of their spiral arms to
a rotating wave pattern that circles the galaxies' centers (Figure 3-4). The
pattern consists of regions of alternating higher density regions (the *arms*)
and lower-density regions (interarm regions). This rotating pattern, in
which the arms have only a slightly greater density than the interarm re-

Fig. 3-3 The central regions of the great spiral galaxy in Andromeda, photographed in visible light (top), show a crowded mixture of stars, gas, and dust. A photograph taken in X-rays (bottom) can penetrate this mass of material to reveal sources of X-rays within the heart of the galaxy. The X-ray photograph shows the innermost one-fifth of the visible-light picture.

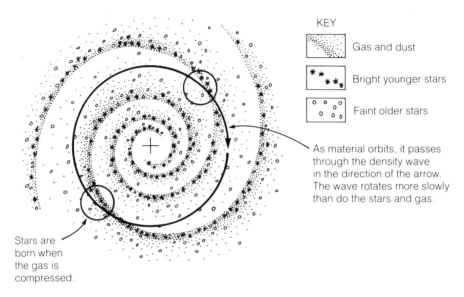

KEY

Gas and dust

Bright younger stars

Faint older stars

As material orbits, it passes through the density wave in the direction of the arrow. The wave rotates more slowly than do the stars and gas.

Stars are born when the gas is compressed.

Fig. 3-4 As the spiral density-wave pattern rotates around a galaxy, it continually compresses clouds of gas, thus starting the formation of a new generation of stars from interstellar matter. The galaxy's spiral structure, outlined by the youngest stars, therefore echoes the recent passage of the density wave.

Fig. 3-5 As a gas cloud passes through the spiral density wave's most compressed part, it can fragment into clumps that become protostars.

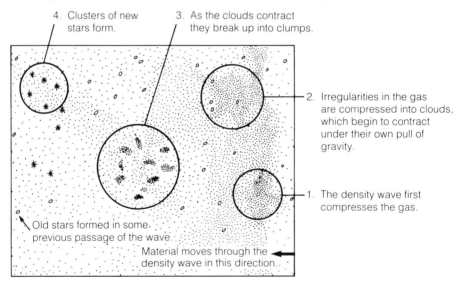

4. Clusters of new stars form.

3. As the clouds contract they break up into clumps.

2. Irregularities in the gas are compressed into clouds, which begin to contract under their own pull of gravity.

1. The density wave first compresses the gas.

Old stars formed in some previous passage of the wave.

Material moves through the density wave in this direction.

An enlarged section of the density wave, showing steps in the sequence of star formation.

gions, resembles the alternating ripples in water waves, except that we replace the water by the stars plus the diffuse gas spread among the stars. Also, the water-wave pattern spreads outward from a disturbance in the water, while the density-wave pattern in a galaxy rotates around and around the galaxy.

Even though the density of matter within the galaxy's spiral arms does not exceed the density outside the arms by much, just this modest increase in density has important effects on the clouds of gas in the disk of the galaxy. The increase in density within the arms means that the average pressure must increase on the gas clouds that wander through interstellar space. When interstellar clouds enter the denser part of the wave pattern, the sudden increase in the pressure around them provokes the fragmentation of the clouds into smaller clumps of matter that can condense into stars (see p. 91). As a result of this process, young stars are born inside spiral arms, which therefore provide a giant cosmic nursery, with the youngest stars located closest to the leading boundary between the arms and the interarm regions (Figure 3-5).

The stars within a spiral galaxy, as well as the gas clouds from which they form, orbit around the galactic center in much the same way as the planets in our solar system orbit the sun (Figure 3-6). Each of these stellar orbits reflects a balance for that particular star between the *gravitational pull* from the material closer to the galactic center and the star's *momentum,* or tendency to move in space in a straight line. Most of the stars in a spiral galaxy have nearly circular orbits around the center, as our sun does in our own galaxy.

Fig. 3-6 Most of the stars in a spiral galaxy orbit the galaxy's center in near-perfect circles, bobbing up and down by a few hundred parsecs several times per orbit. A few stars move in highly elongated orbits.

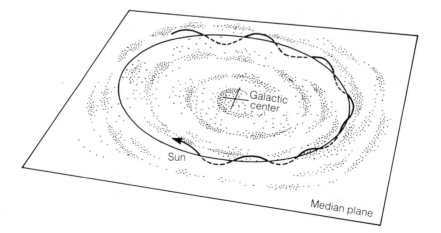

Elliptical Galaxies

In contrast to the spiral galaxies stand the *elliptical galaxies,* the other major galactic type. Elliptical galaxies do not possess a flattened disk of stars; instead, their stars spread into an almost spherical configuration, sometimes elongated into an ellipsoid (see Figure 3-7). Because elliptical galaxies have no structural pattern comparable to spiral arms, one elliptical looks much like another. Elliptical galaxies, however, are almost as numerous as spirals, and a giant elliptical can contain as many or more stars as the largest spiral galaxy.

Elliptical and spiral galaxies differ not only in structure but also in the amount of interstellar gas and dust that they contain. Interstellar gas, mostly hydrogen atoms and hydrogen molecules, represents the remnants of the original matter from which stars have condensed, gradually enriched by matter ejected from exploding stars. In elliptical galaxies, however, star formation appears to be over, for these galaxies contain almost no gas among their stars. Nor do ellipticals show any stars as young as the youngest stars in spiral galaxies. The spirals *do* have 5 to 10 percent of their total mass in interstellar gas clouds, and even now some stars are forming, or have recently formed, from the interstellar gas in spirals such as our own Milky Way.

Fig. 3-7   The giant elliptical galaxy M 87, the largest galaxy in the nearby Virgo cluster of galaxies, has more stars (about 3 trillion) than any other galaxy we know. This galaxy is about 15 million parsecs from us.

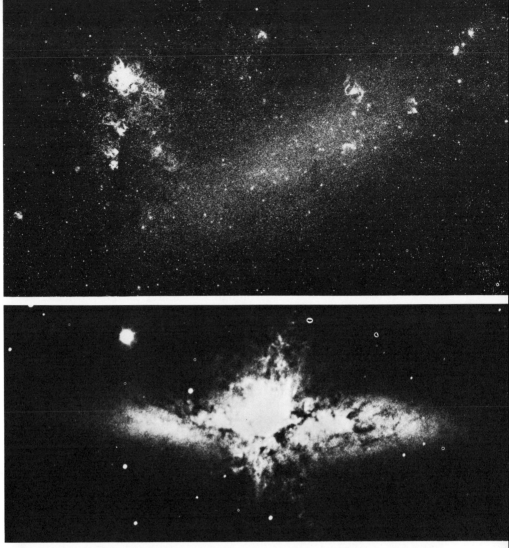

Fig. 3-8 The Large Magellanic Cloud (top), one of the two irregular companions of our Milky Way, is about 60,000 parsecs from us. The irregular galaxy M 82 (bottom), about 4 million parsecs away, shows a barlike arrangement of stars but contains the great amounts of gas and dust typical of irregulars.

Irregular Galaxies

In addition to the spirals and ellipticals, a third type of galaxy exists, the *irregular galaxies*. As their name indicates, irregular galaxies cannot fit into the two major categories, which together account for 90 percent of all the galaxies that we see. Irregular galaxies show neither the disklike flattening of spirals nor the smooth ellipsoidal shape of ellipticals. Two well-studied examples of irregular galaxies appear in Figure 3-8: the spindle-shaped galaxy M 82 and the satellite of our own galaxy called the Large

Magellanic Cloud.[1] Like spiral galaxies, irregular galaxies contain many clouds of gas and dust strewn among their stars, with as much as 20 to 50 percent of their total mass in the form of interstellar gas, not yet formed into stars.

The Formation of Galaxies

Astronomers think that all of these galaxies, save perhaps a few of the irregulars, have existed as galaxies for billions of years. Galaxy formation apparently began soon—less than a few billion years—after the big bang, 15 to 20 billion years ago. Thus, galaxies formed sooner rather than later in the history of the universe, though we cannot tell now whether they formed 1 billion, 3 billion, or even 5 billion years after the big bang. What does seem clear is that elliptical galaxies have always been elliptical and spiral galaxies always spiral (or at least disk-shaped) in their structure.

Galaxies in formation, called *protogalaxies* (*proto* means "earliest form"), contracted in size from vast clouds of gas and dust. We do not know how the original condensations that became protogalaxies ever got

Fig. 3-9 Condensations that formed while a protogalaxy was much larger than the galaxy's present size will have larger, more elongated orbits than the clumps that condensed later on.

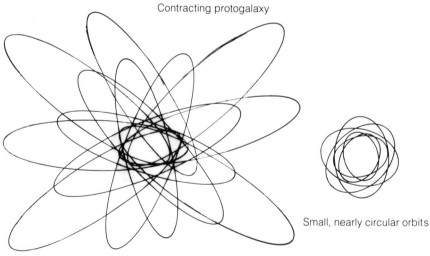

Contracting protogalaxy

Small, nearly circular orbits

Large, elongated orbits

[1]The most famous galaxies and star clusters bear the numbers given to them by Charles Messier, who catalogued galaxies, nebulae, and star clusters during the 1780s. The NGC galaxies have entries in the *New General Catalogue* of galaxies, compiled during the 1890s.

started, but we do know that once condensations form, gravitational forces will tend to contract them further. As a given protogalaxy contracted, it may have acquired some overall spin, which gave each part of the protogalaxy some orbital motion around the center of the contraction (Figure 3-9). Thus, instead of all falling together into one central lump, the individual parts of the protogalaxy, destined to become star clusters and stars, each ended up on orbit around the galactic center. The subcondensations that gained their own identity early on in the contraction acquired the most elongated orbits, while those that condensed later, when the protogalaxy had shrunk further, have almost circular orbits (Figure 3-9).

Star Clusters

The first objects to form as the protogalaxies contracted toward their present sizes were the *globular star clusters,* groups of many thousands of stars within a few parsecs of each other (Figure 3-10). Globular clusters are subunits that belong to galaxies, but because they began to form when the protogalaxy was larger than the galaxy it became, these clusters orbit around the galaxy's center on highly elongated trajectories, sometimes reaching enormous distances from the center (Figure 3-11). The gas in the protoclusters that became globular clusters must have undergone further fragmentation into individual protostars, since today we see only stars, and no gas, within a globular cluster. Most of the stars in globular clusters

Fig. 3-10 This globular star cluster, called M 13, is about 8000 parsecs from the sun, contains about a million stars and has a diameter of 10 parsecs. All of these stars are about 10 billion years old. The Arecibo message (page 402) was sent toward M 13 in 1974, and will arrive in about 25,000 years.

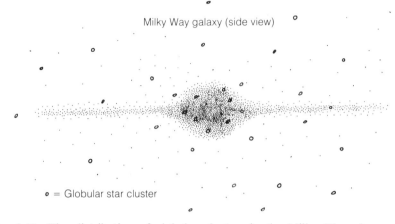

Fig. 3-11 The distribution of globular clusters in the Milky Way shows many clusters much farther out from the center than the other stars in the galaxy. Furthermore, the clusters do not concentrate heavily toward the galaxy's median plane.

rank among the oldest-known stars (see page 93), and none of them is as young as the bright stars that outline the arms of a spiral galaxy. Still, many of these stars have a greater absolute brightness than the sun does. If we lived on a planet that orbited a star in a globular cluster, the concentration of stars would provide us with dozens of stellar neighbors, each shining as brightly as the full moon on Earth!

Within the flat disk of spiral galaxies, enveloped in the web of globular clusters (Figure 3-11), we find a different type of star cluster, the *open cluster*. Open clusters, about the same size as globular clusters (a few parsecs in diameter) contain only a few hundred, or at best a few thousand, stars, rather than the hundreds of thousands of stars that fill a globular cluster. The open clusters have formed, as we can tell from their locations, *after* the galaxy contracted to its disklike shape. (Elliptical galaxies contain no open clusters, though they have many globular clusters around them.) Open clusters do not persist as clusters for the full lifetime of a galaxy (about 12 billion years, so far, in the case of the Milky Way). Unlike globular clusters, which contain enough mass to hold themselves together as separate units for billions of years, open clusters contain too few stars to remain as compact masses over the entire life of a galaxy. After a dozen or so galactic rotations, each of which takes a couple of hundred million years, the stars in an open cluster will merge with the other stars in the disk of the galaxy. Our sun, for instance, probably formed along with several hundred other stars as part of a loose *association* some 4.6 billion years ago, but we remain entirely ignorant of where the sun's brothers and sisters may be now: They are certainly much farther from us than our closest stellar neighbors.

Fig. 3-12 The Pleiades are an open cluster of about a hundred stars, all about 120 parsecs from the sun. Dust among the stars reflects some of the light of the brightest stars. The cluster's diameter equals about 4 parsecs.

A typical open cluster that we can see now, such as the Pleiades (Figure 3-12), will have an age measured in tens of millions of years, and will still show some remnants of the gas and dust from which the individual stars condensed. Since we think that life requires at least a few hundred million, if not billions, of years for its origin and development, young open star clusters do not appear to be good places to search for extraterrestrial life. *All* the stars in the Pleiades are, we think, far too young for life to have begun on any planets that may orbit around them. Our hopes rest not with the young, bright members of the stellar population but with the quieter, older stars that permeate a galaxy such as our own.

Radio Galaxies

Among the millions upon millions of galaxies accessible to our telescopes, we find a few exceptional ones that produce great quantities of radio waves. Since stars emit only a small fraction of their nuclear energy as radio waves, and since galaxies consist mainly of stars, we must be dealing with peculiar galaxies indeed. Astronomers give the name *radio galaxies* to those galaxies that radiate as much (or even more) energy per second in radio photons as they do in visible-light photons. Compare this with the Milky Way, a typical giant spiral galaxy, which produces *one millionth* as much radio energy as visible-light energy each second, and we can see how exceptional the radio galaxies must be.

Radio galaxies apparently owe their tremendous outflow of radio photons to violent events inside them. By studying the details of the photons emitted by these galaxies, astronomers have concluded that the radio waves from most radio galaxies arise from the *synchrotron* process. This name, bestowed in honor of the particle accelerators where scientists first observed the process in detail, describes the fact that *charged particles moving at almost the speed of light in a magnetic field will produce photons whenever the particles change their velocity, or their direction of motion, or both.* The production of photons robs the particles of some energy; without an additional energy input, the charged particles will soon slow down and cease to produce photons by the synchrotron process. From the fact that this process occurs in radio galaxies, we can conclude that the galaxies contain many charged particles (probably electrons) that are moving *at almost the speed of light in magnetic fields.* Thus, some sort of violence within radio galaxies has accelerated the particles to these enormous velocities, even though we cannot specify the cause of this violence.

Fig. 3-13 The distribution of radio-wave emission from the radio galaxy Cygnus A (see Fig. 3-15) shows the most intense emission from regions far outside the galaxy that we see in visible light, which is contained within the box at the center of the radio map.

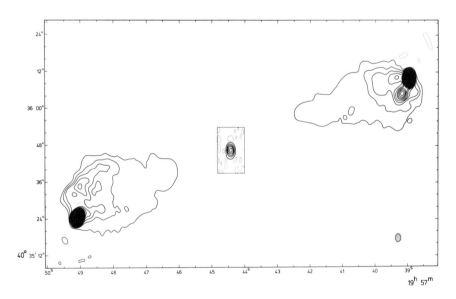

When we study the radio waves from radio galaxies in detail, we usually find that the photons come from large regions far from the center of the galaxy, and indeed often outside the galaxy's visible stars (Figure 3-13). The most widely accepted theory of the cause of radio galaxies suggests that the galaxies' centers are places where particles receive tremendous accelerations. Somehow, certain peculiar galaxies shoot fast-moving particles in opposite directions. The reason that the number of radio photons produced by synchrotron emission often peaks near the outer boundary of the *radio clouds,* as shown in Figure 3-13, is that the particles encounter intergalactic gas. To clear this gas out with fast-moving particles becomes progressively more difficult, so a "snowplow" effect slows the particles, which tend to pile up near the outer edge of the clouds.

Quasars

Quasars, or quasistellar radio sources, burst upon the world of astronomers in 1963, shattering once again any assumption that we were well on our way to explaining all that we observe in the sky. Quasars may be either the most distant objects known and the most powerful sources of photon emission, or they may be satellites of relatively ordinary galaxies, in which case the photons from quasars cannot be explained by the known laws of physics. We shall consider first the appearance of quasars, and then the evidence that has been gathered for the two possibilities of their distances. As the observational evidence has accumulated, most astronomers have concluded that quasars in fact are farther from us than any other objects we can see, but a determined minority has argued that the more compelling observations are those which tend to demonstrate the association in space between quasars and not-so-distant galaxies.

Quasistellar radio sources owe their name to the fact that most of them emit large amounts of radio waves. When astronomers located these radio sources accurately on visible-light photographs, they found points of light similar to stars (Figure 3-14). The light from these quasistellar images, spread into its spectrum of colors, astonished astronomers by showing *none* of the spectral features familiar to them from their study of stars. After a time of confusion, the apparent explanation dawned on the astronomer Maarten Schmidt: The light from the first two quasars to be discovered had such large redshifts that spectral features had been shifted all the way from the yellow part of the spectrum into the red! The redshifts for these two quasars, interpreted as the result of the Doppler effect, indicated recession velocities equal to 15 and 30 percent of the speed of light!

By applying Hubble's Law, astronomers calculated that these velocities imply distances to the two quasars of 900 million and 1.8 billion parsecs. This put the quasars as far away as the most distant galaxies, yet the

Quasi-stellar Radio Sources

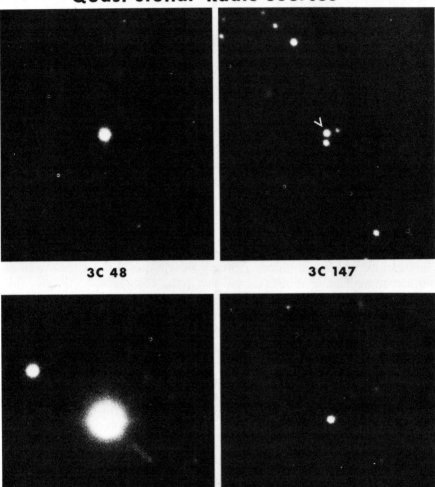

3C 48   **3C 147**

3C 273   **3C 196**

Fig. 3-14 Four well-known quasars look much like stars on visible-light photographs, though we can see a jet of material apparently being thrown out of the
quasar 3C 273.

quasars' apparent brightnesses equaled those of some large galaxies 100
times closer to us. Still more mysterious, quasars must clearly be smaller
than galaxies, for they appear to be point sources of light, while even the
most distant of galaxies appear as fuzzy, extended blobs of light (Figure
3-15).

In recent years, astronomers have been able to detect infrared light and X-rays from quasars, and have found that some quasars emit even more energy per second in these kinds of photons than they do in visible light or radio waves. When we add all the observed forms of photon energy, we find that some quasars produce 100,000 times more energy per second than a giant galaxy does, provided that our estimate of the quasars' distances from us is correct. An additional puzzle about quasars arises from the fact that some of them vary in brightness over a period of days, weeks, or years, while galaxies show no such variation.

No one knows how quasars manage to emit so much energy, or why their tremendous energy output varies in a relatively brief time. Quasars may represent some early stage in the formation of galaxies, since their enormous distances from us imply that we see them as they were billions of years ago. (The most distant quasars known have distances of about 12 billion light years from us!) We can imagine that young galaxies could,

Fig. 3-15 Even a distant galaxy such as Cygnus A, which is more than a billion light years away, appears as a fuzzy blob of light in this enlarged photograph. A quasar at the same distance from us appears point-like rather than fuzzy and must therefore be much smaller than Cygnus A.

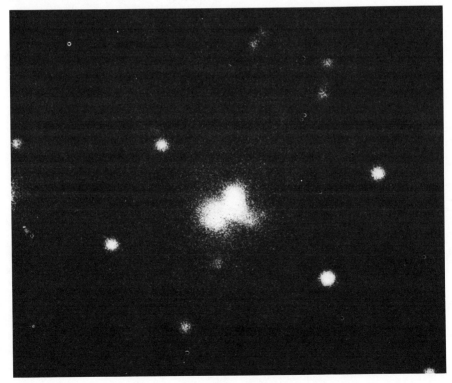

perhaps, produce tremendous numbers of exploding stars, or *supernovae,* that could provide an enormous energy output for some millions of years, but this remains merely an hypothesis, as yet unproven.

Some astronomers have suggested that quasars' rates of energy output lie far below those we have described above, because their distances from us have been grossly overestimated. If quasars are, say, 100 times closer to us than we thought, then we will have overestimated their true brightnesses by 10,000 times in deriving a rate of energy output from the quasars' observed apparent brightnesses. The astronomers who believe that this has been done rest their case on the fact that quasars often appear on the sky close to peculiar galaxies (Figure 3-16). These galaxies, though a bit unusual in their structure, have ordinary redshifts, corresponding to recession velocities of a few percent of the speed of light. Yet the quasars near them on the sky have redshifts 10 or 20 times greater!

How can we explain this apparent contradiction? Do we just happen to see the quasars, far more distant than the galaxies, in almost the same direction? Statistical arguments suggest that this is unlikely. But if the quasars are located in space close to relatively nearby galaxies, what produces their enormous redshifts? The Doppler effect provides the only reasonable explanation, yet how can we then explain that we see only redshifts (recession), and never blueshifts (approach)? And how could matter be shot (always away from *us!*) at such enormous velocities, yet remain a coherent object?

Fig. 3-16 This negative print shows a somewhat peculiar spiral galaxy with a quasar just below it (arrow). Either this is a chance line-up of a quasar and a much closer galaxy, or else the quasar and galaxy are indeed close to one another in space.

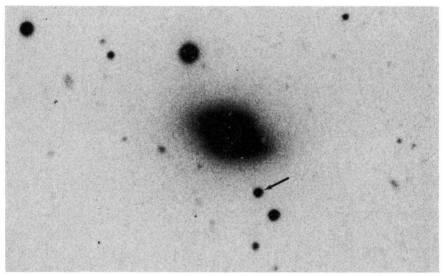

The answer to these questions is: Nobody knows. Fifteen years have not been enough to determine the distances to the quasars, nor to unravel their basic nature, nor their means of producing so much energy. Perhaps the quasar puzzle holds the key to some new part of the domain of astrophysics, some new way to obtain large redshifts in situations as yet unknown to us, or some new way to produce tremendous amounts of energy, if quasars are as far away as their redshifts alone would imply. Another 15 years—less, if we are lucky—may help us resolve our difficulties in understanding the faraway parts of the universe represented by the quasistellar objects. For now, we feel safe in saying only that they seem to represent a natural rather than an artificial astronomical puzzle, since we cannot begin to understand how a civilization would make a quasar.

Summary

Matter no longer spreads evenly through space, as it did in the early universe, but instead appears in clumps called stars, which themselves clump into galaxies and galaxy clusters. Such clumps appear to have bound themselves together through their self-gravitational forces early in the history of the universe and to have contracted toward their present sizes as the rest of the universe continued its expansion.

Spiral galaxies, which number about half of all galaxies, show a disklike distribution of matter, in which the youngest, brightest stars outline the distinctive spiral arms. Elliptical galaxies, the other major type, are far less flattened than spirals and have apparently turned all of their original gas and dust into stars. Spiral galaxies still have 5 or 10 percent of their mass in interstellar gas, while irregular galaxies, with no recognizable structure, have 20 to 50 percent of their mass in interstellar matter.

The small fraction of stars that appear in star clusters represent two distinctly different stellar types. Globular clusters, whose elongated orbits can carry them much farther from a galaxy's center than most stars ever go, represent the oldest parts of the galaxy, the first to form as the protogalaxy contracted. Open clusters, in contrast, are found only in spirals and irregulars, not in elliptical galaxies, and always near the plane of symmetry of spiral galaxies. These clusters usually consist of young stars, such as those in the Pleiades, and have only a few hundred members rather than the million or so stars in a globular cluster.

Quasistellar radio sources, or quasars, appear as points of light on photographs rather than as the fuzzy blobs that show the size of galaxies; they produce vast quantities of infrared and visible light, and of X-rays, in addition to their radio emission. Quasars' visible-light spectra show the largest redshifts yet observed. If these redshifts arise from the expansion of the universe, then quasars must be the most distant objects known, as well as the most powerful sources of photon emission. Many quasars, however, appear on the sky close to peculiar-looking galaxies. If the quasars are in fact close to these galaxies in space, then quasars are about a hundred times closer to us than we thought, and their enormous redshifts then become a mystery as yet unsolved by astrophysics.

Questions

1. Why do galaxies cluster together instead of appearing strewn evenly through space? Why do stars cluster into galaxies?

2. What are the key differences between spiral and elliptical galaxies? How do these galaxy types resemble each other?

3. What importance does the "density-wave pattern" have for the formation of stars in a spiral galaxy?

4. Why do the youngest stars in a spiral galaxy always appear in the spiral arms?

5. Why don't all the stars in a galaxy end up at the center of the galaxy? Why don't all the stars fly off into space?

6. What are the differences between the two major types of star clusters in our galaxy? Do similar star clusters appear in elliptical galaxies?

7. The galaxies M 87 and NGC 7793 have almost the same apparent brightness, but M 87 is four times farther away from us than NGC 7793. How do the two galaxies compare in their *true brightness*?

8. Why do most astronomers believe that quasars are the most distant objects known? Why do some astronomers disagree?

9. The quasar 3C 147 shows a redshift in its visible-light spectrum of 0.55; that is, all the *wavelengths* in the spectrum are 55 percent longer than they would be in a terrestrial laboratory. How have the *frequencies* and *energies* of the light in the quasar's spectrum changed?

10. The mathematical relationship between the observed photon energy and the original photon energy for the Doppler effect is

$$\frac{\text{Observed energy}}{\text{Original energy}} = \sqrt{\frac{1 - (v/c)}{1 + (v/c)}}$$

in which v is the velocity of recession and c is the velocity of light. What recession velocity does the quasar 3C 147 have, if we assume that its redshift arises from the Doppler effect?

11. The largest known radio galaxy, 3C 236, shows two regions of radio emission 5 million parsecs apart. If this radio emission arises from particles that were shot out from the galaxy between them, what is the minimum time interval that must have elasped between the ejection of

particles from the galaxy and the production of the radio waves that we observe?

Further Reading

Burbidge, Geoffrey, and Hoyle, Fred. 1970. The problem of the quasi-stellar objects. In *Frontiers in astronomy,* ed. by Owen Gingerich. San Francisco: W. H. Freeman and Company.

Field, George, Arp, Halton, and Bahcall, John. 1973. *The redshift controversy.* Menlo Park, Calif.: W. A. Benjamin.

Golden, Fred. 1976. *Quasars, pulsars, and black holes.* New York: Charles Scribner's Sons.

Hubble, Edwin. 1936. *The realm of the nebulae.* New York: Dover Books.

Sandage, Allan. 1960. *The Hubble atlas of galaxies.* Washington, D.C.: Carnegie Institution of Washington.

Shapley, Harlow. 1972. *Galaxies.* 3rd ed. Rev. by Paul Hodge. Cambridge, Mass.: Harvard University Press.

CHAPTER 4

Interstellar Gas and Dust

Strewn among the stars in spiral galaxies (and in irregular galaxies as well), a vast quantity of gas and dust, often clumped into *interstellar clouds,* represents the material from which all the stars and planets have formed. Even today stars are still forming in our own galaxy, and astronomers can sometimes see the infrared light from a protostar in its final formation process.

The recent discovery that interstellar clouds contain many types of molecules, with as many as a dozen individual atoms per molecule, has shown that complex molecules can assemble themselves in these clouds in much the same way that similar molecules formed on the primitive Earth. Some astronomers have even suggested that life itself began in interstellar clouds, to be seeded later onto planets. Although this last hypothesis has not received much acceptance, all astronomers agree that the interstellar gas and dust represents the best place to study the birth of stars as well as the formation of a wide variety of molecules in space (Figure 4-1).

Probing the Interstellar Medium

Astronomers first discovered interstellar matter by observing the "milky way," the band of diffuse light that circles the sky (Figure 4-2). This hazy glow arises from the combined contribution of millions of stars in our galaxy, each too distant to be visible as an individual star without telescopic aid. After Galileo first showed that the milky way consists of stars, later astronomers realized that our solar system lies *within* a galaxy of stars, the Milky Way galaxy, so that we see the galaxy's central plane as a band of light. But at certain points around the milky way, most clearly in

Fig. 4-1 An edge-on view of a spiral galaxy shows a dark stripe produced by the absorption of starlight. Such dark lanes reveal the presence of interstellar dust, concentrated toward the plane of the galaxy.

the constellation Cygnus, the band seems to split into two pieces (Figure 4-3). The Milky Way galaxy, however, does not divide at these points; instead, the apparent split arises because invisible dust particles absorb the light from stars behind the obscuring matter. Thus we can "see" interstellar particles by observing the *absence* of light in parts of the milky way.

More detailed studies of the absorption of starlight have distinguished the effects of dust particles, each made of millions of atoms, from those of much smaller atoms or molecules. Dust grains tend to absorb all colors of starlight, though they absorb blue light more efficiently than red light; atoms and molecules absorb only particular frequencies or colors. The analysis of the spectrum of starlight that has passed through clouds of interstellar matter can therefore determine the different kinds of atoms and molecules in the clouds, and also the number of these different sorts of particles, as well as the general properties of the interstellar dust grains.

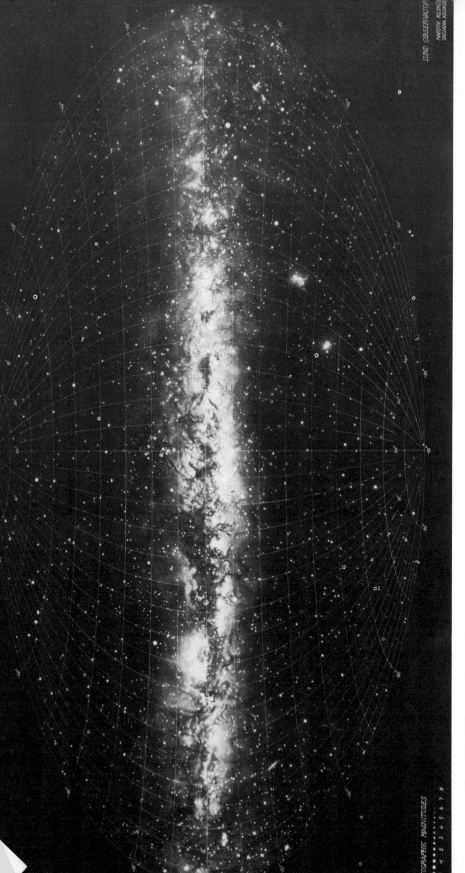

PHOTOGRAPHIC MAGNITUDES
0 1 2 3 4 5 6 7 8

Fig. 4-2 This mosaic drawing, made from many photographs of the night sky, shows the band of light called the "milky way" that seems to circle the skies.

Fig. 4-3 The "milky way" seems to split into two parts as it passes through Cygnus. In reality, dust grains between ourselves and the faraway stars absorb the starlight to produce an apparent splitting.

Fig. 4-4 Young, hot stars inside interstellar gas clouds produce great quantities of ultraviolet light, which can ionize hydrogen atoms close to the stars. As the electrons recombine with the ions, they produce visible-light photons that make the entire ionized region glow with visible light.

Some clouds of interstellar gas and dust do not happen to lie conveniently between ourselves and a bright star. Other clouds are so dense that no starlight can penetrate them, no matter how bright the stars behind them may be. To study these clouds, astronomers use the fact that interstellar matter can *emit* various kinds of photons, so that the clouds can be observed directly. For example, interstellar gas close to young, hot stars will be lit by the starlight energy that the gas absorbs. Still more important, many kinds of molecules, and also hydrogen atoms, can emit radio waves by themselves, without needing any energy input from nearby stars.

The first example, the interstellar gas around young, hot stars, produces the giant, glowing gas clouds called *H II regions*.[1] In these clouds, the atoms of gas have all been ionized—stripped of one or more electrons— by the intense flow of ultraviolet light from the hot stars that the clouds contain (Figure 4-4). As the electrons recombine with the ions, each atom can emit one or more photons of visible light as the electron jumps from a larger orbit into a smaller one. Thus, the entire cloud of gas, whose atoms

[1]The symbols "H II" stand for ionized hydrogen—hydrogen atoms that have each had their electron knocked loose. Neutral or un-ionized hydrogen atoms bear the name "H I" in astronomical language.

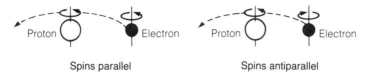

Spins parallel   Spins antiparallel

Fig. 4-5 The electron that orbits around the proton in a hydrogen atom can have its spin either parallel (in the same direction) or antiparallel (spins in opposite directions) to the proton's spin.

are repeatedly ionized, recombined, and ionized again, can glow with the light emitted as part of the recombination process. The stars that power this entire cycle may appear relatively faint in *visible* light, because most of the photons emitted by the star have ultraviolet energies, which in fact are the energies needed to ionize hydrogen and other common atoms in the gas around the star.

The second possibility for studying the interstellar medium embraces the radio waves that many types of molecules can emit. Among atoms, the ability to emit radio waves is rare, but remarkably enough, hydrogen atoms, the most abundant atoms in our galaxy and in the universe, possess this ability. Hydrogen atoms, each with one proton and one electron, can emit radio waves because the two particles each resemble a tiny, spinning magnet (Figure 4-5). The rules of atomic physics dictate that these spinning magnets can be only parallel or antiparallel to one another.

When the proton and electron spins are parallel, the atom has a bit more energy than in the antiparallel case. Hence, a parallel-spin atom can flip the electron spin into the antiparallel position (Figure 4-5). As the atom does this, it emits a radio-wave photon, whose small energy equals the difference in energy between the two spin configurations. This photon's frequency will always be 1420 megahertz, perhaps the most significant frequency the universe produces; the corresponding wavelength of the radio emission equals 21.1 centimeters.

In interstellar space, hydrogen atoms by the trillions upon trillions emit radio waves of 1420-megahertz frequency. Once an atom has reached the antiparallel-spin state, mild collisions between atoms can bump the electron's spin back into the parallel position, from which the atom can again emit a 1420-megahertz photon. With carefully designed antennas and receivers, astronomers can determine the number of hydrogen atoms in a given direction, by measuring the intensity of the 1420-megahertz radio waves from that direction. In addition, these radio astronomers can estimate the distances to various groups of hydrogen atoms by observing the Doppler effect on the radio waves. The amount of change in the frequency of the radio emission can be related to the atoms' velocities relative to ourselves and thus to their distances from us, because of the way that our spiral galaxy rotates.

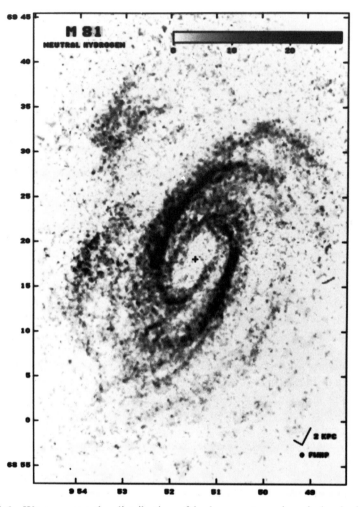

Fig. 4-6 We can map the distribution of hydrogen atoms in spiral galaxies, such as M 81 shown here, by observing the intensity of 1420-megahertz radio emission from different parts of the galaxy. The darkest regions of the map show the location of the most intense radio emission.

By observing 1420-megahertz radio waves, we can map the distribution of hydrogen atoms in our own and other spiral galaxies (Figure 4-6). Since hydrogen is the most abundant element in all galaxies, such maps already tell us a great deal about the distribution of interstellar matter. We should, however, also consider interstellar *dust grains,* as well as those interstellar clouds that are dense enough for the atoms to join into *molecules.*

Dust grains are specks of matter that each include a few million atoms, with diameters of a few millionths of a centimeter. These grains apparently form in the atmospheres of cool stars, which gently blow them out into interstellar space as the stars age. The dust grains' composition, although still a bit of a mystery, embraces mostly silicon, carbon, and oxygen atoms, perhaps with an outer mantle of hydrogen and water molecules.

Interstellar Molecules

The molecules that have formed in the interstellar medium have special interest to the search for life, because the initial stages in forming living creatures must begin with the formation of molecules from the basic mixture of atoms. In interstellar clouds of gas and dust, the most abundant molecules are hydrogen (H_2), each made from two hydrogen atoms, the most abundant type of atom.

The formation of hydrogen molecules occurs by a process that may seem roundabout but that nonetheless appears to be the way in which hydrogen atoms can pair up to make hydrogen molecules. Individual hydrogen atoms that hit dust grains within the interstellar clouds will tend to stick for a while on the surface of the grain (Figure 4-7). During this time, the atoms can combine with other hydrogen atoms that have also stuck to the grain. The resulting molecule will most likely pop off the grain as part of the formation process (Figure 4-7). This process, according to astronomers' calculations, works well for hydrogen molecules, but other forms of molecules will not appear in significant numbers from the grain-surface reactions.

In contrast, other common interstellar molecules, such as carbon monoxide (CO), ammonia (NH_3), and formaldehyde (H_2CO) have apparently formed in a quite different way, as first two atoms combined with another,

Fig. 4-7 The best way to form hydrogen molecules (H_2) in interstellar clouds seems to have hydrogen atoms run into dust grains and stick to them. While an atom is stuck to the surface of a grain, it actually wanders from point to point on the surface, and can meet another atom that has also stuck to the grain. The two atoms will then tend to form a hydrogen molecule, which will pop off the grain and reenter the general mixture of atoms and molecules in the cloud.

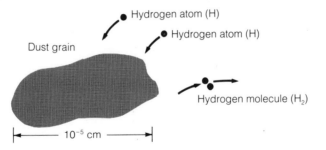

then a third atom joined the first two, and so forth, without any sticking to the surfaces of interstellar dust grains (Figure 4-8). To form molecules in this way, however, requires a fairly large density of matter—that is, a fairly large number of atoms per cubic centimeter. Otherwise, collisions among

Fig. 4-8 Molecules other than H_2 form through a rather complicated set of reactions that often begin with the ionization of a hydrogen atom by a fast-moving particle (misleadingly called a "cosmic ray"). The proton left behind from this ionization can snatch an electron from an oxygen atom, and once this occurs, the ionized oxygen can combine with one of the atoms in a hydrogen molecule to form an ionized molecule of OH. Such an OH ion is likely to interact with a hydrogen atom to make ionized water molecules. Through further interactions with hydrogen atoms and with electrons, these ionized water molecules can produce uncharged molecules of either water or OH. Other sorts of molecules can be formed through similar chains of reactions among the atoms and molecules that already exist in the interstellar medium.

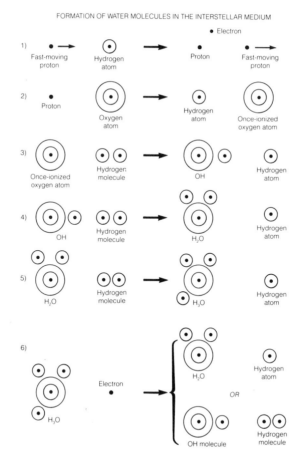

FORMATION OF WATER MOLECULES IN THE INTERSTELLAR MEDIUM

atoms will be so rare that only a few molecules can form, even during the billions of years that an interstellar cloud of gas and dust may exist.

Indeed, when we examine the distribution of interstellar matter within the Milky Way galaxy, our hypothesis about the formation of molecules finds confirmation. Within the general pancake-shaped arrangement of matter, interstellar gas and dust tends to collect into clouds, regions where the density of matter far exceeds the average density. Taking the galaxy as a whole, the interstellar matter averages only about *one atom per cubic centimeter,* while the interstellar clouds each contain at least 10 times this density of matter. But an important difference exists between two types of interstellar clouds, the ordinary clouds and the dense, or *molecular, clouds.*

Ordinary interstellar clouds have densities of matter that range from a dozen atoms per cubic centimeter up to a few hundred atoms per cubic centimeter. Such clouds typically contain dust grains as well, whose total mass may equal about 1 percent of the mass of the atoms. As we might expect from the way that hydrogen molecules form, the ordinary clouds do contain many hydrogen molecules. The ordinary interstellar clouds are the ones that can be studied by the absorption of starlight that we described on page 67. By carefully measuring the absorption produced by particular sorts of atoms, we can determine that these clouds have temperatures between 40 and 250 degrees above absolute zero, colder than Earth's surface but well above the zero degrees that would occur if the matter in the clouds did not receive some photon energy from other regions of space.

Each ordinary interstellar cloud contains a total mass of gas and dust that equals several hundred times the sun's mass, and may span a distance of a few parsecs, perhaps a few dozen parsecs for a large cloud. Thus, the sizes of ordinary interstellar clouds correspond roughly to the average distance between stars, or a little more. On the other hand, the average density of matter within such a cloud falls short of the average density of matter in a *star* by a factor of 10 billion trillion (10^{22})! Clearly, matter that has collected into interstellar clouds has a long way to go before it condenses into something as dense as a star.

Dense Interstellar Clouds

The second type of interstellar cloud, the dense or molecular cloud, presents a different mixture of constituents, testimony to the importance of a greater density of matter in the cloud. Each molecular cloud has a mass of at least a few thousand, and often hundreds of thousands, times the sun's mass, so these clouds far exceed the ordinary clouds in mass. However, because the matter in molecular clouds has a greater *density* than

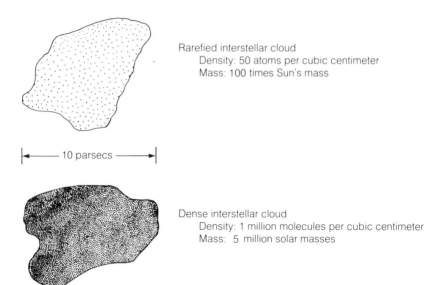

Rarefied interstellar cloud
   Density: 50 atoms per cubic centimeter
   Mass: 100 times Sun's mass

|←——— 10 parsecs ———→|

Dense interstellar cloud
   Density: 1 million molecules per cubic centimeter
   Mass: 5 million solar masses

Fig. 4-9 Dense molecular clouds, which contain many thousands of atoms or molecules per cubic centimeter, may thus contain a total mass hundreds of times greater than that in an ordinary interstellar cloud, which has about the same volume but a density of only a few tens of atoms per cubic centimeter.

that in ordinary clouds, the dense clouds do not occupy greater volumes than the ordinary clouds do (Figure 4-9). The density of matter in a molecular cloud usually equals about 1 million particles per cubic centimeter, about 10,000 times greater than the density in an ordinary cloud. Thus, the far greater mass of a molecular cloud fits into about the same volume of space as that of an ordinary cloud—a few parsecs across in both cases.

The greater density of matter within a molecular cloud gives atoms a better chance to collide with one another to form molecules, and it is precisely in these parts of interstellar space that we find molecules in great abundance. Thus, the molecular clouds have passed through the first, and perhaps the hardest, step on the road from atoms to life: They have already manufactured a great variety of molecules from individual atoms.

Let us take a look at a typical dense cloud, the best-studied example, the Orion molecular cloud. This concentration of gas and dust in the direction of Orion's sword has a million times the mass of the sun. Most of this matter has a temperature just a few tens of degrees above absolute zero, but in part of the cloud complex, the part we call the Orion Nebula, the density of matter has grown so large that stars have condensed and have recently begun to shine (Figure 4-10). The stars inside the Orion Nebula have ages of only a few hundred thousand years, far younger than a typical star, such as our sun, whose age is measured in *billions* of years.

Not only do we find *stars* inside the gas in the Orion Nebula; we also find regions nearby where stars have not yet formed but where they are

Fig. 4-10 The Orion Nebula, about a thousand parsecs away from us, contains several thousand solar masses of gas, illuminated by bright young stars within the gas. A great amount of absorbing dust hides the parts of the nebula below the center from our view. These regions, which emit great amounts of infrared light, contain stars that are forming right now.

now forming or are about to form during the next few tens of thousands, or hundreds of thousands, of years. These regions have densities of billions of particles per cubic centimeter, far more favorable to the formation of molecules than the densities of matter in ordinary interstellar clouds. Within the molecular clouds of the Orion Nebula complex, astronomers have found dozens of different types of molecules, ranging from small molecules such as carbon monoxide and cyanogen (CN) up to molecules as large as ethyl alcohol (CH_3CH_2OH).

The larger molecules draw our attention in the search for life, though they are not as abundant as the small ones, because they take us farther down the road to the complex molecules found in living organisms. With molecules such as methylamine (CH_3NH_2), we seem well on the way to forming the simplest *amino acids* (see page 152). Amino acids, the basic structural units in protein molecules, are certainly not themselves alive; nor has anyone found amino acids in dense interstellar clouds (though in fact the search for amino acids has not really even begun). The discovery of molecules made of as many as 11 atoms in dense interstellar clouds does, however, suggest that amino acids might well have formed there.

The simplest amino acid, glycine, contains 10 atoms; the next simplest, alanine, has 13 atoms; the other amino acids have from 14 to 26 atoms. Almost all these atoms are either hydrogen, carbon, nitrogen, or oxygen, with an occasional sulfur atom in some of the amino-acid molecules. Especially dense clouds, such as those in the Orion Nebula complex and at the center of our galaxy, provide the prime candidate areas in the search for interstellar amino acids. If amino-acid molecules do indeed exist in dense interstellar clouds, then we may well expect that amino acids have managed to form under a variety of conditions throughout our galaxy, and in other galaxies too.

Table 4-1 shows the 50 different types of molecules that have been detected, so far, in dense interstellar clouds. These molecules consist of the most abundant elements in the universe (with the exception of helium and neon, which do not combine with other atoms willingly)—hydrogen, oxygen, carbon, and nitrogen, plus silicon, and sulfur. Notice, in particular, the large number of molecules that contain one or more *carbon* atoms. Carbon atoms form the key to life on Earth, the backbone of our molecular structure. The importance of carbon atoms comes from each atom's ability to combine with as many as four other atoms and to form long chains in which a string of carbon atoms supports the entire molecule. The existence of still larger molecules built on chains of carbon atoms seems a fairly sure bet in dense interstellar clouds, but the possibility of *much* larger molecules, those with many dozens, or many hundreds of atoms, remains a question for open speculation.

The discovery of these fifty varieties of simple molecules in dense interstellar clouds has important implications in our search for life. First

TABLE 4-1

MOLECULES DETECTED IN DENSE INTERSTELLAR CLOUDS

Name of Molecule	Chemical Symbol	Year of Discovery
Methylidyne	CH	1937
Cyanogen	CN	1940
Hydroxyl	OH	1963
Ammonia	NH_3	1968
Water	H_2O	1968
Formaldehyde	H_2CO	1969
Carbon monoxide	CO	1970
Hydrogen cyanide	HCN	1970
Cyanoacetylene	HC_3N	1970
Hydrogen	H_2	1970
Methyl alcohol	CH_3OH	1970
Formic acid	HCOOH	1970
Ionized formyl radical	HCO^+	1970
Formamide	$HCONH_2$	1971
Carbon monosulfide	CS	1971
Silicon monoxide	SiO	1971
Carbonyl sulfide	OCS	1971
Methyl cyanide	CH_3CN	1971
Isocyanic acid	HNCO	1971
Methylacetylene	CH_3C_2H	1971
Acetaldehyde	CH_3CHO	1971
Thioformaldehyde	H_2CS	1971
Hydrogen isocyanide	HNC	1971
Hydrogen sulfide	H_2S	1972
Methanimine	H_2CNH	1972
Sulfur monoxide	SO	1973
Protated nitrogen ion	N_2H^+	1974
Ethynyl	C_2H	1974
Methylamine	CH_3NH_2	1974
Dimethyl ether	CH_3CH_3O	1974
Ethyl alcohol	CH_3CH_2O	1974
Sulfur dioxide	SO_2	1975
Silicon sulfide	SiS	1975
Acrylonitrile	H_2CCHCN	1975
Methyl formate	$HCOOCH_3$	1975
Nitrogen sulfide	NS	1975
Cyanamide	NH_2CN	1975

TABLE 4-1

MOLECULES DETECTED IN DENSE INTERSTELLAR CLOUDS

Name of Molecule	Chemical Symbol	Year of Discovery
Cyanodiacetylene	HC_5N	1976
Formyl	HCO	1976
Acetylene	C_2H_2	1976
Cyanohexatetrayne	HC_7N	1977
Carbon subnitride	C_3N	1977
Methyl nitride	CH_2NH	1977
Ketene	H_2C_2O	1977
Propionitrile	CH_3CH_2CN	1977
Carbon	C_2	1977
Cyanooctatetrayne	HC_9N	1978
Methane	CH_4	1978
Nitric oxide	NO	1978
Butadiynyl	C_4N	1978

of all, these molecules have formed under conditions quite different from those on planetary surfaces, yet in interstellar clouds we find the same sorts of molecules as those which we believe existed on the surface of our planet early in the Earth's history.

The fact that these molecules can assemble themselves in clouds which, even though we call them "dense," are far more rarefied than our atmosphere shows that we may expect to find these types of molecules widely distributed in the cosmos, since many more favorable sites should exist. We might conclude straightaway that we shall restrict ourselves unnecessarily if we look for life only on the surfaces of planets, and should consider the possibility of life in dense interstellar clouds.

Second, we notice that different types of molecules found in dense interstellar clouds have different degrees of relevance to life on Earth. Molecules such as methylamine (CH_3NH_2) have an intimate connection with the kinds of molecules found in terrestrial organisms, while molecules such as sulfur dioxide (SO_2) have far less relevance to the kind of life we find on Earth.

We shall follow the implications of the distinction between life-involved and nonlife molecules in Chapters 8 and 11. For now, we shall pursue the implications of the existence of so many types of molecules in dense interstellar clouds.

First of all, as Table 4-1 reminds us, all but six of the molecules that we have found in the interstellar medium were detected during the past decade. Hence, we may reasonably assume that many more molecular types await our discovery. No doubt exists that the simple molecules are likely to be far more abundant than the more complex molecules; thus, for example, hydrogen molecules outnumber all other types by a factor of more than 1000. The fact, however, that molecules are constantly forming (and coming apart) in interstellar clouds—the continuing *chemical* evolution of molecular clouds—shows that chemical reactions pervade our entire galaxy, even if they occur primarily in localized regions called "dense interstellar clouds."

Second, we must admit that we have no knowledge of how far this chemical evolution has gone in molecular clouds; that is, of what sorts of truly complex molecules may have formed there. Molecules with greater numbers of atoms become progressively more difficult to detect, especially if the abundance of these molecules falls below that of the less complex molecules. The chemical evolution in dense clouds, such as those in the Orion molecular complex or at the center of our galaxy may have proceeded to form much larger molecules than those listed in Table 4-1.

Third, we do not know how the existence of *interstellar* molecules relates to the existence of molecules on *planetary* surfaces. Our own planet, for example, has a great variety of molecular types on its surface, some apparently made by nonorganic processes during the Earth's 4.6-billion-year history, others made by the early stages of what we call life, still others made by human beings in complicated chemical reactions, unlikely to occur elsewhere without planned intervention into natural events. The question of *which* molecules may have existed before the Earth formed remains incompletely answered.

Did Life Begin in Interstellar Clouds?

Two well-known astronomers, Fred Hoyle and Chandra Wickramasinghe, have suggested that complex molecules could form in great quantities in dense interstellar clouds. The typical sorts of molecules that Hoyle and Wickramasinghe have in mind are polysaccharides, long-chain molecules made mostly of carbon, oxygen, and hydrogen atoms. The best-known examples of polysaccharides are cellulose molecules, the principal structural molecule in plants. Hoyle and Wickramasinghe suggest that interstellar clouds are loaded with cellulose, and that we should not ignore the possibility that life has begun in these clouds.[2] But more important still from a human perspective, these astronomers believe that *comets,* the most primitive objects in the solar system, perhaps also contain organic molecules of great complexity, and that comets may indeed contain living cells and viruses, formed from frozen interstellar molecules.

[2]See Fred Hoyle's science-fiction novel, *The Black Cloud* (New York: Signet Books, 1959).

We shall discuss comets in Chapter 12, but we can pause now to see the implications of this theory, if it should prove to be true. (Most astronomers do not think that this theory has much merit as a description of reality, but the crucial experiments await accurate sampling of a comet.) If comets should prove to be a storehouse of interstellar organic molecules—if, in fact, comets do contain primitive forms of life—then the possible interaction of comets such as Halley's comet with our Earth would become extremely important. Hoyle and Wickramasinghe assign the origin of life to the cometary lumps of gas, ice, and dust that condensed in molecular clouds such as those in the Orion complex. Comets then could seed planets with life, or at least with large molecules, as they pass by, and every close pass of a comet would carry the possibility of further seeding. Hoyle has recently suggested that outbreaks of epidemics, such as influenza and smallpox, arise from such close encounters with a comet, and that the age-old tradition that comets bring bad luck has it roots in similar epidemics.[3]

Although the theories proposed by Hoyle and Wickramasinghe await further testing, they serve to remind us that interstellar molecules may have a direct bearing on the origin of life. The molecules that were later incorporated in *planets* could hardly have survived the formation process; instead, the molecules almost certainly broke apart into their constituent atoms at that time. But comets, frozen lumps of old interstellar matter, could preserve the molecules that formed in dense clouds and might later have deposited some of these molecules on the surfaces of planets *after* they had formed. If this hypothesis proves correct, it provides a strong argument in favor of the idea that life should generally be about the same throughout our galaxy, since life should then have arisen from much the same kinds of molecules made in similar molecular clouds. If, however, the opposing hypothesis—that life truly began on Earth—turns out to be correct, then we should expect to find a greater diversity of types of life from planet to planet, since each planet would provide a specialized set of conditions within which life could begin. In either case, the interstellar gas and dust must have been the birthplace of stars and of the planets around them; whether or not life too began in these clouds remains an exciting mystery, to be debated until the time comes when we can resolve the issue by seeing just what comets, and molecular clouds, are made of.

Summary

Spiral galaxies such as our own Milky Way contain clouds of gas and dust, whose properties can be determined by the radio waves that certain atoms and molecules emit, and by the photons of particular energies that will be absorbed when starlight passes through an interstellar cloud. Through ra-

---

[3]See the article by Fred Hoyle, "Astrochemistry, Organic Molecules, and the Origin of Life" in *Mercury* 7:2 (Jan./Feb. 1978).

dio and visible-light studies of the interstellar medium, astronomers have found that the less dense clouds contain few molecules, having their gas (mostly hydrogen and helium) instead in the form of atoms. Young, hot stars within a cloud of gas will ionize most of the atoms to produce the H II regions that shine brightly as the atoms temporarily recombine.

In dense interstellar clouds, more than forty types of molecules have been discovered, ranging from hydrogen (H_2), the simplest and by far the most abundant, to molecules as complex as HC_9N, which contains a chain-like structure of nine carbon atoms. These more complex molecules, in many cases, resemble the basic building blocks of living matter on Earth. The fact that such molecules appear in "dense" interstellar clouds, which are far less dense than our atmosphere, tells us that molecules of at least this complexity seem to form naturally in relatively difficult situations. This conclusion in turn suggests that the basic molecules required for life may be widely distributed in interstellar space as well as on planets.

Questions

1. What do interstellar clouds consist of? Why do molecules appear only in the denser clouds?

2. What does the appearance of simple organic molecules such as formaldehyde in dense interstellar clouds imply for the origin of life on Earth?

3. Should the formation of molecules on Earth be more difficult than their formation in interstellar clouds? Why?

4. The ability of a given receiver of photons to resolve small angles varies in proportion to the diameter of the receiving antenna divided by the wavelength of the photons. How does the human eye, with an antenna diameter of 3 millimeters, compare with the Arecibo antenna, with a diameter of 300 meters (30,000 cm), if the human eye observes visible-light photons of 5×10^{-5} cm wavelength, and the Arecibo telescope observes radio photons of 50 cm wavelength? How large would the Arecibo telescope have to be to achieve the same angular resolution, when observing photons of 50 cm wavelength, that the human eye achieves in observing photons of 5×10^{-5} cm wavelength?

5. The ability of any antenna to gather photons for detailed study varies in proportion to the *area* of the antenna. How many times more photons can the Arecibo antenna detect each second than the human eye, if its effective diameter exceeds that of the human eye by 100,000 times?

6. Interstellar dust scatters photons of all visible-light frequencies, but the dust scatters blue-light photons more effectively than red-light photons. Will light from a star that passes through a thin cloud of interstellar dust emerge redder or bluer than before its passage? Why?

Further Reading

Bok, Bart, and Bok, Priscilla. 1976. *The milky way.* 4th ed. Cambridge, Mass.: Harvard University Press.

Gammon, Richard. 1977. *Chemistry between the stars.* Washington, D.C.: NASA, U.S. Government Printing Office.

Lovell, Bernard. 1973. *Out of the zenith.* New York: Harper & Row.

Sanders, R. H., and Wrixon, C. T. 1975. The center of our galaxy. In *New frontiers in astronomy,* ed. by Owen Gingerich. San Francisco. W. H. Freeman and Company.

Turner, Barry. 1975. Interstellar Molecules. In *New frontiers in astronomy,* ed. by Owen Gingerich. San Francisco: W. H. Freeman and Company.

Verschuur, Gerrit. 1974. *The invisible universe.* New York: Springer-Verlag New York.

Science Fiction

Hoyle, Fred. 1959. *The black cloud.* New York: Signet Books.

CHAPTER 5

Energy Liberation in Stars

As protogalaxies contracted to form galaxies, they must have developed subcondensations within them that became protostar clusters and protostars. Although we do not know in any detail how such smaller clumps of gas and dust ever formed, we can see the results of this process throughout the universe: Stars form the basic units of observable matter, shining with almost constant brightness for millions or billions of years.

Because we think planets offer the likeliest places for life to develop in the universe, and because stars provide the most abundant sources of energy for life, we have a direct interest in how stars evolve when we consider the chances for life elsewhere. The most important characteristics of a star, so far as life nearby is concerned, are the star's intrinsic *brightness* and its total *lifetime*. Many stars have too little absolute brightness to keep their planets warm, if the planets have orbits resembling those in the solar system, while a few stars shine so brightly that life would have a hard time surviving the stars' heat. By a significant irony of stellar evolution, *the stars with the greatest true brightness have the shortest lifetimes*. These stars burn themselves out long before life would have a chance to evolve on planets in orbit around them.

When we study the brightnesses and lifetimes of various stars, we conclude that our hopes for finding life on planets around them rest with the average stars, bright enough to keep planets warm, but not so bright that the star burns out before life can originate or evolve in a significant way. Our own sun furnishes a perfect example of such an average star: Brighter than 80 percent of all stars, it will nevertheless last for a total lifetime of 10 billion years. Half this time has already passed, during which life on Earth has appeared, evolving into a fantastic complexity that continues to increase.

Stellar Lifetimes

Why do some stars continue to shine for billions of years, while others have their energy-liberating lifetimes measured in mere millions of years? Why do some stars explode violently as supernovae, while most stars simply fade calmly into white dwarf obscurity? Astronomers have slowly resolved the answers to these questions by generations of patient observations of stars and by calculations that attempt to determine what happens deep inside stellar interiors.

The most far-reaching result of this research into stellar structure and evolution connects the total lifetime of any star with the star's *mass*: Stars with higher masses burn themselves out far more rapidly than stars with lower masses do. Large-mass stars turn out to be spectacular fireworks, roman candles with lifetimes that fall a thousand times below the sun's 10 billion years. These high-mass stars, which last for only a few million or perhaps a few hundred million years, are prime candidates for *supernova* explosions, violent catastrophes that put the stars out with a bang, not a whimper.

To understand why stars of different masses have widely different lifetimes, we must understand how all stars liberate energy through nuclear fusion reactions. We must also consider the options open to a star as it runs out of its basic nuclear fuel, which will lead us to discover that most of the elements present in the Earth and in our bodies were made in fiery stellar furnaces that later exploded to seed their ashes through the universe.

How Stars Liberate Energy

All stars that shine owe their energy output to the liberation of energy of motion from energy of mass. This conversion of one form of energy (*energy of mass*) into another (*energy of motion* or *kinetic energy*) follows Einstein's most famous equation, $E = mc^2$. This equation describes the amount of energy of mass E that any object with a certain mass m contains: We multiply the mass by the square of the speed of light, c. Every object with some mass therefore has an equivalent energy of mass, m times c^2, that can, at least in theory, be converted into energy of motion. For example, the energy of mass in a 5-gram nickel, 4.5×10^{21} ergs, equals the amount of kinetic energy consumed in the United States each minute.[1] Hence, we could supply the energy needs of the United States with just half a million nickels each year, *if* we could find a simple way to convert energy of mass into kinetic energy with complete efficiency.

---

[1] One erg is about equal to the energy of a housefly in full flight.

On Earth, this problem remains unsolved. Even in the universe at large, we rarely find complete efficiency in converting energy of mass into kinetic energy. Deep inside stars, however, where the temperature rises to tens of millions of degrees, nature has created billions of natural nuclear fusion reactors, shielded by thousands of kilometers of the stars' outer layers. Even though the conversion of energy of mass into energy of motion proceeds with only 1 percent efficiency in these natural fusion reactors, the resulting energy produces the starlight that sprinkles the sky as far as we can see. One of these reactors, inside our sun, provides the basic energy source that allows life to flourish on Earth.

Humans have not yet built a nuclear fusion reactor that would mimic the insides of stars, and discussions of "nuclear power" still center around the rare element uranium, of which some isotopes fall apart, or *fission,* with the release of energy of motion, leaving a dangerous, radioactive waste product. Uranium, which ranks among the least common elements both on Earth and in the universe, would never serve as the basic energy source for all the stars. Instead, stars use the most abundant element, hydrogen, as their basic nuclear fuel. The lightest as well as the most common of all elements, hydrogen has the great virtue that the *fusion,* or melding together, of hydrogen nuclei *(protons)* releases energy of motion from energy of mass. Human beings have used this fact to make immensely powerful (by our standards!) "hydrogen bombs," but we have not yet learned how to obtain a *steady* release of energy in the way that stars do.

Stars rely on the fact that if two protons (hydrogen nuclei) fuse together, the fusion products have *less mass* than the two colliding particles (Figure 5-1). This result may contradict what we expect to happen in particle collisions, but it lies at the heart of energy liberation through nuclear fusion: Some energy of mass *disappears* in the fusion process, and turns into new kinetic energy. This newly liberated kinetic energy adds to the total, so that the particles that emerge from the fusion reaction have a combined kinetic energy greater than that of the original colliding particles. The increased kinetic energy can be shared among the particles that surround the nuclear fusion region through many collisions that heat the star's interior. The liberated energy of motion will in turn generate photons, and these photons will eventually emerge from the star's surface in the form of ultraviolet, visible, and infrared light.

The Proton-Proton Cycle

The fundamental series of nuclear reactions that occurs in most stars is called the proton-proton cycle, because in the first of three nuclear reactions in the series, two protons collide (Figure 5-1). This collision causes the two protons to fuse together, producing a *deuteron,* a *positron,* and a *neutrino.* We can think of a deuteron as a proton and neutron combined

THE THREE STEPS OF THE PROTON-PROTON FUSION CYCLE

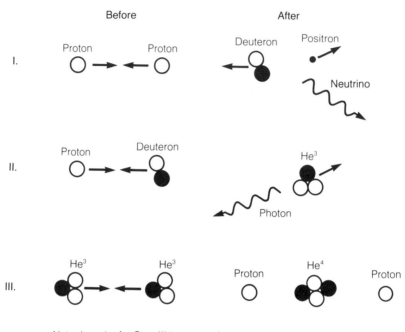

Note: In order for Step III to proceed,
we require <u>two</u> of Steps I and II.

Fig. 5-1 In each of the three steps of the proton-proton cycle, fusion reactions among elementary particles produce new particles with a total mass, and therefore a total energy of mass, that is less than the total before the fusion. In each step, the total kinetic energy increases by the same amount as the total energy of mass decreases.

into one nucleus, but the most important fact for us is that a deuteron has less mass than two protons do, and still less mass than a proton mass plus a neutron mass. The very act of binding together causes the mass of the deuteron to be less than the mass we would expect from the fact that a deuteron can be considered to be a proton bound to a neutron.

The sum of a deuteron's mass, a positron's mass, and a neutrino's mass (zero) falls below the mass of two protons. Therefore, *some mass disappears* in the fusion of two protons. This missing mass amounts to $m = 7 \times 10^{-28}$ gram, and corresponds to an energy of mass equal to $m \times c^2 = 6.3 \times 10^{-7}$ erg. Precisely this amount of energy appears as new kinetic energy, which adds to the kinetic energy that the colliding protons had (Figure 5-1).

The positron produced in the first step of the proton-proton cycle is an antielectron, which will soon meet an electron; the resulting mutual annihilation of the positron and electron will turn *all* of their energy of mass into the energy of motion of the photons, neutrinos, and antineutrinos that result from matter-antimatter collisions. The neutrinos that appear in the first step of the cycle, and in electron-positron annihilations, can, amazingly enough, escape directly from the center of the star! Neutrinos are so reluctant to interact with matter that most of them can pass through hundreds of thousands of kilometers of matter in a straight line, as easily as visible-light photons pass through air.[2]

In the second step of the proton-proton cycle, a proton collides with a deuteron. The two particles fuse together to produce a nucleus of helium-3 (He^3) and a photon (Figure 5-1). Once again, some energy of mass disappears during the fusion process, and the total kinetic energy increases by the same amount.

The third and final step of the proton-proton cycle liberates the greatest amount of kinetic energy. In this step, *two* nuclei of He^3 collide, and this collision produces a nucleus of helium-4 (He^4) and two protons. The total mass before the collision again exceeds the total mass after the collision; once again, the decrease in the total energy of mass matches the increase in the total kinetic energy.

The three steps of the proton-proton cycle are summarized at the bottom of Figure 5-1. Because *each* nucleus of He^3 comes from the fusion of a proton and a deuteron, the first and second steps of the cycle must occur twice for the third step to occur once. The three steps in the cycle, along with the positron-electron annihilations, liberate a grand total of 4.25×10^{-5} erg of kinetic energy. This does not seem like much energy, since even a bumblebee uses a million times more energy as it moves. Inside a star such as our sun, however, about 10^{38} proton-proton fusions occur each second, so that a grand total of 4×10^{33} ergs of kinetic energy appears each second, 10,000 times more energy than the human race has consumed during the past 5000 years!

Strewn throughout the universe in star after star there exist controlled nuclear fusion reactors, as perfect as nature can make them, in which the rate of energy liberation hardly varies over most of the stars' lifetimes. The tremendous power of the proton-proton cycle arises from the enormous number of individual fusion reactions that occur, and this number in

---

[2]This same property of neutrinos—their unwillingness to interact with matter—makes them extremely difficult to detect on Earth. The best attempts at such detection use 100,000 gallons of fluid in a giant tank, buried a kilometer underground to avoid interference from other particles. With this apparatus, physicists have detected about one neutrino each day! Oddly enough, the sun produces so many neutrinos that the experimenters expected to detect five neutrinos per day from the sun. The reason for this discrepancy remains unsettled at the present time.

turn arises from the tremendous abundance of protons (hydrogen nuclei) in every star.[3] It is a fact of the greatest cosmic significance that the fusion of the most abundant nuclei, hydrogen, into the next most abundant, helium, liberates kinetic energy from energy of mass, providing the fundamental means by which kinetic energy appears in the universe.

The liberation of energy through nuclear fusion inside stars provides a good way to see the interplay of the four types of forces that exist in the universe. *Gravitational forces* hold the entire star together. *Weak forces* and *strong forces* are responsible for the fusion reactions among nuclei, but they act only over extremely small distances. At larger distances, the *electromagnetic forces* among charged particles generally repel the nuclei from one another. Consider, for instance, the two protons that fuse together in the first step of the proton-proton cycle. Since each proton has a positive electric charge, electromagnetic forces between the two protons produce a mutual repulsion. Only if the two protons can approach within about 10^{-13} centimeter can they fuse together through the effect of strong forces. But how can the protons ever achieve this close approach, since they repel one another?

The Importance of Temperature inside Stars

Temperature provides the answer, because the temperature (in degrees Kelvin) measures the average energy of motion per particle.[4] At low temperatures, particles have little energy of motion and low velocites. At higher temperatures, each particle's kinetic energy, and thus its velocity, increases. Protons need enormous velocities to overcome their mutual electromagnetic repulsion and fuse together. For this reason, the proton-proton cycle of fusion reactions *cannot begin* inside stars until the temperature reaches about 10 million degrees Kelvin.

How do these enormous temperatures arise in stellar interiors? The temperatures owe their rise to the gravitational forces that hold each star together. Gravity pulls each part of the star toward every other part, and the total result makes each part of the star feel an attractive pull from the star's center (Figure 5-2). In a contracting protostar, the strength of this gravitational attraction increases as the particles approach one another, because gravitational forces increase as 1 over the square of the distance. As the particles move toward each other, they tend to collide more often and to move more rapidly. Squeezing molecules into a smaller volume

[3]A minority of stars, those with the largest masses, fuse protons into helium nuclei not through the reactions of the proton-proton cycle, but rather through a different set of reactions called the *carbon cycle*. Both these sets of nuclear reactions, however, have the same net result: Four protons fuse into one helium nucleus, liberating energy of motion as a result of the fusion process.

[4]The absolute or Kelvin scale of temperatures begins at absolute zero, the coldest possible temperature. Water freezes at 273.15 K and boils at 373.15 K.

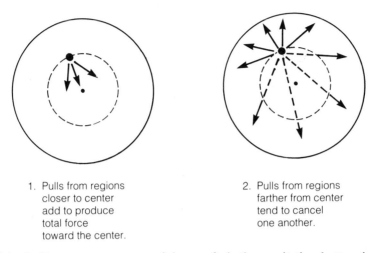

1. Pulls from regions
closer to center
add to produce
total force
toward the center.

2. Pulls from regions
farther from center
tend to cancel
one another.

Fig. 5-2 Inside a star, every part of the star feels the gravitational attraction of all the other parts. As a result of these combined forces, each part of the star feels a net force that pulls toward the star's center.

raises the average kinetic energy per particle. If we pump a bicycle tire, for example, we find that both the air inside the pump and the pump itself grow warmer as we compress the air to pump it into a tire.

Stars represent agglomerations of so much matter that each entire star, although completely gaseous, holds together solely by gravitation. Likewise, gravity provides the force of compression as a protostar contracts. Inside such a contracting protostar, the increasing gravitational forces produce an increase in the gas pressure and gas temperature during the millions of years of slow contraction. Even though protostars radiate away some of the additional heat (greater kinetic energy in particles) that results from their contraction, the temperature inside a protostar must continue to rise—first to thousands of degrees at the center, then to hundreds of thousands, finally to millions of degrees—as the protostar contracts to an ever-smaller size. Finally, when the temperature at the center of the protostar reaches about 10 million degrees Kelvin, the proton-proton cycle of nuclear reactions will begin to liberate kinetic energy from energy of mass, and the contraction will be halted.

Only at temperatures of tens of millions of degrees can protons begin to fuse together *despite* their mutual electromagnetic repulsion, and only then can kinetic energy be created from energy of mass. The newly liberated kinetic energy pushes outwards, opposing the inward pull of gravity. This opposition allows the protostar to cease contracting and to become a *star*, capable of supporting itself against its own self-gravity by liberating kinetic energy, energy which eventually flows from the star's center to its surface, from which it can radiate outward into space.

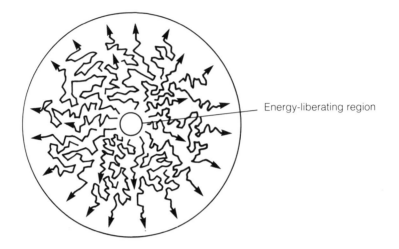

Energy-liberating region

Fig. 5-3 The kinetic energy that is liberated through fusion reaction in a star's central regions spreads outward through collisions among the protons, electrons, and helium nuclei within the star. Since the energy spreads into an ever-larger volume, each cubic centimeter outward from the center receives progressively less energy.

The liberation of kinetic energy in a star's central regions means that the particles emerging from nuclear fusion reactions there have more kinetic energy than the particles that went into the reactions. Collisions among particles near the center can then spread this additional energy to all of the particles in the star, with those near the center gaining a larger share than those near the surface (Figure 5-3). Hence, the temperature inside a star decreases steadily from the center (about 10 million degrees) to the surface (3000 to 50,000 degrees Kelvin). The amount of kinetic energy that is liberated near the center supplies just the right amount to keep the particles in the star zipping to and fro rapidly enough to prevent the star from contracting further. If this were not so—if, for example, the energy liberated each second were not enough to halt the contraction entirely—then the additional contraction would raise the star's central temperature and thus would increase the rate of nuclear fusion reactions. Higher temperatures imply larger rates of fusion reactions, because the particles have more kinetic energy and therefore can more easily overcome their mutual repulsion and fuse together upon close approach. Hence, any increase in a star's central temperature must produce an additional liberation of kinetic energy which would balance the star's tendency to contract.

The Influence of Mass upon Stellar Lifetimes

More massive stars have higher central temperatures and shorter lifetimes than less massive stars. A star with 10 times the sun's mass, for example, will last for only 100 million years, instead of the sun's 10 billion years, before expending its stock of protons and exhausting the fusion reaction's potential. Such a star, bright as it may be while it lasts, does not furnish a good environment for life, because we think (using our Earth as a guide) that life needs something like a billion years to start, and billions of years to evolve intelligence.

Even though massive stars begin with a greater supply of hydrogen nuclei, these stars liberate kinetic energy at such large rates that they manage to burn through their fuel reserves at relatively great speeds. Roughly speaking, the rate at which a star liberates kinetic energy from energy of mass will vary as the *cube* of the star's mass, so a 10-solar-mass star will turn energy of mass into kinetic energy at least 1000 times more rapidly than the sun. Such a star, even with 10 times the sun's original supply of protons, can last for less than a hundredth of the sun's lifetime. In general, the stars' lifetimes vary approximately as 1 over the square of the stars' masses, so that a star with half the sun's mass should have a total lifetime four times longer than the sun's.

Why do the more massive stars liberate kinetic energy at such enormous rates? The reason is that their central temperatures are higher, so that protons and other nuclei fuse together much more rapidly. If we double the temperature at the center of a star—say, from 10 million to 20 million degrees absolute—the rate of energy liberation there will increase, not by a factor of 2 but by *50* times. The particles' increased ability to overcome their mutual repulsion produces a vast increase in their rate of fusion, and thus in the star's ability to turn energy of mass into kinetic energy.

And why do more massive stars have higher temperatures at their centers? Simply because the total weight of the overlying layers presses down more firmly at the centers of more massive stars. The Earth's atmosphere, by analogy, exerts 1 kilogram of pressure per square centimeter on any object at the Earth's surface. This pressure arises from the weight of the atmosphere, held down by the Earth's gravity and compressed into a much smaller volume than it would occupy if the force of gravity were much less. If we doubled the total amount of gas in our atmosphere, we would increase the pressure at the Earth's surface, because more gas would weigh down on each square centimeter.

Similarly, if we double the mass of a star, we increase the mass that tends to concentrate toward the star's center. The increased concentration (increased self-gravitation) will raise the temperature by the extra compression of the gas. To oppose this tendency to compress, the more massive star must liberate more kinetic energy each second, because a greater mass of gas must be prevented from further contraction. Each star has just the right central temperature for its individual mass. That is, the rate of energy liberation in the star's center (governed by the central temperature) provides just the amount of energy of motion needed for each star to resist

TABLE 5-1

ESTIMATED MASSES, CENTRAL TEMPERATURES, LUMINOSITIES, AND LIFETIMES OF CERTAIN STARS

Star	Estimated Mass (Sun = 1)	Central Temperature (degrees Kelvin)	Rate of Energy Liberation per Second (Sun = 1)	Estimated Main Sequence Lifetime (Years)
Rigel	10	30,000,000	50,000	2,000,000
Sirius	2.3	20,000,000	23	1,000,000,000
Procyon	1.8	18,000,000	7.6	2,400,000,000
Alpha Centauri A	1.1	15,000,000	1.5	7,000,000,000
Sun	1.0	13,000,000	1.0	10,000,000,000
61 Cygni A	0.63	8,000,000	0.079	80,000,000,000
Proxima Centauri	0.1	6,000,000	0.00006	16,000,000,000,000

its own self-gravitation. Table 5-1 shows the central temperatures that have been calculated to exist inside stars of various masses, along with the stars' rates of energy liberation and their expected lifetimes. The table shows that the less massive, far less energy-liberating stars will outlast their more massive, energy-prodigious cousins by tremendous margins.

When we look at the stars, we cannot, of course, observe their masses or their central temperatures directly. The largest telescopes can hardly show stars other than the sun as anything but points of light—clear proof of the enormous distances to even the nearest stars. What astronomers *can* determine with relative ease are the stars' *surface temperatures* and *apparent brightnesses*. By using these two measured quantities, astronomers have developed a fruitful science of classifying the various types of stars.

Types of Stars

To determine the temperature of the surface layers of a star, astronomers use a *spectrograph* to spread the starlight into the spectrum of colors; that is, of the light's various frequencies or wavelengths. Figure 5-4 shows the spectrum of the light from our sun, by far the best-studied star. In this figure, the frequency (and thus the photon energy) decreases from top to bottom and also from left to right. The brightness of the photograph at a particular place in the spectrum shows how many photons of that particular frequency or wavelength are present in the sun's light. We can see at once that at certain frequencies, the sun emits few photons or none at all. These particular photons have been absorbed in the sun's surface layers, as the light from below (which contains photons of all the frequencies of visible light) escapes into space.

The sun's surface layers have a temperature of 5800 degrees Kelvin, cool enough to allow many familiar types of atoms—for example, hydrogen, helium, nitrogen, oxygen, and sodium—to exist, with all their electrons in orbit around the atomic nuclei. These atoms, along with many singly or doubly ionized ions of carbon, calcium, magnesium, iron, and other elements, absorb photons of certain precisely known frequencies. Only these photons can excite the atoms or ions into states of larger electron orbits. Photons with other energies will pass right by the atoms and ions with little interaction, filling in the spectrum between the frequencies of photon absorption.

The temperature of the sun's surface layers determines which atoms will be ionized and the degree of their ionization. Thus, the temperature has a direct influence on the spectrum of the light that emerges from the sun or from any other star, since the presence or absence of a given type of atom or ion will determine whether or not light can emerge from the star

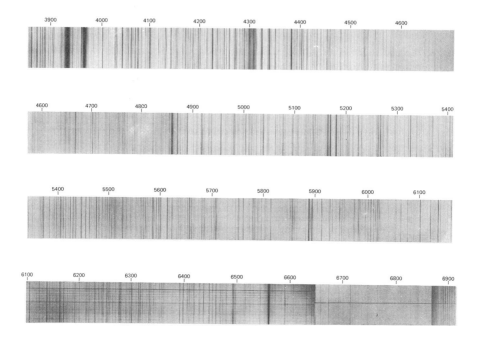

Fig. 5-4 When we separate the photons that form sunlight into the various photon energies (colors), we see dark lines at those energies where relatively few photons emerge. The sun's spectrum shows two especially dark regions (top left) which mark the absence of photons with the right energy to be absorbed by calcium ions. The numbers on the spectrum give the photon wavelengths in 10^{-8} centimeter units.

at the particular frequencies at which those atoms or ions will absorb photons.

Astronomers use stellar spectra to classify stars on the basis of their surface temperatures. These *spectral types,* shown in Table 5-2 and in Figure 5-5, acquired their names and definitions before astronomers recognized that the differences in stars' spectra arise primarily from differences in the stars' surface temperatures. Hence, we find a notable lack of consistency in the names of the spectral types: The hottest stars are type O; next comes type B, then A, F, G, K, and M. The O stars have surface temperatures of 30,000 degrees Kelvin, 10 times those of the M stars. In addition to these basic categories, astronomers subdivide each class of stars by placing a number (0 through 9) after the letter, so that, for example, a G9 star has a slightly greater surface temperature than a K0 star. A helpful memory guide for recalling the spectral types from hottest to coolest is ''Our Baby Asks for Grandma's Kitchen Matches,'' in which the initial letters of the words give the spectral types.

TABLE 5-2

SURFACE TEMPERATURES AND SPECTRAL FEATURES OF DIFFERENT TYPES OF STARS

Spectral Type	Average Surface Temperature (degrees absolute)	Outstanding Spectral Features
O	30,000	Absorption lines from ionized helium, weak hydrogen absorption
B	20,000	Absorption lines from un-ionized helium; stronger hydrogen absorption features
A	10,000	Hydrogen absorption lines strongest; still some absorption from un-ionized helium atoms
F	7000	Hydrogen lines still dominate spectrum; absorption features from "heavy" elements
G	5500	Hydrogen lines weak; many absorption lines from once-ionized and un-ionized "heavy" elements
K	4000	Increasing number of absorption features from un-ionized atoms of heavy elements; no hydrogen lines visible
M	3000	Absorption features of un-ionized atoms and of simple molecules (in particular, of titanium)

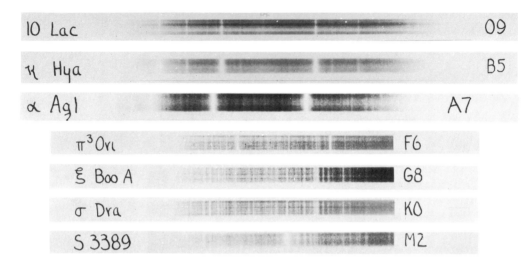

Fig. 5-5   We can classify stars on the basis of their surface temperatures, which can be determined from studying the differences in the stars' spectra. Such differences arise from the ways that different temperatures change the ionization of the atoms in the stars' outer layers.

The Magnitude Scale of Brightness

We have described how astronomers determine and classify the surface temperatures of stars by examining the spectrum of starlight. The other key observational parameter, the star's apparent brightness, can be measured on a scale of *magnitudes,* or sizes, that goes back two thousand years in time. Just as astronomers carry on an old tradition of naming the hottest stars first and the coldest stars last in their classification of spectra, so too they call the brightest stars *first-magnitude stars* and the faintest stars visible without a telescope *sixth-magnitude stars,* thus specifying that *larger magnitudes imply fainter stars.*

The basis of comparison among the stars' apparent magnitudes—that is, among the apparent brightnesses—is the principle that *five magnitudes imply a brightness ratio of 100.* Thus, a first-magnitude star has 100 times the apparent brightness of a sixth-magnitude star, and 1 million times the apparent brightness of a sixteenth-magnitude star. Individual magnitudes are separated by a factor of 2.512 (the fifth root of 100), so that a second-magnitude star has 2.512 times the apparent brightness of a third-magnitude star, and 6.310 (2.512 × 2.512) times the apparent brightness of a fourth-magnitude star.

When astronomers managed to measure the distances to the nearest stars, they could finally determine the true brightness or *absolute brightness* of these stars. By absolute brightness, we mean the brightness that stars would have if they all were at the same distance from us. For convenience, astronomers choose this standard distance to be 10 parsecs, and define a star's absolute magnitude as the apparent magnitude the star would have at a distance of 10 parsecs from us.

By using their distance measurements, astronomers have found that stars' true brightnesses vary over a tremendous range from star to star. The absolute magnitude of the sun, +4.8, places it near the middle of the total range of brightness. Some stars have 100,000 times the true brightness of the sun, or even more, while other stars (the faintest white dwarfs) have 100,000 times *less* true brightness than the sun does.

The Temperature-Luminosity Diagram

When we look at the true brightnesses of stars together with the stars' surface temperatures, a remarkable pattern emerges. Figure 5-6 shows a

Fig. 5-6 When we plot the surface temperatures and absolute luminosities of a large number of stars, we find that most stars' characteristics place them on the "main sequence" of this diagram. Stars above and to the right of the main sequence are "red giants"; those below and to the left are "white dwarfs."

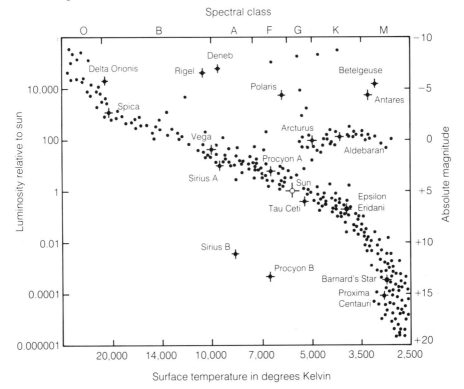

graph of the absolute luminosities and the surface temperatures of a representative group of stars, including some of the best-known ones. The temperature-luminosity diagram is also called the Hertzsprung-Russell (H-R) diagram for its originators. Figure 5-6 shows that most stars have surface temperatures and absolute magnitudes that place them in a definite part of the graph, the *main sequence*. Stars on the main sequence have surface temperatures that range from a few thousand degrees to 30,000 degrees or more. These stars show a clear correlation between their surface temperatures and absolute brightnesses: Stars with higher surface temperatures have much greater true brightnesses or luminosities. This result could be expected if all of the stars on the main sequence have roughly the same size. The hotter stars radiate much more energy per second from each square centimeter of their surfaces, and the number of square centimeters on each star's surface would be about the same.[5]

The main sequence of stars, which includes most of the stars that we can observe, represents the essential part of a star's energy-liberating lifetime. All of the stars whose surface temperatures and true brightnesses place them somewhere on the main sequence are turning energy of mass into kinetic energy at a steady rate. These stars can hold themselves to an almost constant size as they liberate kinetic energy, thanks to the balance between their own self-gravitation and the outward push of the liberated energy of motion.

Red Giants and White Dwarfs

But what of the stars that do *not* appear along the main sequence of the temperature-true brightness graph? Such stars, as we shall discuss in the next chapter, have finished the stage of the steady liberation of kinetic energy. These stars, which have moved farther toward the end of their lives than the main-sequence stars, appear either *above* the main sequence (the *red giants*) or *below* it (the *white dwarfs*).

Red giant stars have relatively low surface temperatures (2000 to 6000 degrees Kelvin), but shine with great luminosities. For the true brightness of a low-temperature star to be large, the star itself must be enormous, since a low-temperature surface will radiate relatively little energy per second from each square centimeter. Thus, we require an enormous surface area, hence an enormous radius for the star, to produce a true brightness

---

[5]Life is a bit more complicated than this, and in fact the stars with larger surface temperatures also have somewhat greater sizes than the stars with lesser surface temperatures. The primary effect that determines the stars' true brightnesses, however, remains their surface temperatures, because the hotter stars radiate *much* more energy each second from each square centimeter of their surfaces than the cooler stars do.

large enough to account for the energy output of red giant stars. The largest such stars, the red supergiants, have sizes thousands of times larger than the sun's radius, large enough to contain the orbits of the Earth and Mars! Since these stars contain at most only a few times the *mass* of the sun, however, they must be extremely tenuous, and consist mainly of a gauzy near-vacuum that surrounds a smaller, dense core.[6]

White dwarf stars, on the other hand, have relatively high surface temperatures (5000 to 15,000 degrees Kelvin) but extremely small true brightnesses. Therefore, the white dwarf stars must have tiny surface areas and minuscule radii. Most white dwarfs have sizes similar to the *Earth's,* yet contain almost as much mass as the sun! Since the sun's radius exceeds the Earth's by 100 times, the sun's volume is 1 million times the Earth's. Thus, a white dwarf that gathers the mass of the sun into a region as small as the Earth must have a density of matter 1 million times the density of matter in the sun. A cupful of white dwarf matter would weigh 1000 tons at the Earth's surface!

The classification of stars by their surface temperatures and true brightnesses has helped to reveal the important ways in which stars resemble and differ from one another. By making calculations that simulate the evolutionary history of stars, astronomers have established the fact that all of the stars on the main sequence fuse protons into helium nuclei to liberate energy of motion. The stars' *masses* determine whether a given star will appear high up on the main sequence (large true brightness, high surface temperature), in the middle of the sequence (average true brightness and surface temperature), or at the low end (small true brightness, low surface temperature).

During the contraction of the protostars that formed the stars that now appear on the main sequence, the stars did not occupy the same points in the diagram that they do now (Figure 5-7). But since the time when these stars began to liberate kinetic energy from energy of mass, they have maintained nearly the same surface temperatures and the same true brightnesses, and thus almost the same positions in the main sequence. In short, stars evolve on to the main sequence, and later evolve off it, but they do not evolve *along* the sequence; instead, they remain in almost the same position on the main sequence throughout their lifetimes of steady energy liberation.

Most of the matter in any star on the main sequence consists of hydrogen nuclei (protons), with helium nuclei providing about 25 percent of the total mass. *All the elements heavier than hydrogen and helium form no*

---

[6]Astronomers also use the terms blue giant stars and blue supergiants. These stars are merely the hottest, brightest stars on the main sequence; they do not represent a category as separate as those of red giants and red supergiants.

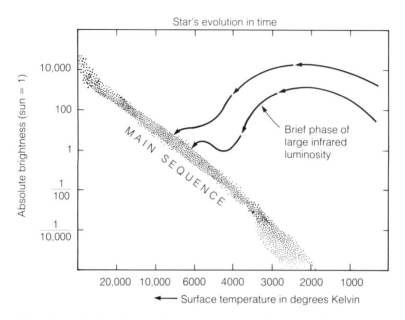

Fig. 5-7 Stars farther down the main-sequence (lower-luminosity stars) have smaller masses and lower surface temperatures than the brighter stars on the main sequence. These low-mass stars also last much longer than stars high up on the main sequence.

Fig. 5-8 As stars contract to the sizes that they maintain throughout their main-sequence lifetimes, they pass through a brief phase of low surface temperature and large luminosity. In this phase, while the star's size far exceeds its main-sequence size, the star radiates great amounts of infrared light.

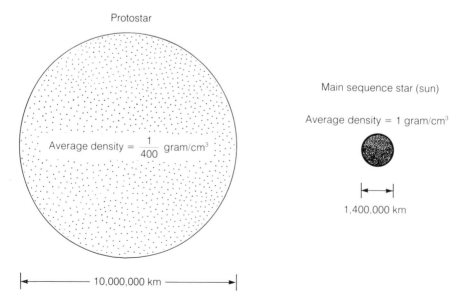

more than 1 or 2 percent of a star's mass. This fact should not surprise us when we consider that the big bang left behind mostly hydrogen and helium, and that most of the physical processes within a star have as their result the production of helium nuclei from hydrogen nuclei. In fact, the question that must be answered turns out to be: Why do stars contain *as much as* 1 or 2 percent of their mass in the form of heavy elements— carbon, nitrogen, oxygen, neon, and so forth? How were these elements produced? The answer to these questions lies in the final stages of stars' lifetimes, when the familiar ways of liberating kinetic energy become no longer available. In the next chapter we shall see what happens inside stars that have exhausted their basic energy-liberating processes and thus have passed through middle age to enter a glorious senility.

Summary

Most stars seem to have condensed out of collapsing gas clouds billions of years ago, although some stars in our own galaxy, in other spiral galaxies, and in irregular galaxies, are much younger than others, and new stars continue to appear even today. In spiral galaxies, the passage of a density wave triggers a burst of star formation in interstellar gas clouds, such as the nearby Orion Nebula.

Stars shine by turning some energy of mass into kinetic energy, in accord with Einstein's formula $E = mc^2$. To do this, most stars use a series of nuclear fusion reactions called the proton-proton cycle. In these reactions, four protons combine to form one helium nucleus, with the release of some kinetic energy as the energy of mass decreases slightly. Huge numbers of reactions occur in the hottest, innermost regions of stars each second. This energy, shared among all the particles inside the star through collisions, gives the particles enough kinetic energy to resist the star's tendency to collapse under its own self-gravity. Because the outward flow of liberated kinetic energy balances the inward pull of self-gravitation, most stars can regulate their rate of energy liberation accurately. Stars that achieve a steady rate in this manner appear on the main sequence of a graph of stars' true brightnesses and surface temperatures. Large-mass stars have larger surface temperatures, and *much* greater true brightnesses, than small-mass stars. As a result, the large-mass stars burn out their supply of protons, by fusing them into helium nuclei to release kinetic energy, much more rapidly than small-mass stars do.

The total lifetime of a star, and particularly the length of time that the star lasts on the main sequence, has great importance in the search for life. *Only those stars whose main-sequence lifetimes exceed a few billion years are likely to permit life to develop on any planets they may have.*

Life on Earth required billions of years first to appear and then to evolve beyond the stage of single-celled, primitive organisms. If our planet's history represents any sort of guideline, we must look to the lower-mass stars, those with less than about 1.5 times the sun's mass, to provide the long lifetimes needed for the development of life. Luckily for us in the search for life, such stars far outnumber the more spectacular high-mass but short-lived stars that form most of the stars we can see in the night skies.

Questions

1. How do stars liberate kinetic energy from energy of mass? Which part of a star has a temperature high enough for this process to work? Why are high temperatures necessary?

2. Why do no stars shine simply by using the annihilation of matter and antimatter to release a steady stream of photons? Why do no stars use the fission of uranium nuclei to liberate their kinetic energy?

3. Which part of the proton-proton cycle produces a particle of antimatter? What happens to this antiparticle?

4. Why did protostars stop contracting? What makes the transition from a protostar to an actual star? Why doesn't the liberation of kinetic energy inside stars make them expand?

5. Why do more massive stars exhaust their nuclear fuel in less time than less massive stars?

6. Alpha Centauri, the closest star to the sun, has an apparent magnitude of −0.2, while the star Wolf 359 (the fifth closest star to the sun) has an apparent magnitude of +9.8. How many times greater is Alpha Centauri's apparent brightness than Wolf 359's?

7. Wolf 359 has a distance from us about twice Alpha Centauri's. By how many times does the *absolute brightness* of Alpha Centauri exceed that of Wolf 359?

8. Why do most stars lie close to the main sequence line in the graph of true brightness versus surface temperature? Which stars do *not* lie along the main sequence?

9. Why are red giants red? What are red supergiants?

10. Antares and Barnard's star (the third closest star) have the same surface *temperature,* but the surface *area* of Antares exceeds that of Barnard's star by 100 million times! How do the absolute brightnesses of these stars compare?

Further Reading

Baade, Walter. 1963. *The evolution of stars and galaxies.* Cambridge, Mass.: Harvard University Press.

Bok, Bart. 1975. The birth of stars. In *New frontiers in astronomy,* ed. by Owen Gingerich. San Francisco: W. H. Freeman and Company.

Gamow, George. 1961. *A star called the sun.* New York: Mentor Books.

Herbig, George. 1975. The youngest stars. In *New frontiers in astronomy,* ed. by Owen Gingerich. San Francisco: W. H. Freeman and Company.

Jastrow, Robert. 1969. *Red giants and white dwarfs.* New York: New American Library.

Schatzman, Evry. 1974. *The structure of the universe.* London: George Weidenfeld and Nicolson.

Page, Thornton, ed. 1968. *The evolution of stars.* New York: Macmillan.

CHAPTER 6

How Stars End Their Lives

If our own existence provides any guide, stars keep life going on planets that may orbit around them. Therefore, both for ourselves and in our search for life, we have more than a passing interest in the fate of stars as they begin to exhaust their supplies of protons. The changes that a star undergoes as it ages have certain fantastic and far-reaching ramifications that do not correspond directly to our Earth-formed intuition of what they should be. In particular, it comes as a surprise to most people to learn that every molecule in our bodies contains matter from exploded stars. The air we breathe, the ground on which we live, the seas around us, and the flowers that bloom in the spring all came from the cinders of long-vanished stars that first collapsed, then erupted, as they exhausted all the processes that could liberate more kinetic energy in their interiors.

Nuclear Fuel Consumption in Stars

Because a star's self-gravitation never ceases, any star will contract if it cannot liberate enough kinetic energy each second to overcome this tendency and if it cannot find some other way to support itself against gravity. Once most of the protons in a star's central regions have fused into helium nuclei, the star cannot liberate much more energy from the fusion of protons. Helium nuclei provide a natural back-up source for further fusion reactions, but with a catch: Helium-4 (He^4) nuclei are extremely reluctant to fuse together. To fuse He^4 nuclei requires a temperature not of millions of degrees, but rather *hundreds of millions* of degrees. Only at these enormous temperatures can He^4 nuclei overcome their mutual repulsion and

approach closely enough in collisions to fuse together. These fusion reactions first make two He^4 nuclei fuse into a nucleus of beryllium-8 (Be^8)
plus a photon. Soon afterwards, the Be^8 nucleus will fuse with another He^4
nucleus to produce a nucleus of carbon-12 (C^{12}) and another photon (Figure
6-1). The combination of the two reactions fuses three He^4 nuclei into a
C^{12} nucleus plus two photons, and liberates 1.2×10^{-6} erg of kinetic energy. Notice that this amount of energy falls almost four times below the
4.2×10^{-5} erg of energy liberated in the proton-proton cycle: The star can
liberate far less kinetic energy from the fusion of He^4 nuclei than from the
fusion of protons.

The Evolution of Stars

Consider, then, what happens as a star exhausts the protons in its center
through fusion reactions. First, the dwindling supply of protons must continue to supply enough energy to keep the star from collapsing. The star
can solve this problem by contracting its central core, the natural result of
self-gravitation insufficiently opposed by energy liberation. This raises the
temperature in the core, causing the protons to fuse together more rapidly,
with a greater rate of energy liberation and a greater rate of exhaustion of
the proton supply. The star's central regions, which contain about half the
star's mass, keep on contracting and growing hotter until finally, as almost

Fig. 6-1   When helium nuclei start to fuse together, they first produce a nucleus of
beryllium-8, plus a photon, from two nuclei of helium-4. The next step fuses the
beryllium-8 nucleus with another helium-4 nucleus, producing a nucleus of carbon-
12, plus another photon.

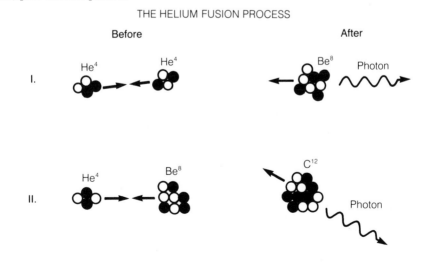

all of the protons have fused into helium nuclei, the helium nuclei themselves start to fuse together because the temperature in the core has risen to 200 million degrees absolute.

The onset of nuclear fusion reactions among the helium nuclei provides a sudden rush of liberated kinetic energy, which, for a short time, makes the star much brighter. This *helium flash* marks the high point of a star's true brightness; afterwards, the slide towards obscurity continues. Figure 6-2 shows the changes in the true brightness and surface temperature of stars at various points along the main sequence. As most of the protons in the stars' inner regions fuse into helium nuclei, the stars must leave the main-sequence locations on the graph that typify steady proton fusion. During this phase, the stars' inner cores become almost pure helium, while

Fig. 6-2 As stars age, they leave their main-sequence positions on the temperature-luminosity diagram and become much cooler in their outer layers than main-sequence stars of the same true brightness.

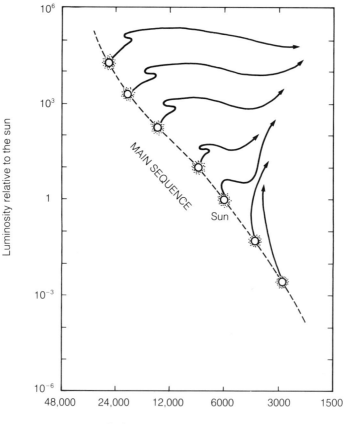

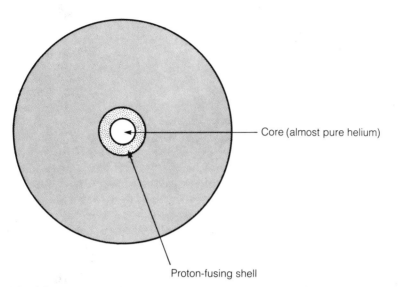

Fig. 6-3   After a star has fused almost all of the protons near its center into helium nuclei, a shell of matter around the helium core continues to fuse protons.

protons continue to fuse in a shell surrounding the central core (Figure 6-3). Some stars actually increase their true brightnesses while contracting their insides; they all raise their central temperatures to fuse their remaining protons more and more rapidly in the thin shells around their pure-helium cores.

The helium flash liberates so much kinetic energy in such a short time that the core *expands* slightly. This in turn lowers the temperature and the rate of helium fusion. Nonetheless, the helium fusion reactions continue, and the shrinking core liberates kinetic energy from a diminishing supply of helium nuclei. Because the rate of fusion reactions depends on the temperature in an extreme manner, the star can, for a brief time, liberate more kinetic energy per second than before from a depleted stock of nuclei. As the star does so, the increased amount of newly liberated kinetic energy pushes outward on the star's outer layers, expanding them beyond their original size. The expansion cools the gases, and the star creates a cool, highly extended outer shell at the same time that it produces an extremely dense and hot central core.

The star thus becomes a red giant, cooler and more rarefied on the outside and hotter and denser within than a main-sequence star. Eventually, the outermost layers of the star will evaporate into space, exposing the hotter subsurface layers beneath. For a time, the star may shine with a large true brightness and a high surface temperature—so large, in fact, that the star emits most of its energy in the ultraviolet part of the spectrum, invisible to the human eye. The discarded outer layers may, however, trap

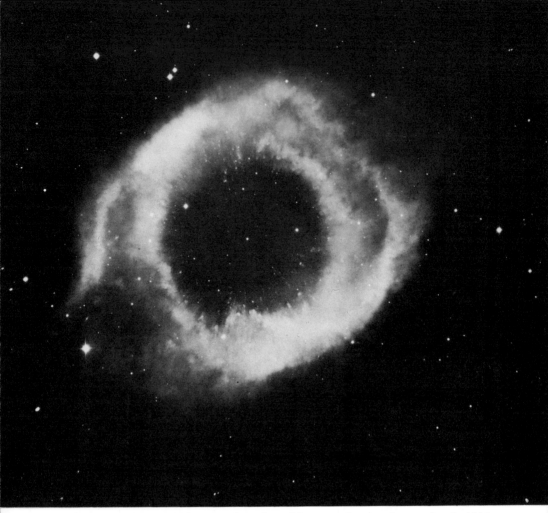

Fig. 6-4 After a red giant star has ejected its outer layers, the previously hidden lower layers become the star's new surface, much smaller and hotter than the previous one. The star at the center of this "planetary nebula" in the constellation Aquarius produces far more ultraviolet light than visible light, and powers the entire glowing shell around it.

some of the ultraviolet photons. These photons will ionize the atoms in the shell around the star. When the atoms recombine, they can produce photons of visible-light frequencies so that we see a *planetary nebula* from the star's discarded mantle (Figure 6-4). Such planetary nebulae draw their energy from a nearly invisible (to us) star, at the center of a spherical shell of gas that used to be part of the central star.

How long do the various phases of a star's evolution take? If we imagine the entire lifetime of a star to be equal to a human lifetime (70 years), then the initial contraction and the protostar phase represent the star's childhood (say, 15 years), while the final stages of contraction and the onset of energy liberation pass much more rapidly (1 year). The main-sequence lifetime lasts for perhaps 50 years, and the evolution from the main sequence up to the time of the helium flash takes just a year more. The helium flash itself must be reproduced by a billionth of a second (in real life it takes less than a second[1]), while the red giant phase that follows lasts for perhaps three years. The subsequent planetary nebula phase and all that follows will require less than a year before the star either explodes or becomes a white dwarf (see pp. 112–115). We use the analogy with a human lifetime, rather than naming specific numbers of years in a star's true history, because stars of different masses require quite different times for each phase of their lifetimes. Thus, for example, the sun will spend 10 billion years on the main sequence and 100 million years from there to the helium flash, while a star with 5 solar masses will spend just 70 million years on the main sequence and 5 million years from there to the helium flash. These times, estimated from detailed calculations, underscore the key fact that *most of the star's lifetime passes on the main sequence.*

We have followed a star's evolution from the onset of proton-proton reactions through the stages of helium-to-carbon fusion. What, then, happens as a star exhausts its helium nuclei by fusing them into carbon nuclei? Each star will follow one of two basically different life stories, depending on the mass that the star has. Low-mass stars will become white dwarfs, while high-mass stars will not, and are likely candidates for supernova explosions.

The Exclusion Principle of White Dwarf Stars

White dwarf stars are the highly condensed remnants of once-normal stars that have shed their outer layers to expose their inner cores, now mostly carbon nuclei and electrons. What distinguishes white dwarfs from other stars, and supports the white dwarfs against their self-gravitation, is the fact that the electrons in white dwarf stars obey the *exclusion principle.* This scientific term is a fancy name for an apparently fanciful concept: Certain types of particles (in particular, electrons, protons, and neutrons) simply cannot be squeezed together indefinitely. This resistance to further squeezing acts in addition to, and much more powerfully than, the general repulsion electrons exert on one another through electromagnetic forces. The exclusion principle becomes important for electrons, however, only when the density of matter exceeds about 1 million grams per cubic centimeter, a million times the density of water. At lower densities, the exclusion principle has no role to play in the way that electrons in a star behave,

---

[1]The helium flash takes place in less than a second *in the star's core.* For the energy to diffuse outward to the star's surface takes much longer, so if we were observing a star at the time of its helium flash, we would see a brightening that lasted for several days.

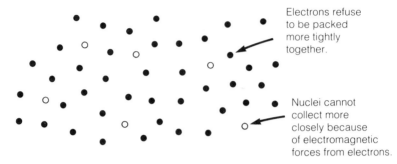

Electrons refuse to be packed more tightly together.

Nuclei cannot collect more closely because of electromagnetic forces from electrons.

Fig. 6-5 In a white dwarf star, the exclusion principle prevents the electrons from being squeezed beyond a density about a million times the density of water. If the electrons cannot be packed more tightly than some density, then neither can the nuclei, which are held back by electromagnetic attractive forces from the negatively charged electrons.

so we do not encounter its effects until we consider the huge densities of matter inside white dwarf stars.

Consider a star that keeps contracting its central regions, thus raising the temperature there. Once most of the helium in the star's core has become carbon nuclei, the carbon nuclei could themselves fuse into heavier nuclei, liberating still more kinetic energy. If, however, the star's interior has contracted so far that the matter there has reached a density of 1 million grams per cubic centimeter, this fusion of carbon will not occur. Why not? The electrons, obeying the exclusion principle, will refuse to compress to a greater density. If the electrons resist further compression, so do the nuclei, because the electromagnetic attractive forces between the negatively charged electrons and the positively charged nuclei will drag the nuclei back as they try to obey their mutual gravitational pull toward the center of the star (Figure 6-5).

Because the electrons in a white dwarf star hold the carbon nuclei apart rather than allowing them to rush toward the star's center, the nuclei cannot fuse together, for they can never collide with enough speed to overcome their mutual electromagnetic repulsion. Main-sequence stars have a lower density of matter in their centers than white dwarf stars, but in the interiors of main-sequence stars occasional head-on collisions drive the cycle of proton-proton fusion reactions. In contrast, white dwarf stars maintain extremely high densities without the possibility of fusion reactions, because the nuclei can never collide with enough kinetic energy to fuse the nuclei together.

A white dwarf star, the Earth-sized remnant core of a once-ordinary main-sequence star, can slowly conduct heat from its interior and radiate it away into space, but the star can liberate no more kinetic energy. Strangely enough, because white dwarf stars radiate relatively little energy (their true brightnesses reach at most one thousandth of the sun's), they can shine for billions of years, growing slowly but steadily dimmer as they expend the stored energy from their earlier years. These white dwarfs would provide a stable energy source for any living creatures that orbited close around them, content with a tiny fraction of a percent of the sun's energy output. When we consider, however, the red giant and helium-flash phases that preceded the white dwarf, not to mention the planetary nebula expulsion of the star's outer layers, we may easily conclude that life might not *reach* this stable configuration. Any civilization on a planet orbiting an aging star would have to deal with tremendous changes in the star's energy output. Only if a civilization foresaw what would happen to their parent star, managed to ride out the ups and downs of the red giant phase, and then moved close to the white dwarf remnant would the possiblity of "lowered expectation" living become a reasonable strategy for the next billion years!

Supernova Explosions

So much for the numerous but modest white dwarfs. What of the other pathway of stellar evolution, the supernovae that made the elements within us? Why do some stars explode, seeding the universe with useful nuclei, rather than fading away as compact dwarfs? Were it not for the fact that a small proportion of stars *do* explode at the end of their lives, we would not be here to think about them, though we are fortunate that our own sun is not one of the fast-evolving, soon-to-explode supernovae that prove so essential for life.

Stars end their lives differently because they have different masses, which in turn imply different densities at the stars' centers. More massive stars have *lower* central densities, because the more massive stars can reach a given central temperature with a lower density of matter in their cores. Why? Because the more massive stars have more matter pressing down on their centers, weighing more heavily on the matter there. The material at the center responds to this greater weight by raising its temperature more readily; that is, without requiring so large a density of particles to produce a given temperature, or even to produce the higher central temperatures that characterize large mass stars. In the lower-mass stars, the lesser gravitational force cannot raise the star's central pressure and temperature as easily, so the star must achieve this result (in order to balance the gravitational force) by drawing the matter more closely together in the star's interior. Hence, less massive stars always have a greater density of matter at their centers than more massive stars at the same point in their evolution. In the less massive stars (masses less than about 1.4 times the sun's mass), the exclusion principle becomes important by the time their helium nuclei have fused to form carbon nuclei. All white dwarf stars, so far as we can tell, have masses less than 1.4 times the sun's mass, and most have masses that are less than the sun's.[2]

The more massive stars never become so dense in their interiors, even after helium-to-carbon fusion, that the exclusion principle is important. These stars can proceed to fuse carbon nuclei into heavier nuclei. Such fusion reactions liberate a bit more kinetic energy, but each successive type of reaction liberates less and less energy. If two C^{12} nuclei fuse together to form a nucleus of magnesium-24 (Mg^{24}), the fusion process liberates only one tenth as much kinetic energy as the fusion processes in the proton-proton cycle. But the star must continue to liberate kinetic energy as rapidly as always or face ever-increasing contraction. In fact, massive stars pass through a series of fusion reactions that yield less and less kinetic energy, producing all the elements from carbon to iron before they completely exhaust their ability to liberate more kinetic energy through nuclear fusion.

Iron marks the end of the line, because to make nuclei heavier than iron-56 (Fe^{56}, with 26 protons and 30 neutrons per nucleus), we must *add* kinetic energy. In other words, fusion reactions that make the nuclei lighter than iron end with *less* energy of mass and *more* kinetic energy, while fusion reactions that make nuclei heavier than iron finish with *more* energy of mass and *less* kinetic energy.[3] A star that has fused most of the nuclei in its core into Fe^{56} can no longer come up with a new, even if less effective, fusion reaction to liberate more kinetic energy. Such a star, if it has not yet achieved a central density high enough for the exclusion principle to be significant, has exhausted all means of supporting its core against collapse.

So the core collapses under its own weight, shrinking in size by thousands of times in less than a second. Why doesn't the exclusion principle prevent this collapse? The reason is that at the fantastically high temperatures inside the collapsing core, the electrons begin to fuse together with protons broken off the colliding nuclei. A star that has fused its nuclei all the way to iron will have an internal temperature of a few *billion* degrees absolute. As the core of such a star collapses, the collapse itself provides the energy needed to break the iron nuclei apart into protons and neutrons as the nuclei collide with each other. The protons can then fuse with elec-

---

[2]Detailed calculations show that if a star has a mass greater than 1.4 times the sun's mass, it simply cannot exist as a white dwarf star. That is, the exclusion principle cannot support an indefinite amount of mass against collapse under its self-gravitational forces. The fact that, of those white dwarf stars whose masses can be determined, none has a mass greater than 1.4 times the sun's mass tends to confirm the theory that predicts that more massive white dwarfs simply cannot exist.

[3]Since we gain kinetic energy as we fuse more and more complex nuclei until we get to iron-56, we must supply kinetic energy to break apart these nuclei. In contrast, since it *takes* kinetic energy to make nuclei more complex than iron, we can gain kinetic energy when these nuclei break apart (for example, from the fission of uranium nuclei).

Before   After

Proton   Electron   Neutron   Neutrino

Fig. 6-6 At densities as enormous as those inside a collapsing stellar core, protons and electrons fuse together in a weak reaction, the result of weak forces, to produce a neutron and a neutrino from each proton and electron.

trons to produce neutrons and neutrinos, in a reaction that is almost the opposite of the decay of individual neutrons (Figure 6-6).[4]

Such fusion reactions between protons and electrons are *weak* reactions, governed by weak forces. Since electrons do not feel strong forces, they do not ordinarily fuse with nuclei as protons and neutrons do. Furthermore, the electromagnetic forces, which do attract the oppositely charged electrons and nuclei, always get the electrons into orbits *around* the nuclei and never make them fuse together. This leaves the weak forces as the means of producing the fusion of electrons and protons (since gravitational forces are insignificant at the size level of elementary particles). As their name implies, weak forces do not usually have much success at producing such fusion. If, however, the temperature in stellar cores rises to billions of degrees and the density rises to hundreds of thousands of grams per cubic centimeter (still not high enough for the exclusion principle to have an important effect on the electrons), then the electrons will indeed begin to fuse with the protons. As the electrons disappear, they take with them the star's chances of using the exclusion principle's effect upon electrons to hold itself up. Thus, the fusion of protons with electrons allows the star's core to continue its sudden, violent collapse—which itself provides the temperatures and densities high enough to make the protons and electrons fuse together.

The Production of Heavy Elements in Supernovae

The collapse of the core of a massive star occurs within about 1 second, and it is an important second for the star and for the rest of the universe. During the collapse, fusion reactions by the trillions occur as the nuclei smash into one another. Such collisions mostly have the effect of stripping

[4]Because of weak forces, any isolated neutron will fall apart (decay) after about 15 minutes into a proton, an electron, and an antineutrino. If we imagine the exact reverse of this process, we must bring a proton, an electron, and an antineutrino together to form a neutron. A near-duplicate of this reverse process brings a proton and an electron together to make a neutron and a neutrino.

the nuclei (iron-56 and similar nuclear species) into their constituent protons and neutrons. In addition, however, such collisions can produce nuclei
heavier than iron, such as mercury, silver, lead, gold, platinum, and uranium. In effect, the star's collapse uses the last energy source available—
the energy that held up the core—to make these heavy elements.

When we look at the elements that make up the Earth and comprise
living organisms, we find an important distinction in the *abundance* of
elements. The elements whose nuclei were made during the millions of
years before the star's core collapsed, such as carbon, nitrogen, oxygen,
aluminum, silicon, and iron, are far more abundant than those nuclei made
during the single second of collapse, such as molybdenum, silver, platinum, gold, and mercury. Living organisms, in fact, consist almost exclusively of elements lighter than iron; that is, of the more abundant elements,
made during the long periods before stellar collapse. Trace amounts of the
elements heavier than iron do, however, play a role in most living creatures
on Earth. In a similar fashion, more than 99.9 percent of the Earth consists
of elements no heavier than iron, but the tiny amounts of silver, gold, and
uranium made during the final second of a star's life have played a key role
in human history, partly because of their very scarcity, the result of the
short amount of time during which they were made.

How did these elements ever blast outwards from the collapsing star?
Besides the production of heavy nuclei, the rush of fusion reactions inside
the collapsing star creates enormous numbers of neutrons, the result of
proton-electron collisions at tremendous energies. These neutrons quickly
form a stellar core called a *neutron star,* a tiny, compact object made almost
entirely of neutrons. Like electrons, neutrons obey the exclusion principle
and cannot be squeezed past a certain density much higher than the corresponding density for electrons. As the neutrons refuse to be squeezed
further, matter from the outer layers falls onto the surface of the newly
made neutron star and actually bounces off at enormous velocities in an
explosion, simply the reversal of the infall at tremendous speeds about a
second earlier. This last gasp of outward energy can light up the sky with
a sudden outburst that signals the *collapse* of the star's inner regions to
produce an incredibly dense neutron-star core (Figure 6-7).

Such supernova explosions would, of course, be fatal for any life that
inhabited planets around the star. Even if we suppose that living creatures
on these hypothesized planets had managed to survive the pre-supernova
evolution of the star, when the star radiated energy at a greater and greater
rate, the final outburst must surpass all preparation. Most supernovae
release hundreds of millions or even billions of times more energy each
second than the sun does, and they do so for several weeks or months.
The effect on a planet like the Earth would exceed what would happen here
if the sun were suddenly brought to the *moon*'s distance from us. In fact,
this comparison is far too mild; it would be more accurate to compare the

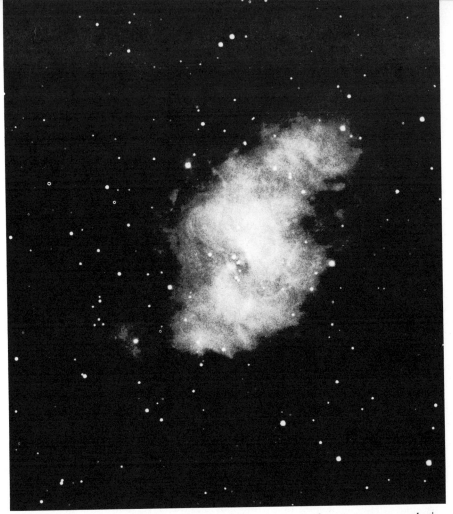

Fig. 6-7 The Crab Nebula in Taurus is the remnant of a supernova explosion observed on Earth in the year 1054, though the star had in fact exploded some 6500 years before. Matter from the star exploded outwards at great velocity; even today we find gas moving at about 1000 kilometers per second. Many charged particles within this expanding mass of gas are moving at almost the speed of light and produce photons of synchrotron emission.

supernova's energy release on nearby planets to the experience of a hydrogen bomb close up.

From a distance, supernova explosions are spectacular but safe. In a large spiral galaxy—our own Milky Way, for example—one star becomes a supernova every 50 or 100 years (Figure 6-8). Astronomers then see a new star appear where they could see no star before, since the presupernova star typically has too little brightness to be visible. The supernova, however, can be a billion times brighter than the star that preceded it (Figure 6-9). After a few months, the supernova fades from being by far the brightest star in its galaxy to invisibility. Thus, the total energy output from the supernova, immense though it seems, adds up to less than the energy

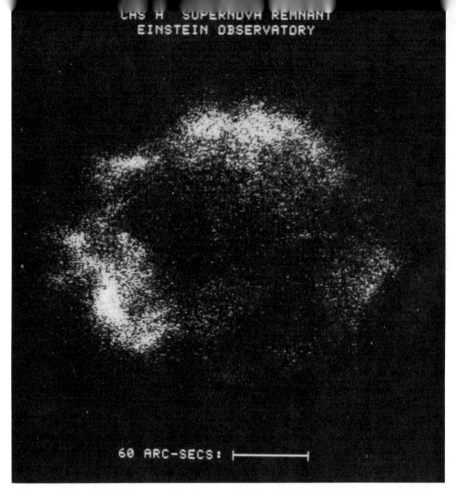

60 ARC-SECS:

Fig. 6-8 An exploding star in the constellation Cassiopea has left behind almost no traces in visible light, but this X-ray photograph clearly shows the ring of material blasted outwards. Radio maps also reveal the presence of an expanding shell, and this supernova remnant is the strongest source of radio waves in our skies.

that flowed during the hundreds of millions or billions of years in which the presupernova star shone fairly steadily.

Astronomers observed supernovae within our own galaxy most recently in the years 1572 and 1604. The supernova shown in Figure 6-8 probably appeared during the early 1700s, but escaped detection because absorption by interstellar dust prevented it from reaching a great apparent brightness. By the laws of probability, we should expect to see another supernova in our galaxy soon. In view of the fact that we are probably thousands of light years from the next supernova that we shall see—as, indeed, we were from the last three—we may be almost positive that the next supernova to appear to us has *already* exploded, but its light has not yet reached the Earth. Every year brings the discovery of several new supernovae in relatively nearby galaxies, and the last supernova to be observed in our giant neighbor, the Andromeda galaxy, appeared in the year 1885.

We have seen how supernova outbursts arise from the collapse of old stellar cores, which turn part of the energy of collapse (simply the stored energy that opposed the collapse) into the explosion of matter away from the core. This outward explosion produces the light that suddenly appears in our familiar skies. Before we examine the remnant cores that are left after the explosion has developed, let us pause to consider the other effects of supernova outbursts. These explosions have the useful and important effect of spreading elements heavier than helium through space. In a galaxy such as our own, the first generation of stars formed from the matter that emerged from the big bang. Such matter consisted almost entirely of hydrogen and helium nuclei, electrons, neutrinos, and antineutrinos, with *less than one millionth* of the matter made of nuclei heavier than helium. The first generation of stars, apparently contained relatively little of the heavier elements. These stars have all vanished. The oldest stars we see now, called Population II stars, must have formed somewhat later. These stars

Fig. 6-9 This photograph shows a supernova in a galaxy about 40 million parsecs from us; at its peak brightness, the supernova shone with about one thousandth of the brightness of the total galaxy.

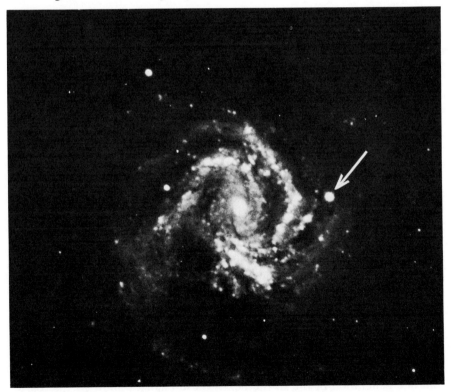

have far less of the heavier elements (carbon, oxygen, neon, and so forth) than the one percent that characterize stars like our sun, called Population I stars, which therefore must have formed still later.

It seems likely that a high proportion of the early stars had relatively high masses, several times the mass of our sun. These stars therefore evolved quickly (in less than a few billion years) through their energy-liberating lifetimes and exploded as supernovae, flinging the heavier elements (all the way from carbon through uranium) out into the rest of the galaxy. The third and following generations of stars, which include most of the stars now in our galaxy, resemble our sun in the fraction of heavy elements they contain. One or two percent (by mass) of these stars consists of nuclei heavier than helium, formed in other stars and seeded through space when the stars exploded. Planets such as the Earth represent a collection of these cinders from burnt-out stars, from which the two most abundant elements in the universe—hydrogen and helium—have evaporated to leave a heavy residue of elements, many of which are essential to life.

Cosmic Rays

As if this were not enough, supernovae seem to have made another contribution to the evolution of life. We have seen that the outer layers of a supernova explode into space at enormous velocities as they bounce off the neutron-star core. Because the presupernova star had progressively less density of matter at greater distances from its center, the blast wave produced by the catastrophic infall and subsequent rebound will encounter progressively less matter to accelerate as it moves outward. As a result, the matter in the outermost layers of a supernova will acquire fantastic speeds, 99.9 percent of the speed of light and even more. These layers quickly turn into individual *cosmic ray* particles—electrons, protons, helium nuclei, and heavier nuclei traveling at nearly the speed of light. Such cosmic rays (actually fast-moving elementary particles) permeate our galaxy, and they arrive at the top of the Earth's atmosphere by the billions every second. Luckily for us, our atmosphere prevents most of the cosmic ray particles from reaching the Earth's surface, thus saving us from a deadly rain that would tend to destroy life as we know it within several generations. Some of these cosmic rays can penetrate molecular clouds in interstellar space, where they ionize hydrogen atoms to begin the process of forming molecules shown in Figure 4-8.

The cosmic ray particles that do reach the Earth's surface may play a vital role in our existence. Such particles, along with similar fast-moving particles from radioactive minerals (also the result of supernova explosions!) are one cause of *mutations,* sudden changes in individual genes from one generation of life to the next. These changes allow for the basic pattern of evolution, the difference in reproductive ability among members of the same species. Most mutations reduce an individual's ability to survive and to reproduce, so the characteristics typical of such mutations quickly disappear from the general population, but successful individuals owe their success in part to the specific mutations that comprise their genes. If cosmic ray particles do somehow induce mutations, then supernova explosions provide not only the raw material of which we are made, but also some of the changes that allow biological evolution to occur. A relatively nearby supernova would then increase the influx of cosmic rays, and thus the rate of mutations, so that we would tend to evolve more rapidly. Some scientists have speculated that just such a nearby supernova was the cause of the extinction of the dinosaurs about 70 million years ago, as new species of mammals competed with them for survival with ever-increasing success.

Summary

As stars grow older, they exhaust the supply of protons at their centers, depriving themselves of their former ability to liberate kinetic energy from energy of mass. For a time, a star can compensate for the increasing shortage of protons by contracting its central regions, thereby raising the temperature to fuse its remaining protons at an ever-increasing rate. The additional kinetic energy liberated in this way expands the star's outer layers, cooling them slightly in the process and producing a red giant star that has a huge outer surface and a small, fast-burning inner core. More massive stars will run out of protons in less time than the less massive stars, so in a cluster of stars all of the same age, the more massive stars will leave the main sequence of the temperature-luminosity diagram to become red giants *before* the less massive stars do. A red giant star will eventually contract its core enough to start fusing the helium nuclei into carbon nuclei. This helium flash temporarily expands the core, but as helium nuclei fuse completely into carbon, the core again grows denser and denser as the star tries to maintain itself against self-induced gravitational collapse.

Smaller stars will stop contracting their cores after they have fused helium nuclei into carbon nuclei. The star's interior will be so dense at this point that the exclusion principle keeps the electrons from packing any tighter. The electrons hold the nuclei from further compression through electromagnetic forces, so the exclusion principle can support the entire star against its own gravitation. After the star's outer layers evaporate, the inner core persists as a white dwarf star that begins a long, slow fade into dimness.

No white dwarf, however, can exist if its mass exceeds about 1.4 times the sun's mass. Stars with more mass are likely to produce supernova explosions as they age. Supernovae arise from the fact that more massive stars will not become so dense in their centers as less massive stars as they fuse helium nuclei into carbon nuclei. Instead, these stars will fuse carbon into heavier and heavier nuclei to keep on liberating kinetic energy, but at the stage of iron nuclei, no more energy-liberating reactions exist, so the star's core collapses. The collapse produces a tiny, incredibly dense neutron star, and infalling matter from the star's outer layers bounces off the neutron star to produce an outward-directed explosion. This explosion makes the star exceedingly bright for a short while. More important, supernovae seed heavier nuclei throughout the galaxy in which they are located, and they probably produce most of the fast-moving cosmic rays, nuclei and electrons traveling at nearly the speed of light, which may be responsible for some of the mutations that allow evolution on Earth to proceed.

Questions

1. What happens to the central region of a star as most of the protons there fuse into helium nuclei? What happens to the outer layers of the star?

2. What keeps white dwarf stars from collapsing? How does the exclusion principle prevent all the carbon nuclei in a white dwarf star from rushing to the star's center?

3. Do nuclear fusion reactions occur inside a white dwarf star? What makes a white dwarf keep on shining?

4. Which stars do not end their lives as white dwarfs? Do these stars pass through their main-sequence lifetimes more rapidly or more slowly than other stars do? Why?

5. What happens to the stars that do not become white dwarfs? What importance does their fate have for the formation of later generations of stars?

6. Vega has three times the sun's mass and an absolute brightness 30 times the sun's. How does Vega's main-sequence lifetime compare with the sun's?

7. The supernova that appeared in our galaxy in the year 1604 had a distance from us of about 5000 parsecs. In approximately what year did the explosion actually occur?

8. The supernova that appeared in the Andromeda galaxy in 1885 had a distance of 600,000 parsecs, about 100 times the distance of the 1604 supernova. In what year had the 1885 supernova exploded? If both supernovae reached the same absolute brightness at maximum, and if the apparent magnitude of the supernova in Andromeda was +6, what was the apparent magnitude of the 1604 supernova?

9. Why do Population I stars contain a larger fraction of their mass in the form of heavy elements than Population II stars do? Why do we distinguish heavy elements from light elements? To which population does the sun belong?

10. What are cosmic rays? Which part of a supernova may have produced them?

Further Reading

Cameron, A. G. W., 1976. Endpoints of stellar evolution. In *Frontiers of astrophysics*, ed. E. H. Avrett. Cambridge, Mass.: Harvard University Press.

Greenstein, Jesse. 1970. Dying stars. In *Frontiers in astronomy,* ed. by Owen Gingerich. San Francisco: W. H. Freeman and Company.

Oort, Jan. 1970. The Crab Nebula. In *Frontiers in astronomy,* ed. by Owen Gingerich. San Francisco: W. H. Freeman and Company.

CHAPTER **7**

Pulsars, Neutron Stars, and Black Holes

In 1972 and 1973, two spacecraft, called Pioneer 10 and Pioneer 11, left Cape Kennedy, accelerated away from Earth, and later coasted by the planet Jupiter after almost two years of travel. Seven years after the first launch, each of these spacecraft has covered about 3 billion kilometers; Pioneer 10 has passed the distance of Uranus from the sun, while Pioneer 11 was re-directed to pass by Saturn in September 1979. In another 100,000 years, these spacecraft will cover a distance equal to that from the sun to the nearest star, though they are not directed toward Alpha Centauri but simply out into the depths of interstellar space.

· Each of the two Pioneer spacecraft carries a gold-anodized plaque, 15 by 25 centimeters, with an intriguing picture etched on it (Figure 7-1). The right-hand portion of the figure, which drew the most attention on Earth, shows a man and a woman in front of a stylized drawing of the spacecraft (to give the correct sizes). The top of the drawing shows the spin flip of a hydrogen atom (see page 71), while the bottom represents schematically the path of the spacecraft out from the Earth, the third planet from the sun.

But what of the left-hand part of the picture, a sort of spider-web arrangement of lines with dashes along them? This diagram represents the best way that astronomers Carl Sagan and Frank Drake could think of to show our sun's location in the Milky Way galaxy, in case some far-distant civilization might find the plaque after millions of years and wonder from what outlying corner of the Milky Way it might have come. Each of the lines in the spider-web figure shows the direction from the solar system to a *pulsar*, a strange sort of cosmic object that emits pulses of radio waves, and sometimes of light, at precise intervals. The exact intervals between pulses are written in binary notation along each of the 14 lines, using the spin-flip frequency of hydrogen atoms (hence the top part of the drawing) to define the basic interval of time.

124

Even though the Pioneer plaques will only be messages to ourselves for at least the next few thousand centuries, they do show how pulsars can serve as cosmic beacons, lighthouses that differ from one another in the interval between successive flashes. Pulsars may prove extremely useful when one civilization wants to tell another its precise location in space, because they each have a characteristic interval between flashes, recognizable at great distances throughout the galaxy.

What produces these precisely timed beacons, the pulsars? Once again, we meet supernova explosions in a role important to the search for life. Pulsars arise from *neutron stars*, the collapsed inner remnants of stars that have exploded. To see how a stellar collapse can produce a pulsating source of radio waves, flashing on and off at precise intervals of time, we

Fig. 7-1 The plaque that rides the Pioneer 10 and Pioneer 11 spacecraft shows male and female humans with a line drawing of the spacecraft and (bottom) a representation of the spacecraft's trajectory from the solar system. The left side of the picture locates the solar system by means of 14 pulsars, using binary notation to specify the pulsation intervals at the time of launching, in terms of the natural frequency of the 1420-megahertz radio waves emitted by hydrogen atoms (symbolized at top left).

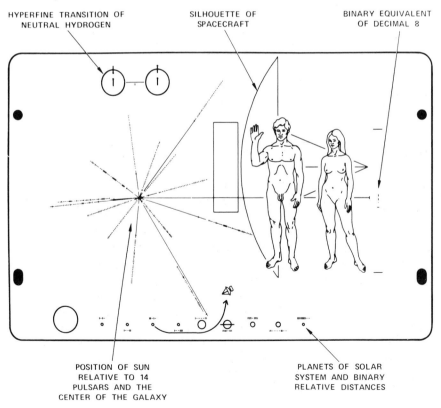

HYPERFINE TRANSITION OF NEUTRAL HYDROGEN

SILHOUETTE OF SPACECRAFT

BINARY EQUIVALENT OF DECIMAL 8

POSITION OF SUN RELATIVE TO 14 PULSARS AND THE CENTER OF THE GALAXY

PLANETS OF SOLAR SYSTEM AND BINARY RELATIVE DISTANCES

must study the process of stellar collapse and the neutron stars that can arise from such a catastrophe.

Collapsing Stellar Cores

When a star's core collapses because it can no longer oppose its own gravitation, the particles within the core rush at one another with immense fury. In the inner regions of the core, the individual nuclei collide with such enormous impact that they break apart into protons and neutrons. Immediately afterward, the fusion of protons with electrons forms neutrons and neutrinos out of the charged particles. The neutrinos can escape, with some difficulty because of the tremendous increase in density, but the neutrons stay behind and the core becomes a neutron star.

Neutrons, like electrons, obey the exclusion principle: They cannot be forced into almost the same place for any length of time. Just as the exclusion principle acts on the electrons in white dwarf stars to hold the electrons apart, so too it acts on the neutrons in neutron stars, but only at much greater densities of matter. A white dwarf star, with a radius close to the Earth's, has a density of matter a million times the density of water. A neutron star, with about the same mass as a white dwarf (say, one solar mass), will have a radius of only 6 kilometers, a thousand times less than the Earth's! Thus, the *volume* of a neutron star will be one billion times less than that of a white dwarf, and the density of matter (mass divided by volume) will be a billion times greater, or 10^{15} grams per cubic centimeter. Such neutron-star matter would weigh a million tons per cubic centimeter at Earth's surface![1]

Neutron Stars and Pulsars

In addition to their characteristically tiny sizes and enormous densities of matter, neutron stars have two other important properties: rapid rotation and strong magnetic fields. Neutron stars rotate rapidly precisely because they have such small sizes. Any spinning object that is relatively free from outside influences will spin more rapidly as it contracts. This fact, which scientists call *conservation of angular momentum*, affects our performance whenever we make acrobatic high dives (Figure 7-2). When we contract, our bodies rotate more rapidly than when we open to full length to enter the water without spinning. The conservation of angular momentum describes the fact that the rate of spin (number of rotations per second) must

---

[1]An object's *weight* measures the force of the Earth's gravity on the object, while the object's *mass* is the amount of matter in it. Neutron-star matter is packed so tightly that each cubic centimeter contains at least 10^{15} times more mass than the same volume of water!

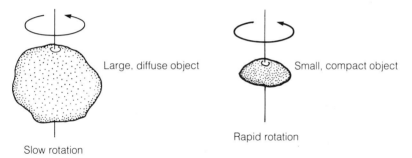

Fig. 7-2   Spinning objects that are relatively free from outside influences will spin more rapidly as they contract. This fact applies to high divers on Earth and to collapsing stars in space, both of which increase their rates of rotation as they decrease their sizes.

vary in inverse proportion to 1 over the *square* of the object's size.[2] Thus, an object that contracts to half its original size will spin four times more rapidly than before. In the case of a neutron star, the collapsing core might shrink 20,000 times, from a radius of 120,000 kilometers to 6 kilometers. If the core has been rotating, the collapse will increase the core's rotation rate by the square of 20,000, or 400 million times! Hence the collapsed neutron star will spin 400 million times more rapidly than before. If the original, uncollapsed core rotated once each day, then the collapsed neutron star will rotate 46 times each second!

The neutron stars' rapid rates of spin have significant effects because the stars whirl their magnetic fields as they rotate. Most stars have some initial magnetic field; our sun, for example, has a relatively weak field that has a north and a south magnetic pole, like a bar magnet. Even the sun's weak magnetic field can affect the motion of charged particles near its surface (Figure 7-3), because the magnetic field produces electromagnetic forces on the charged particles nearby. If the magnetic field had a greater strength, then the particles' motions would respond still more to the presence of the field.

In a collapsing star, the strength of the magnetic field at the star's surface increases as 1 over the square of the star's radius. Thus, the strength of the magnetic field at the surface of a star that shrinks in radius 20,000 times will increase 400 million times, so the entire neutron star becomes a super-dense, spinning magnet, whipping the field around and around at 46 times each second (Figure 7-4). Any charged particles that remain near the star's surface will be accelerated by the motion of the magnetic field and will spiral along the field lines. Some of them will even-

[2]To be precise, we should note that when we speak of the object's *size,* we must really consider only its size in the direction perpendicular to the object's axis of spin.

tually escape into space with quite impressive energies, adding to the cosmic ray flux from the original supernova explosion.

More important, the charged particles accelerated by the rotating magnet will emit photons by the *synchrotron* emission process we described on page 58. These are the photons detected at visible light and radio frequencies that reveal the pulsar's existence.

Near the surfaces of neutron stars, some charged particles constantly appear as neutrons decay into protons, electrons, and antineutrinos. (In the star's interior, such decays are quickly compensated by the additional fusion of protons and electrons into neutrons and neutrinos.) The charged particles are rapidly accelerated almost to the speed of light by the rotating magnetic field. Thus, the particles produce photons of synchrotron emission, which escape outwards from the immediate vicinity of the star's surface. This radiation process decreases the star's rotational kinetic energy by transferring this energy first to the accelerated particles and then to the synchrotron photons. As a result, the neutron star's kinetic energy

Fig. 7-3 The sun's magnetic field, like Earth's, has a north and a south magnetic pole, and basically resembles a giant bar magnet in rotation. Charged particles tend to follow the lines of force that indicate the direction in which the field points at any location near the sun.

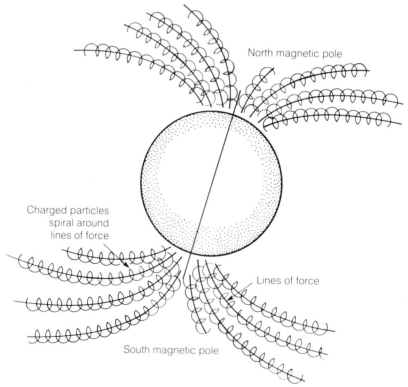

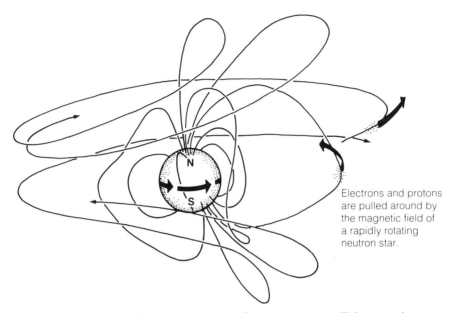

Fig. 7-4 A star that collapses to a dozen kilometers across will increase its magnetic field strength, and its rate of spin, by several hundred million times. Hence, the entire star will resemble a supermagnet that rotates many times per second. As the magnetic field sweeps through space, it accelerates charged particles near the star's surface to enormous velocities.

must decrease, so the neutron star gradually slows its rotation to ten rotations per second, to four, to two, and to even fewer, but at an extremely slow rate, perhaps by one part in a thousand each year.

Astronomers are fairly certain that pulsars are neutron stars. Each pulsar, of which a few hundred have been found so far, produces pulses of photons, rapidly spaced bursts that recur with remarkable regularity anywhere from once every 4 seconds (for the slowest) to 33 times per second (for the fastest). These photon pulses are usually detected at radio frequencies, but two of the best-studied pulsars show photon bursts at gamma-ray, X-ray, and visible-light frequencies, timed in precise synchrony with their radio pulses. According to the best theories now available, the reason that pulsars emit photons in pulses rather than at a steady rate stems from the fact that the pulsars' magnetic fields do not align perfectly with the pulsars' rotation axes (Figure 7-5). Photons emitted by the synchrotron process tend to appear preferentially away from the direction of the magnetic field. Thus, as the neutron stars spin, we sometimes get a full blast of photons and sometimes only a few, with the pattern repeating over and over again as the seconds tick by.

The photons from pulsars arrive with great regularity, but not a perfect one, because the neutron stars' rates of spin must all be slowing down.

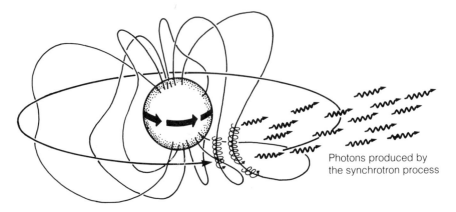

Photons produced by
the synchrotron process

Fig. 7-5 Pulsars presumably emit more photons in some directions than in others. As the regions of greater emission sweep by us, we see a flash of photons. The reason for greater emission in some directions lies in the fact that the magnetic field strength varies in different directions.

For example, at the center of the Crab Nebula, the supernova remnant left behind by the outburst seen in the year 1054, we find a pulsar that blinks on and off 33 times each second, the fastest of all known pulsars (Figures 7-6 and 7-7). Owing to the rapidity of the pulsar's rotation, we may conclude that this pulsar must be extremely young—a conclusion clearly supported by our prior knowledge of the fact that just over 900 years have passed since it first appeared. Precise timing reveals that the period between pulsar flashes is increasing: Each year, the pulse interval increases by about one hundred thousandth of a second.

In 1967, when astronomers detected the first pulsar, they passed through a phase of wondering whether they had not in fact found the first artificial interstellar beacon, another civilization's equivalent of our lighthouses. Each terrestrial lighthouse has its rate of sweep set at a precise interval of time, so that sailors can determine which lighthouse they see simply by timing the interval between flashes. Pulsars would serve admirably in the same manner, as the Pioneer plaques demonstrate, but they seem to be completely natural, though impressive, cosmic timekeepers.

Because pulsars slow down gradually, the interval between pulses grows a bit longer each year, though it takes thousands of years for this interval to change very much. Therefore, if another civilization did find the Pioneer plaque (or if we found a similar picture from them), a comparison of the time intervals given for the pulsars on the plaque with the time intervals at the time of discovery would show how much the pulsars have slowed down.[3] Since we can observe the rate of slowdown over even a few

---

[3]Since pulsars slow down only gradually, another civilization could recognize which pulsars we were describing, even though the intervals between pulses had lengthened.

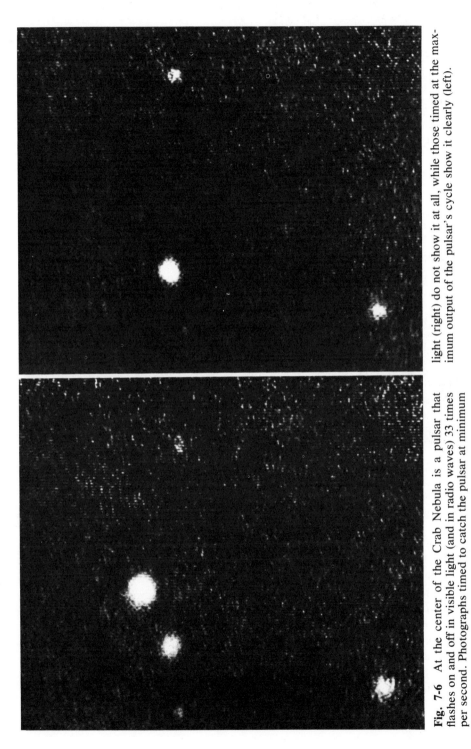

Fig. 7-6 At the center of the Crab Nebula is a pulsar that flashes on and off in visible light (and in radio waves) 33 times per second. Photographs timed to catch the pulsar at minimum light (right) do not show it at all, while those timed at the maximum output of the pulsar's cycle show it clearly (left).

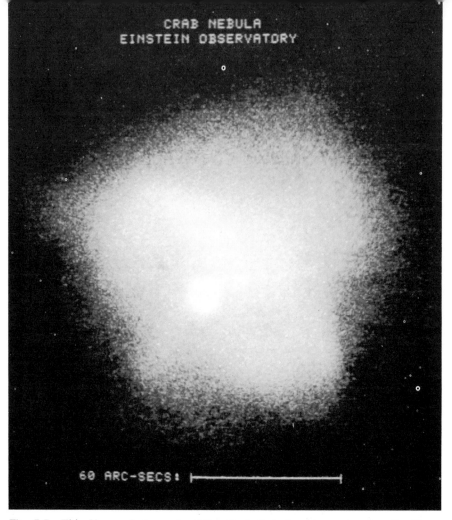

CRAB NEBULA
EINSTEIN OBSERVATORY

60 ARC-SECS:

Fig. 7-7 This X-ray photograph of the Crab Nebula pulsar, showing the X-ray output averaged over all of the pulsar's cycle, shows X-ray emission from the pulsar's immediate surroundings.

years' time (using the incredibly accurate timers now available to astronomers), we could determine just when the spacecraft was launched—and so could the other civilization that we imagine receiving it. Thus, the plaque tells not only where it came from, but when it left as well.

In 1979, astronomers announced the discovery of a strange object in the constellation Aquila. This starlike source of light, called SS 433, has an unusual spectrum that apparently shows the Doppler effect at work on a scale previously unknown in our galaxy. Measurements of the Doppler shift of the spectral features in SS 433 reveal that the object appears to be ejecting matter in two jets that emerge in opposite directions, with each jet traveling at 40,000 km/sec, more than 10% of the speed of light! Although many objects are known to accelerate small numbers of particles (relatively speaking) to almost the speed of light, this is the first object in the Milky Way that has been found to make entire streams of matter move at a significant fraction of the velocity of light.

The most likely explanation of SS 433 is that the source of light is a neutron star about 10,000 light years from us. The neutron star is apparently in mutual orbit with another object, whose gravitational force makes the axis of rotation of the neutron star wobble in space, with a 164-day period. It is interesting to note that an advanced civilization near the neutron star might be able to use the jets of matter to accelerate objects to 40,000 km/sec (see page 370)! But for now we have neither a "use" for SS 433, nor a good explanation of why it is squirting out jets of matter in opposing directions.

Black Holes

Some supernovae completely destroy the star that produced them; others leave a neutron star core behind. But not every supernova explosion leads to one of these two alternatives. Some collapsing stellar cores go beyond any neutron-star possibility to produce the remarkable objects we call *black holes,* regions of space so compact that they almost—but not quite!—disappear from the universe. Black holes are so fantastic to our experience that for decades physicists considered them merely a theoretical construct, unlikely ever to exist in reality. Now, however, we must deal with the possibility that most of the matter in the universe could be in black holes, and that the universe *itself* could be a black hole.

Black holes, the ultimate product of gravitational attraction, represent agglomerations of matter that exert such strong gravitational forces that nothing ever escapes from them. Any object with mass exerts gravitational force on all other objects, and the strength of this force between any two objects varies as the product of their masses divided by the square of the distance between their centers. Let us then consider an object that rests on the surface of a collapsing star. As the star collapses, the distance between the center of the object and the center of the star decreases dramatically (Figure 7-8). As a result, the *square* of the distance decreases still more dramatically, so the gravitational force between the star and the object must *increase* by the same factor. If, for example, the star shrinks in size by a factor of 100,000 (10^5), then the gravitational force will increase by the square of 100,000, or 10 billion (10^{10}) times.

Such an increase in the gravitational force will make the object's escape from the star's surface enormously difficult. We all know that for humans to escape from the Earth's gravity, we must build large, expensive rocket

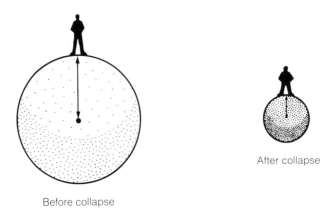

After collapse

Before collapse

Fig. 7-8 As a star collapses, the gravitational force at the star's *surface* increases dramatically, because the distance between the surface and the rest of the star decreases. The gravitational force at the surface varies as 1 over the square of the distance to the star's center, so a decrease of 5 times in radius will produce an increase of 25 times in the gravitational force.

motors to accelerate them. Such rockets must generate an enormous quantity of kinetic energy to overcome the Earth's gravitational force.[4] What is not so obvious is that even particles with no mass—photons, for example—must also expend some energy in leaving the Earth's surface. We have concentrated our attention on Newton's law of gravitation, which tells how particles *with some mass* attract each other. Particles that have no mass at all—photons, neutrinos, and antineutrinos—also feel the effects of gravitational forces, but not in precise accordance with Newton's law, which applies only to particles with some mass.

A famous experiment, performed first in 1919 and with ever-increasing precision since then, shows that the sun's gravitational force will *bend* the trajectories of light rays passing by it (Figure 7-9.)[5] Einstein was the first scientist to predict that gravitational forces actually cause massless particles to deviate from straight-line motion, and a recent experiment aboard the Viking spacecraft on Mars showed that Einstein correctly calculated the amount of this bending, with an experimental error of less than 1 percent.[6]

[4]If we consider a particular object—say, a particular rocket—the amount of energy needed for escape varies in proportion to the mass of the body it seeks to escape from divided by the radius of the body.

[5]Some physicists like to think of gravity as bending space itself, with greater bending near more massive objects. Then photons travel in straight trajectories through bent space!

[6]This result has great importance, because other physicists since Einstein's time have suggested modifications to Einstein's theory of gravitation that predict an amount of bending different from Einstein's prediction. Thus, the verification that Einstein had the correct answer tends to vindicate Einstein's theory as the right one to apply to the real universe.

If photons do feel the force of gravity, then we would expect that photons must use some energy in escaping from a source of gravitational force, just as a rocket must expend energy to overcome this attraction. Indeed, the loss of photon energy has been measured not only on Earth (where the gravitational force is far weaker than it is on highly compressed stars), but also for the sun and for the light from white dwarf stars, where the loss of energy rises as high as $1/200$ of a percent of the original photon energy. This fractional loss was considered enormous when it was first detected (in 1928) but pales to insignificance in comparison with the effect from truly condensed objects.

The theory of gravitational forces shows that if a photon escapes from a source of gravitational force, the photon must lose a certain fraction of its original energy. This fraction will be the same for all photons, so that if, for example, the energy loss were 10 percent, then a radio-wave photon would lose 10 percent of its original energy, while visible-light photons would also each lose 10 percent of their tremendously greater kinetic energies. The fraction of energy lost is proportional to the *mass* of the object divided by the photon's original distance from the object's center. If we look at photons that originate at the object's surface, then *the key ratio is the object's mass divided by its radius*. If this ratio keeps on increasing, then any photon must lose a greater and greater fraction of its energy in escaping from the object's surface. We can imagine the point at which the photon must lose *all* of its energy, which means that the photon will not escape at all. Then the object, by definition, becomes a black hole: No photons can escape. And it turns out that if no photon, no massless particle, can escape, then certainly no particle *with* mass can escape. The

Fig. 7-9 The sun's gravitational force bends the light rays that pass close by from the straight-line paths they would take if the sun did not exist. We can verify this fact by comparing photographs of the stars close to the sun at the time of a solar eclipse with photographs of the same stars several months later, when the sun is no longer in front of those stars.

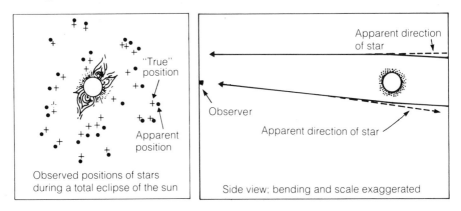

Observed positions of stars during a total eclipse of the sun

Side view; bending and scale exaggerated

object becomes truly black, for it can emit no light, no radio waves, nor anything else.

We can easily calculate the point at which no light can escape from the surface of a collapsed stellar core. Since the energy loss varies as the mass divided by the radius, we find that any object will become a black hole if this ratio rises above a certain number.[7] For an object with a mass equal to the sun's, the critical radius will be 3 kilometers; for an object with 10 times the sun's mass, the critical radius will be 10 times larger, or 30 kilometers. Any object that shrinks within its critical radius thereby becomes a black hole. But which objects are likely to do so?

The answer turns out to be those collapsed stars with more than a few times the sun's mass. We have discussed the fact that only stars with less than 1.4 times the sun's mass can exist as stable white dwarfs. Similarly, calculations show that only those neutron stars with less than three to five times the sun's mass can exist for long. (The fact that we cannot name an exact mass as the possible upper limit reflects the difficulty of making these calculations.) Collapsed stars with masses greater than three to five times the sun's simply cannot support themselves by any means against their self-gravitation. These collapsed stars will undergo the final collapse, falling into themselves until they become black holes.

Can We Find Any Black Holes?

Black holes should "appear" where the most massive, collapsing stellar cores have disappeared. Hence, we could look for black holes where supernovae have flared, since we believe that supernova outbursts are the result of collapsing cores within stars. Unfortunately, black holes are hard to spot; their very nature renders them invisible. But not all chance of detecting their presence vanishes with the catastrophic collapse through the critical radius. First, black holes continue to produce gravitational force. Second, and as a result of the first fact, matter falling into black holes can emit light *before* the matter passes through the critical radius into invisibility (Figure 7-10).

To take the first part first, consider the fact that the gravitational force from a black hole never ceases. In fact, if the mass of a collapsing star does not change; neither will the gravitational force from it on objects that *keep the same distance* from its center. The sun will almost certainly never become a black hole, but if it *did* suddenly collapse to a radius of 3 kilometers (while losing no mass), then the Earth and the other planets would continue to move along their familiar orbits, unaffected in their motion by

---

[7]If we measure mass in units of the sun's mass and radius in kilometers, then this critical number is just about equal to $\frac{1}{3}$.

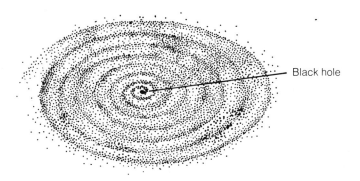

— Black hole

Fig. 7-10   As matter falls into a black hole, it will eventually become unable to emit any light whatsoever. Before the matter falls this far in, however, it can acquire a considerable kinetic energy and can produce large amounts of light from collisions among the particles it contains. A rotating black hole will tend to drag particles into ever-tightening spirals, so the region that emits light should resemble a disk with an invisible center.

the sun's collapse into a black hole. The sun's mass, the planets' masses, and the distances between the sun's center and the planets' centers would all remain the same as the sun collapsed. Of course, we would miss the sun's output of photon energy as we circled the newly formed black hole in sudden darkness, but we could content ourselves with our knowledge that the laws of gravitation remain correct.

Since black holes continue to exert gravitational force, they continue to attract nearby matter, and if, for example, they are surrounded by diffuse gases, they will attract those gases into the black hole, forcing them to accelerate.[8] The particles in the gases will reach quite impressive velocities as they fall toward the black hole, which means that collisions among the particles will occur with large kinetic energies. Such collisions could produce X-ray photons that would carry away some energy from the infalling particles, provided that the photons were produced farther away from the center of the black hole than the critical radius.

When we consider a black hole that forms one component of a double star, we have two chances to detect the existence of the black hole. First, the motion of the other, "normal" star can reveal that another object of starlike mass must be nearby (Figure 7-11). If we can calculate the mass of the unseen companion by studying the star's motion in space, and if this mass exceeds three to five times the sun's mass, then we can conclude with some confidence that the companion must be a black hole. Second,

[8]If the black hole is rotating, as we think it ought to be, then the infalling gas will form a disk of matter around the black hole. The greater the rate of rotation of the black hole, the thinner the disk will be.

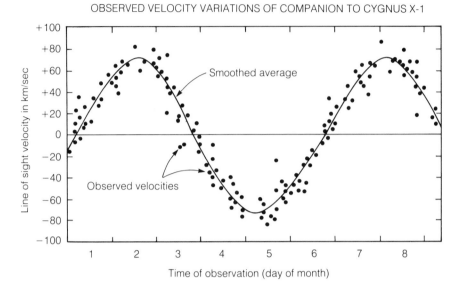

OBSERVED VELOCITY VARIATIONS OF COMPANION TO CYGNUS X-1

Fig. 7-11 The star located at the same position on the sky as the X-ray source Cygnus X-1 shows periodic changes in its *velocity* along our line of sight. These changes repeat in a cyclical fashion, over a time interval of 5½ days. We can calculate from the time interval, and from the size of the velocity changes, that the unseen companion has a mass of at least four solar masses, and more likely equals eight solar masses.

the normal star's outer atmosphere provides a natural source of matter to be attracted by the black hole. Thus, we may expect that the chance of observing photons from the infalling matter would be far greater than in the case of a black hole alone in space.

We have considered the black holes that might form from collapsing stars and would therefore have masses that are at least several times greater than the sun's mass. Black holes with smaller masses *could* exist, but we would not expect them to form under the present conditions in the universe. The Earth, for example, seems entirely capable of maintaining its present size indefinitely, supported against its self-gravitation by the electromagnetic forces that hold rocks together. If, however, some force could compress the Earth, or any object with an equal mass, to a radius of *3 centimeters,* then the object would be a black hole. So far, Earth has endured.

What about black holes with masses much greater than the sun's? Could such huge black holes exist? Figure 7-12 shows the sizes and masses of various objects that occur in the universe. Half of this graph includes familiar objects such as galaxies, stars, planets, and people, while the other half consists of black holes, all objects for which the ratio of mass to radius exceeds ⅓, the critical number given in the footnote on

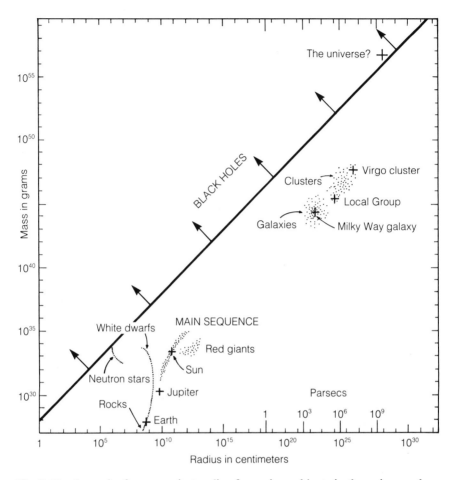

Fig. 7-12 A graph of mass against radius for various objects in the universe shows the critical line that separates black holes from everything else. Any object whose mass and radius place it above and to the left of the critical line must be a black hole.

page 136. The line that separates black holes from other objects represents just those objects for which the ratio of mass to radius exactly equals the critical number.

Some astronomers have suggested that *every* giant galaxy, including our own Milky Way, contains a black hole with at least several hundred million solar masses at its center, and that matter continuously falls into this black hole. This infall would produce great numbers of infrared, visible-light, ultraviolet, and X-ray photons, some of which might penetrate the veil of gas and dust near the galaxy's center, to be observed by ourselves, who (in this picture) would calmly orbit the central black hole at a distance of 10,000 parsecs. This conjecture remains unproven, but if such a massive black hole does exist, it could well have an "ordinary" density of matter inside it, though it would still possess the formidable *event horizon,* or critical radius, that can be crossed in only one direction—inwards. Intriguingly enough, during the past few years, astronomers have deduced from the motions of stars in the giant elliptical galaxy M 87 (Figure 3-7) that this galaxy's central regions seem to contain far more matter than can be accounted for in the form of stars, interstellar gas, or anything else except an invisible black hole. This fact has led to speculation that M 87 does indeed possess a giant black hole at its center, one with about 5 billion times the sun's mass! Such a black hole would span only a few times the size of the solar system but would contain more mass than all but the largest of galaxies!

Is the Universe a Black Hole?

As long as we are thinking about supermassive black holes, we might pause to ask ourselves a fascinating question: Is the universe itself a black hole? If so, then of course we would be living *inside* a black hole, while speculating on the possibility of black holes within a black hole. Notice that since the universe, by human definition, contains everything that exists, nothing can leave the universe, just as nothing can leave a black hole. On the other hand, there are some reasons to believe that the universe is not what we "usually" regard as a black hole. First, the average density of matter in the universe seems to be remarkably constant from place to place, taking the broad view (that is, comparing regions the size of galaxy clusters). Inside a black hole, we would expect a greater density toward the center of the black hole because of the self-gravitation of the matter that forms it. Second, the lack of a "center" in the universe (see p. 37) reminds us that to visualize something while remaining *outside* (either physically or mentally) must be an easier task than to conceive of what something is while remaining inside it. The question of whether what we call the universe is indeed a black hole is not likely to be resolved with certainty in the near future.

Summary

When stars' cores collapse at the ends of their lifetimes, the collapsed remnant of what was once a real star can sometimes become a neutron star. Neutron stars have sizes slightly greater than 10 kilometers, and such tiny sizes make them rapidly rotating, highly magnetic objects. Such spinning magnets apparently form pulsars, sources of photon pulses that repeat with impressive regularity, and the photons themselves are thought to originate from charged particles accelerated almost to the speed of light by the neutron star's rapidly rotating magnetic field.

If a stellar core shrinks too far, it will become a black hole, an object with such strong gravitational forces that neither matter nor photons can escape from it. Matter can fall into a black hole, but nothing that passes inside the critical radius of a black hole will emerge again. We can hope to deduce the existence of black holes if we find them in orbit with ordinary stars; the study of the star's motion implies the presence of a companion and provides information about its mass. Calculations suggest that no neutron star nor white dwarf star can exceed certain limiting masses. Therefore, if we can find an unseen companion and can prove that it has too large a mass to be either a neutron star or a white dwarf, then we may have found evidence of a black hole. We may also detect black holes in orbit with stars from the X-ray photons produced by gas falling into the gravitational "well" of the black hole.

Questions

1. What produces neutron stars? What do neutron stars produce?
2. Can a neutron star expand to become a white dwarf star? Can a white dwarf star shrink so much that it becomes a neutron star? Will neutron stars eventually become black holes?
3. What is a black hole? What makes it work?
4. If the sun decreased its radius about 200,000 times, it would become a black hole. How many times stronger would the gravitational force at the sun's surface become as a result of such a contraction?
5. Is the sun likely to become a black hole? What types of stars are most likely to collapse so far that they become black holes?
6. Can the entire "universe" be inside a black hole? What are some arguments against this view?
7. Why do pulsars blink on and off with an almost constant interval between photon pulses?
8. Why can we identify pulsars within our galaxy simply by specifying the period of pulsation?
9. Can black holes of any mass whatsoever exist? Which masses appear to be the most likely for black holes?
10. Why do some astronomers think that galaxies might have a black hole at their centers? Would most stars in the galaxy fall into these black holes?

Further Reading

Asimov, Issac. 1977. *The collapsing universe*. New York: Doubleday and Company.

Gursky, Herbert, and van den Heuvel, Edward. 1975. X-Ray emitting double stars. In *New frontiers in astronomy,* ed. by Owen Gingerich. San Francisco: W. H. Freeman and Company.

Ostriker, Jeremiah. 1975. Pulsars. In *New frontiers in astronomy,* ed. by Owen Gingerich. San Francisco: W. H. Freeman and Company.

Sagan, Carl. 1973. *The cosmic connection*. New York: Doubleday & Company.

Shipman, Henry. 1976. *Black holes, quasars, and the universe*. Boston: Houghton Mifflin Company.

Stephenson, F. Richard, and Clark, David. 1976. Historical Supernovas. *Scientific American 6:100.*

Taylor, John. 1974. *Black holes*. New York: Avon Books.

Thorne, Kip. 1975. The search for black holes. In *New frontiers in astronomy.* ed. by Owen Gingerich. San Francisco: W. H. Freeman and Company.

More Speculative

Gribbin, John. 1977. *White holes: Cosmic gushers in the universe*. New York: Delta Books.

Science Fiction

Niven, Larry. 1962. *Neutron star*. New York: Ballantine Books.

Niven, Larry. 1975. The borderland of Sol. In *Tales of known space*. New York: Ballantine Books.

PART THREE

Life

When he came wholly forth
I took him up in my hands and bent
over and smelled
the black glistening fur
on his head, as empty space
must have bent
over the newborn planet
and smelled the grasslands and the ferns.
                              —GALWAY KINNELL

    Life on Earth provides us with the single example we know of life in the universe, an example of extraordinary complexity that nonetheless relies on a basic structure of impressive simplicity. The complexity arises from billions of years of evolution; the simplicity from the limited number of atomic species and the rules that govern how they can join together. Both the complexity and the simplicity deserve close examination if we hope to derive an estimate of how life might have arisen elsewhere, and of the stages to which life might have evolved in other situations than our own. In this effort, we must rely on a double extrapolation: first to determine the past course of evolution on our own planet, then to apply this knowledge to the imperfectly known sets of circumstances that occur throughout the universe. Although this extrapolation remains unavoidable, we can rely on well-tested principles of science as we form our estimates of what life must have been like on Earth and what it may be like in the universe at large.

Shiva, the god of destruction and rebirth, appears here trampling upon Asura, the demon of ignorance. Since death, in Hindu belief, is a transition to a new form of life, Shiva is often called the Bright One or Happy One.

CHAPTER 8

The Nature of Life on Earth

Now that we have obtained a general understanding of how the universe fits together, we can proceed to look around us to try to determine the best places to find life. This search would logically begin with a study of the different types of life in the universe, and the various habitats in which we find them. Unfortunately, we cannot use this approach, because we know of only *one* example of life, life on Earth. Therefore, we cannot use the comparative methods that have proven so helpful in studying stars and galaxies. We must proceed as best we can from our single example to try to determine how life begins, why life is what it is and where it is, and how often the conditions occur that lead to the origin and development of life in our galaxy and in the universe at large.

As we study these problems, we must be aware that many characteristics of life on Earth, especially the most visible ones, such as the size and weight of living creatures, are undoubtedly the product of the specific environment offered by the particular planet on which life has developed. We shall be more interested in any "fundamental" properties that we can identify for use in our efforts to determine the probability that life has arisen many times in the universe and is not confined to the single example we know on Earth.

What Is Life?

Since life seems to be a property of matter, we can inquire about life as we might ask about magnetism. What *is* this mysterious property that distinguishes "living" matter from all other combinations of atoms and molecules? Can we describe life in terms of conventional physics and chemistry, or are we missing some special insight that will enable us to solve

TABLE 8-1

The Most Abundant Elements in the Sun, in the Earth, in the Earth's Crust and Atmosphere and in Living Organisms[a]

Sun		Earth		Earth's Crust		Earth's Atmosphere		Bacteria		Human Beings	
Hydrogen	93.4%	Oxygen	50%	Oxygen	47%	Nitrogen	78%	Hydrogen	63%	Hydrogen	61%
Helium	6.5	Iron	17	Silicon	28	Oxygen	21	Oxygen	29	Oxygen	26
Oxygen	0.06	Silicon	14	Aluminum	8.1	Argon	0.93	Carbon	6.4	Carbon	10.5
Carbon	0.03	Magnesium	14	Iron	5.0	Carbon[b]	0.011	Nitrogen	1.4	Nitrogen	2.4
Nitrogen	0.011	Sulfur	1.6	Calcium	3.6	Neon	0.0018	Phosphorus	0.12	Calcium	0.23
Neon	0.010	Nickel	1.1	Sodium	2.8	Helium	0.00052	Sulfur	0.06	Phosphorus	0.13
Magnesium	0.003	Aluminum	1.1	Potassium	2.6					Sulfur	0.13
Silicon	0.003	Calcium	0.74	Magnesium	2.1						
Iron	0.002	Sodium	0.66	Titanium	0.44						
Sulfur	0.001	Chromium	0.13	Hydrogen	0.14						
Argon	0.0003	Phosphorus	0.08	Phosphorus	0.10						
Aluminum	0.0002			Manganese	0.10						
Calcium	0.0002			Fluorine	0.063						
Sodium	0.0002			Strontium	0.038						
Nickel	0.0001			Sulfur	0.026						
Chromium	0.00004										
Phosphorus	0.00003										

[a] Abundances are given as the percent of the total number of atoms in each case, rounded off to two significant figures (after A. G. W. Cameron, C. W. Allen, and B. Mason).

[b] Of the carbon in the Earth's atmosphere, 99.5% is in carbon dioxide molecules and 0.5% is in methane molecules.

this puzzle? Does there exist a "vital force," an essence that can make ordinary matter come alive?

Instead of struggling with the philosophical implications of these questions, we shall use a descriptive approach, trying to identify the characteristics that distinguish life from inanimate matter. *The most striking characteristics of matter that we call alive are the abilities to reproduce and to evolve.* A flame can ignite other flames (thus "reproducing itself"), but cannot evolve into other sorts of flames, such as iron rust. But we have extensive evidence from fossils that life has been able to transform itself continuously, so we are presently surrounded by abundant examples of living matter from whales and redwoods to amoebae and blue-green bacteria. To understand how these transformations could occur, and to gain an appreciation of the unity that underlies the apparent diversity of life, we must examine the atoms and molecules that form living organisms, and the basic chemical reactions that enable matter to remain alive.

We can begin by asking what life is made of. This apparently simple question immediately leads to a useful perspective: Life, as we know it, is made with a recipe dominated by a remarkably small number of ingredients. Eighty-five stable elements exist in nature, ranging from hydrogen (the lightest) to uranium (the heaviest). Just four of these elements—hydrogen, oxygen, carbon, and nitrogen—comprise 95 percent of the matter that we call alive. These four elements are the most abundant elements in the universe, except for the inert gases (chiefly helium and neon), which do not form chemical compounds. In other words, *the composition of living matter resembles the composition of the stars more closely than the composition of the planet on which we find ourselves* (see Table 8-1).[1]

The large amount of hydrogen and oxygen in living organisms follows naturally from the high percentage of water that all life contains, and this percentage seems natural because water is so common at the Earth's surface. But carbon and nitrogen, though abundant in stars, are relatively *rare* on Earth, far less abundant than silicon or aluminum, for example. The concentration of carbon and nitrogen in living matter demands additional explanation, which hinges on the chemical properties of carbon and nitrogen atoms.

The unique ability of carbon atoms to form complex molecules, which can store the information necessary for the continuation of life, clearly makes carbon an important element for biology. But complexity in molecules has no use if the molecules are either unstable or too stable to combine with other molecules. The unusual ability of nitrogen and oxygen atoms to *share more than one electron* with a carbon atom produces strong

---

[1]Does this explain why many cultures believe that their ancestors descended from the sky and that they will return to the stars when they die? Do we see a longing of our atoms to return to their primordial wombs?(!)

but breakable bonds that hold together stable molecules (Figure 8-1). Nitrogen has the additional ability to form a stable gas that aids the cycling of this element between organisms and their environment, while oxygen atoms easily combine with other atoms and molecules in chemical compounds that release energy as they form (Figure 8-2). Given the unusual chemical properties of carbon, oxygen, and nitrogen, we can see that these atoms, together with hydrogen, probably dominate life *not* by chance, but because of the special nature of these particular atomic elements.

Aside from establishing our kinship with the stars, this coincidence of elemental abundances between the universe and life on Earth has reassuring implications in our search for life elsewhere in the universe. If life on Earth were made primarily from such rare elements as hafnium and holmium, we might conclude that our kind of life would be exceedingly rare. But in actuality, at least the *elemental composition* of life as we know it poses no barrier to the prevalence of life in the universe.

Let us continue with the list of ingredients that form life. If we add calcium and phosphorus to our four basic elements, we can account for 98.6 percent of living matter (by weight). The remaining 1.4 percent usually consists of chlorine, sulfur, potassium, sodium, magnesium, iodine, and iron, plus tiny amounts of *trace elements,* which include manganese, molybdenum, silicon, fluorine, copper, and zinc. This completes the list of elements in a typical organism, but great variations in the abundance of minor elements exist from one organism to another, and even the composition of a single cell within an organism can change with time. We can point, however, to a single important relationship: The concentrations of

Fig. 8-1 Oxygen and nitrogen atoms can each share more than one electron with a carbon atom. This produces relatively strong chemical bonds, though the bonds can be broken with an extra input of energy.

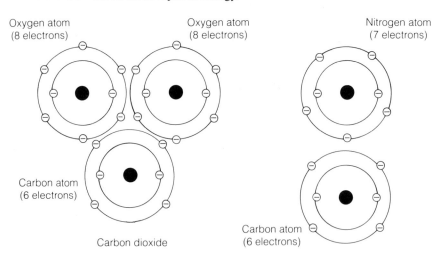

Oxygen atom
(8 electrons)

Oxygen atom
(8 electrons)

Nitrogen atom
(7 electrons)

Carbon atom
(6 electrons)

Carbon dioxide

Carbon atom
(6 electrons)

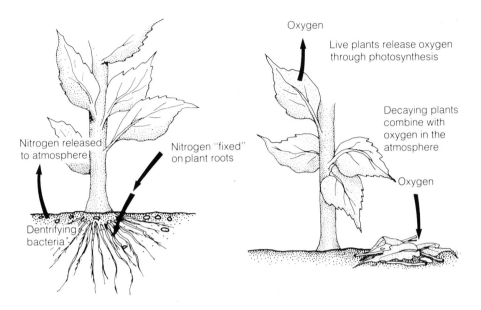

Oxygen

Live plants release oxygen
through photosynthesis

Decaying plants
combine with
oxygen in the
atmosphere

Oxygen

Nitrogen released
to atmosphere

Nitrogen "fixed"
on plant roots

Dentrifying
bacteria

Fig. 8-2 Nitrogen gas can be easily exchanged between an organism and its environment, while oxygen gas easily combines with other molecules to release energy in oxidation.

trace elements in bacteria, in fungi, in plants, and in land animals show a strong correlation with the concentrations of these elements in *sea water.* This correlation not only suggests that life began in the seas, but also indicates that life as we know it indeed arose on Earth rather than arriving here from some other environment (for example, the surface of Mars or an interstellar cloud of gas and dust), where the relative abundances of trace elements would be quite noticeably different.

Thus, the elements that add to the basic four also have much to tell us about living matter. Some of the trace elements, rare as they are in living creatures, are vital to the organism that contains them. Copper appears with great enrichment in certain marine animals, such as oysters and octopi.[2] Zinc forms a component of insulin, while manganese is an essential ingredient in some enzymes. Certain kinds of southwestern "locoweed," a plant that induces hallucinations in animals that eat it, contain as much as 1.5 percent by weight of the rare element selenium. We ought to remember that even though four elements provide the bulk of all living creatures, as many as three dozen additional elements may be essential to such complicated organisms as human beings, proof that life has evolved in complex interaction with its environment.

[2]For these animals, hemocyanin (made with copper) plays the same role that hemoglobin (made with iron) does in vertebrate mammals, such as mice and men.

L-alanine   D-alanine

Fig. 8-3 A typical amino acid has the form shown here schematically. Each amino acid, such as the one called alanine, may come in two forms, which are basically the left-handed (*levo-*, or L) and right-handed (*dextro-*, or D) versions of the same chemical compound.

Biologically Important Compounds

The list of elements in living creatures does not tell us what life is, any more than ingredients describe a cake. We must know how the elements fit together into simple molecules, and how simple molecules join into more complex molecules, in order to see how living organisms work.

When we examine various forms of life at the molecular level, we find that most of them consist of a *small* number of types of rather simple molecules called *monomers*. Among the most important of these monomers are the amino acids that form proteins; other well-known types of monomers are called sugars and fatty acids. Of particular interest is the fact that some of these simple monomers come in two forms, which we can characterize as left-handed and right-handed (Figure 8-3). The two forms are identical in their composition but differ in the way that the atoms that form the monomers fit together, so that one form is basically the mirror image of the other. Except for some proteins found in cell walls of certain bacteria, all the amino acid monomers found in life on Earth are of the left-handed, never the right-handed, variety. This distinctive property of life on Earth apparently arose by chance. The use of just one of the two possible forms increases the efficiency of chemical reactions required to sustain life, but *either* structure would serve this purpose.

The way in which individual types of monomers form larger molecules ultimately distinguishes the matter that is alive from the inanimate matter that forms the bulk of the universe. Living matter consists mainly of long, chainlike molecules, monomers strung together into *polymers* in which a given pattern repeats over and over again, sometimes with small variations (Figure 8-4). In these polymers, ringlike structures and side chains often occur, and the chains themselves sometimes fold into elaborate, highly complex shapes. This ability to assume specific shapes, along with ability to change shape at appropriate times, allows some protein polymers to act as catalysts, sites where chemical reactions can occur much more rapidly than they could otherwise. These catalysts are called enzymes.

An immense variety of chemical compounds can be made by constructing various polymers from the available monomers and linking them together in various ways. An average protein molecule consists of a few hundred amino acid monomers; each type of protein differs from others in the selection of amino acids and in the order in which the amino acids are strung along the polymer chain. Of the tremendous number of *possible* amino acids, only 20 commonly occur in life as we know it (Table 8-2). An average protein molecule, with, say, 100 amino acids, could be made in at least 20^{100} different ways. This enormous number, far greater than the number of atoms in our galaxy, would imply a correspondingly astronomical variety of proteins. Yet most living organisms make, and use, less than 100,000 kinds of protein molecules.

In other words, life shows a tremendous selectivity in the kinds of molecules that it uses. The ability to form highly specific compounds, and to reject a far greater number of molecular types, provides one of the defining characteristics of life as we know it.

In all of these complex molecules, carbon is the element that allows the structure to exist; without carbon, life as we know it would not occur. The complexity of molecular structure *must* exist in order to store, and to transmit, the information that permits one configuration of matter to reproduce itself, or to choose from many chemical compounds just the ones an organism needs to keep alive.

Fig. 8-4 Glycogen, the chief carbohydrate used to store energy in animals, is a polymer made of a long chain of glucose molecules, each of which contains only 22 atoms. The polymeric chain made of such glucose monomers may repeat itself thousands of times.

Part of a glycogen molecule

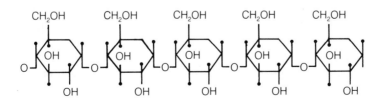

TABLE 8-2

THE 20 AMINO ACIDS FOUND IN LIVING ORGANISMS

Amino Acid[a]	Chemical Formula	Number of Atoms
L-Alanine	$C_3H_7O_2N$	13
L-Arginine	$C_6H_{15}O_2N_4$	27
L-Asparagine	$C_4H_8O_3N_2$	17
L-Aspartic acid	$C_4H_6O_4N$	15
L-Cysteine	$C_3H_7O_2NS$	14
L-Glutamic acid	$C_5H_8O_4N$	18
L-Glutamine	$C_5H_{10}O_3N_2$	20
Glycine	$C_2H_5O_2N$	10
L-Histidine	$C_6H_9O_2N_3$	20
L-Isoleucine	$C_6H_{13}O_2N$	22
L-Leucine	$C_6H_{13}O_2N$	22
L-Lysine	$C_6H_{15}O_2N_2$	25
L-Methionine	$C_5H_{11}O_2NS$	20
L-Phenylalanine	$C_9H_{11}O_2N$	23
L-Proline	$C_5H_9O_2N$	17
L-Serine	$C_3H_7O_3N$	14
L-Threonine	$C_4H_9O_3N$	17
L-Tryptophan	$C_{11}H_{12}O_2N_2$	27
L-Tyrosine	$C_9H_{11}O_3N$	24
L-Valine	$C_5H_{11}O_2N$	19

[a]For those amino acids that have both a left-handed (L) and a right-handed (D) form, we have indicated that only the left-handed stereoisomer appears in living organisms. Only glycine, the simplest of the amino acids, has its left-handed and right-handed forms identical, so that no L or D designation is needed.

The Capacity to Reproduce

The most basic property of life, the essence of its existence and persistence, rests in life's ability to reproduce itself. Cells divide; plants make seeds that grow into new plants; birds and reptiles lay eggs; mammals have babies. Despite all the diversity we see, reproduction at the *molecular* level occurs in all organisms along the same basic plan. In all creatures, a certain kind of long, skinny polymer called DNA (deoxyribonucleic acid) governs the process of reproduction. Furthermore, DNA molecules, together with their close relatives RNA (ribonucleic acid) tell the new organism, as well as the old one, how to function. The general principles by which DNA and RNA molecules work deserve some attention, because although we do not expect identical molecules to appear in other forms of life, we may well expect that molecules of similar *function* do. Thus, it is worth our while to understand how DNA and RNA molecules can do so many things for life on Earth.

DNA molecules store the genetic code that tells the next generation of organisms how to carry out metabolism, grow, and reproduce. These all-important messages reside in the sequence of small molecules that runs along the inside of the DNA (Figure 8-5). In living organisms, DNA has two strands, wound around each other like the handrails of a spiral staircase, the famed double helix of molecular biology. Joining one spiral with the other are *pairs* of four kinds of small molecules (adenine, cytosine, guanine, and thymine) called bases, which *cannot pair at random*. Instead, adenine pairs only with thymine to form one kind of link, and guanine pairs only with cytosine to form the other kind (Figure 8-5). The *order* in which these bases occur along one strand of a DNA molecule provides the genetic code, which consists of a sequence of A's, T's, G's, and C's, standing for adenine, thymine, guanine, and cytosine molecules.

Notice that the order in which the code occurs along one strand of the molecule completely determines the code along the other strand, because each kind of base can pair with just one other kind. During the replication, or copying, of DNA, each DNA molecule separates into two halves along its entire length. As each of the two strands then assembles the other half out of available molecules, the new bases must align themselves so that the pieces fit together exactly as before (Figure 8-5). Thus, each half of a

Fig. 8-5 DNA molecules consist of two spirals, linked by pairs of nitrogen-bearing bases. The long strands, which consist of sugar and phosphate molecules, keep the same distance from one another throughout the length of the molecule. Of the four types of bases, adenine can pair only with thymine, and guanine can pair only with cytosine. Therefore, when the DNA molecule divides lengthwise during replication, each separate half of the molecule can determine precisely which bases can link up to reconstitute the missing strand. In this way, the genetic code, carried by the sequence of bases along each strand, can be reproduced in each of the two "daughter" molecules.

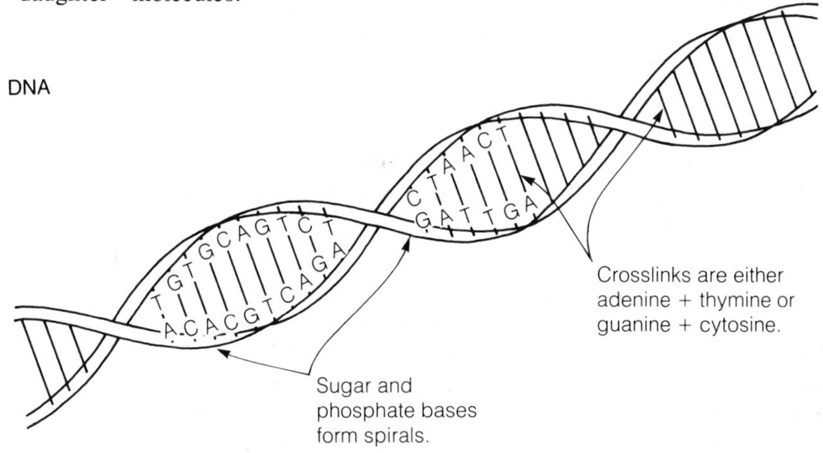

DNA

Crosslinks are either adenine + thymine or guanine + cytosine.

Sugar and phosphate bases form spirals.

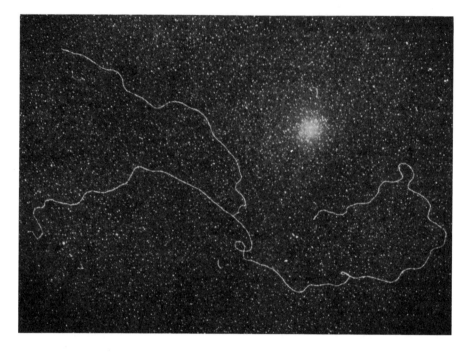

Fig. 8-6 This photograph of a DNA molecule shows the double-stranded spiral in the process of self-replication. As the molecule duplicates itself, a region with two double helices appears until finally two complete double spirals have formed from one original.

DNA molecule can reproduce its partner exactly, to create *two* double strands of DNA, each identical with the "parent" molecule (Figure 8-6).

In addition to reproducing themselves and thus guiding the reproduction of entire organisms, DNA molecules also govern the formation, or synthesis, of protein molecules. This synthesis occurs in several stages, in which RNA molecules, much like DNA, are involved. The DNA molecules that direct the entire synthesis must be able to specify which of 20 types of amino acids should be selected at a particular step in the process. To make this specification, DNA molecules use the sequence of bases—the genetic code—that runs along each of their strands.

How can DNA molecules specify a particular amino acid? If we considered the A's, T's, G's, and C's taken two at a time, we would not be able to specify among 20 different amino acids, because we can pick a sequence of two bases from four types in only 16 different ways. But if we take the bases three at a time, we have 64 possibilities, more than the 20 we need. Careful experiments have shown that DNA and RNA *do* use a triplet sequence for the code that specifies how proteins are to be made.

The sequence of letters (A's, T's, G's, and C's) serve as the letters in the code, and each three bases, taken in sequence, make a "word" (technically, a codon). In effect, each word is the name of a particular amino acid (Figure 8-7). A sequence of 100 to 500 triplets along a strand of DNA can specify a protein or an RNA; if they specify some product of significance to the organism, the sequence is called a *gene.*

As living organisms assemble protein molecules, each DNA molecule serves as a master blueprint. RNA molecules read this blueprint by assembling themselves along a strand of DNA; these messenger RNA molecules then react with another kind of RNA, small RNAs, to which amino acids attach themselves, linking up in the proper order to make a protein. Thus, information flows *from* the nucleic acids to the proteins, but never in the opposite direction.

Impressive as nucleic acids, DNA and RNA, may seem, they are not all-powerful. DNA itself is not "alive" in the strict sense, for DNA molecules need a large and complex environment to reproduce themselves. At the present time, such an environment occurs naturally only within living

Fig. 8-7 The genetic code consists of the 64 possible triplets of three bases in sequence. Each triplet codes an amino acid type, except for three triplets that make a chain come to an end. (See Table 8-2 for the names of the amino acids given here in abbreviation.)

Second letter

First letter	U	C	A	G	Third letter
U	UUU, UUC phe / UUA, UUG leu	UCU, UCC, UCA, UCG ser	UAU, UAC tyr / UAA stop / UAG stop	UGU, UGC cys / UGA stop / UGG try	U C A G
C	CUU, CUC, CUA, CUG leu	CCU, CCC, CCA, CCG pro	CAU, CAC his / CAA, CAG gln	CGU, CGC, CGA, CGG arg	U C A G
A	AUU, AUC ile / AUA, AUG met	ACU, ACC, ACA, ACG thr	AAU, AAC asn / AAA, AAG lys	AGU, AGC ser / AGA, AGG arg	U C A G
G	GUU, GUC, GUA, GUG val	GCU, GCC, GCA, GCG ala	GAU, GAC asp / GAA, GAG glu	GGU, GGC, GGA, GGG gly	U C A G

cells.[3] DNA molecules need special "enzymes" to help them uncoil and to determine which part of the information they encode will be read into an RNA message.

All life as we know it contains DNA and its close relative RNA. These molecules provide the basis for replication, which can occur only within cells. Hence, the smallest, simplest systems that live—that can grow and reproduce—are cells. All forms of life contain proteins, made from amino acids in response to the information carried by the nucleic acids. Proteins are responsible for much of the structure, and for most of the functions, of living cells. Other polymers serve as food (carbohydrates), store and transport energy (fats), and form the basic components of the membranes that organize molecules into cells (lipids).

The Arrow of Time

Let us look more closely at the property of life that we have called most distinctive: the capacity to reproduce and evolve. At the molecular level, life's ability to reproduce arises from the replication of the double spirals of DNA, which can each make two new spirals from one old one. Sometimes a change in the sequence of nucleotide bases, called a *mutation,* occurs in the DNA polymer, a change that causes different amino acids to be selected and thus produces an altered protein. Inheritable changes in DNA sequences can arise from the impact on the DNA molecules of high-energy gamma rays or of fast-moving cosmic rays, or from exposure to various chemical agents called mutagens. The new DNA configuration may be just as stable as the old one; succeeding generations will then carry this mutation if the mutation occurs within the part of the DNA that governs cellular reproduction. Differential reproduction—greater or lesser success of various organisms in producing offpsring—will then determine how well a given mutation becomes established in the general population. The effects of differential reproduction are what we call the *evolution* of species: New species appear, others disappear.

The evolution of living creatures seems to violate our commonsense knowledge that *disorder tends to increase.* All around us, we see configurations of matter move from order to disorder, from improbable states to more probable ones. Paint weathers; rocks crumble; iron rusts; wood decays; stars radiate away their energy over billions of years. But here on Earth, life continues to combine elements into specific molecules and monomers into lengthy polymers, making ever-greater complexity and order from simplicity and disorder. A single DNA molecule, a million times

---

[3]Viruses, which are often called the simplest living organisms, feel this same restriction. Made only of a nucleic acid surrounded by a coat of proteins, viruses cannot reproduce themselves outside of cells. (If only they could, we would be spared many diseases!) The same restriction holds true for viroids, still smaller objects made of naked RNA molecules.

longer than it is wide (but still only a millimeter long at best), represents an exceedingly nonrandom bit of matter. Such molecules store the tremendous amounts of information needed to carry out life's activities, information that can be preserved undamaged through thousands of replications. Hence, a great degree of order must go into keeping matter alive. This order not only persists but has actually *increased* as life has evolved to ever more complex forms on Earth.

How can we explain this apparent contradiction—the maintenance, and even the increase, in the order and complexity of life in a universe that inexorably heads to increasing disorder? The resolution lies in a consideration of the total system, of which life is just one part. Life on Earth does not form a closed system. Instead, life can maintain its highly improbable configuration only at the expense of its environment; that is, life can become highly organized only by increasing the disorganization of its surroundings. The disorder of the total system increases, while the disorder of living creatures within it decreases.

Energy

The various chemical reactions that occur in living organisms require some source of energy. From a chemical standpoint, we need to add kinetic energy to make the reactions proceed. Billions of years ago, living organisms may have drawn some of this energy from the heat of the Earth itself, as a few organisms in hot springs still do. The *dominant* source of energy throughout the lifetime of our planet has been the sun. Thus, most forms of life on Earth have linked their destiny to the energy that arises from thermonuclear fusion deep within our star. Over the past few billion years, many forms of life have developed the ability to obtain energy directly from sunlight through the extraordinary process of *photosynthesis*. In photosynthetic activity, sunlight helps to convert carbon dioxide and water into organic (carbon-bearing) molecules, releasing oxygen molecules as a by-product (Figure 8-8).[4] We can represent photosynthesis by the following equation:

Six carbon dioxide molecules   one organic molecule

  + six water molecules   =   +

   + SUNLIGHT ENERGY   six oxygen molecules

If we prefer chemical symbols, the equation of photosynthesis is

$$6 \ CO_2 + 6 \ H_2O \xrightarrow{\text{light}} C_6H_{12}O_6 + 6 \ O_2$$

[4]Bacterial photosynthesis, which developed before the more familiar plant photosynthesis, does not always involve the release of oxygen molecules (see page 187).

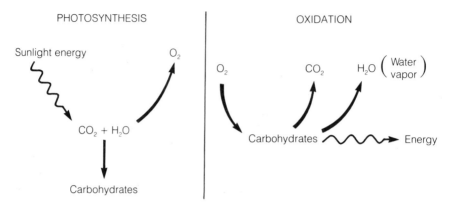

PHOTOSYNTHESIS

OXIDATION

Sunlight energy

O_2

O_2

CO_2

H_2O $\left(\begin{array}{c} \text{Water} \\ \text{vapor} \end{array} \right)$

$CO_2 + H_2O$

Carbohydrates

Energy

Carbohydrates

Fig. 8-8 In plant photosynthesis, sunlight energy helps to make carbon dioxide and water molecules combine, forming carbohydrate molecules with the release of oxygen molecules. The carbohydrates store energy in their chemical bonds, which can be released upon oxidation, when oxygen combines with the carbohydrate molecules.

As we pass from the left to the right side of this equation, photon energy from the sun becomes chemical energy, energy that is stored in the bonds of the organic molecules. On Earth now, this activity occurs predominantly in the chloroplasts of the organisms we call plants (Figure 8-8), allowing the plant to store energy for future use. Photosynthesis produces relatively complex carbon compounds from atmospheric carbon dioxide and releases oxygen molecules into the atmosphere.

In this way, life has solved the need for energy acquisition and storage, allowing some forms of life to feast on the energy that plants have stored in a manner that is complex in structure but simple in operation. The organization of highly specialized plants and animals has such solidity that the *order* can persist while the matter of which life consists constantly changes. We eat, but we do not become what we eat (despite popular sayings to the contrary). Our bodies assimilate the "dead" matter we ingest and use its constituents to remain alive; in this sense, the "dead" food regains "life." Like all other living creatures, we constantly exchange the matter within ourselves for new atoms and new molecules, yet we remain much the same.

A similar pattern holds true for life itself. We tend to be impressed by the fragility of life, for we are surrounded by organisms busily going through their cycles of birth, reproduction, and death. But if we take a longer view, we can be impressed that this exceedingly improbable state of matter that we call life has remarkable continuity. It has persisted on Earth for at least 3.4 billion years, a span that is greater than the lifetimes of many stars. How has it managed to do this? There are many answers, but the key to continued survival is a steady source of energy. The development of photosynthesis has assured terrestrial life a chance to survive on a cosmic time scale.

During this long period, life has changed the face of our planet. Consider the most important side effect of plant photosynthesis, the release of oxygen molecules. This oxygen enters the Earth's atmosphere, where it provides an additional source of energy for any form of life that can use it in *oxidation*, the process of combining oxygen with other molecules. We can see from this why the animal kingdom can be called parasitic, since without plants, animals could neither develop nor live. We may not be able to tell whether the chicken or the egg came first, but we do know that oxygen-producing organisms came before oxygen-using animals as life evolved on Earth.

Let us look once again at our equation for photosynthesis. If we reverse the process and go from right to left in the equation, we are describing oxidation or respiration, in which energy is liberated rather than stored. The released energy may appear in the form of work, heat, or even light. The most familiar examples of oxidation are the burning of plants, wood, coal, or oil, all of which liberate the solar energy stored by photosynthesis. The fact that large deposits of energy-rich organic material exist on our planet indicates that photosynthesis has dominated respiration over past ages; that is, the *net* direction of the arrow in our equation has been as drawn, from left to right. Human activities, however, often have exactly the opposite goal in mind, and we are rapidly depleting the richest deposits of energy stored through photosynthesis: petroleum, natural gas, and coal.

The Unity of Life

We have now identified the elements of which life consists, the principal molecules formed from these elements, the polymers made of smaller molecules, and some of the basic reactions that occur in any configuration of matter that we call alive. We have seen that despite the extraordinary diversity of life shown on our planet, an underlying unity appears at the molecular level of operation. Just 20 amino acids, out of the thousands that could exist, are used to make proteins, and these same 20 appear in all forms of life. The different kinds of life all use the same genetic code, with the same four bases, taken three at a time to specify the particular amino acids. Organisms differ in the sequence of base pairs—which causes them to be different types of organisms—but the code remains identical.

The unity of life's chemistry suggests that all the life we see around us arose in much the same way. This conclusion in turn leads to the expectation that life on some other planet could all be quite different from life on Earth, even if that life uses the same elements that we do. We might expect extraterrestrial life to use a different set of amino acids, for example, or different types of bases in its equivalent of DNA. But life might, of course, differ in still more fundamental ways from ours, perhaps by using a completely different chemical system for the processes of replication, information storage, and energy transport.

We would have a far better perspective for assessing these possibilities if we understood the *essence* of life: what happens to matter to bring it to the level of complexity where reproduction can occur. To this question we still have no answer; no one has yet made a self-replicating, helical polymer from simple monomers without using products furnished by living cells. Then how *did* nonliving matter assemble itself into a self-replicating configuration? We shall deal at some length with the fundamental question of life's origin in the next chapter without reaching any detailed conclusions about the exact steps by which life began. If we discovered another kind of life somewhere else in the universe, the two examples together would surely tell us much more than our single example about the origin of life from inanimate molecules.

Summary

Life on Earth has a composition that more closely resembles that of the sun and other stars than the composition of the matter that forms the Earth. We may conclude from this fact, and from more detailed studies of the ways that life behaves, that the basic elements that appear in living organisms—carbon, oxygen, hydrogen, and nitrogen—appear to have a special role that could make them essential for life throughout the universe.

We must, however, be cautious in the conclusions we draw from our *single* example of life, life on Earth. We refer to all of life as a single example because every organism that reproduces does so in the same way, by the doubling of long, double-stranded, spiral molecules of DNA. DNA molecules carry the genetic information that determines what the next generation will look like, as well as the additional information needed to tell the various parts of an organism how to function. When DNA molecules split, each half of the spiral can reform the missing half from smaller molecules, for two reasons: Only certain molecules can fit into the niches created by the splitting, and the order in which they fit is uniquely specified by the "parent" molecules.

We can define living organisms through their capacity to reproduce themselves and to evolve, but the basic activity of an individual organism—when it is not reproducing itself—consists of continuing its metabolic activities through the direct or indirect use of sunlight energy. Plants can take direct advantage of sunlight through the chemical reactions involved in photosynthesis, which stores sunlight energy in the chemical bonds of carbohydrate molecules formed from water and carbon dioxide. Animals (and some rare plants) depend on this stored energy, which they release through the digestion and assimilation of plant matter, or of other animals that have eaten plants. In addition, the oxygen molecules released in photosynthesis provide another important contribution from plants: The metabolic processes in animals consist of combining these oxygen molecules with organic matter from plants.

In other words, plants make carbohydrates and oxygen (more precisely, they release oxygen molecules into the atmosphere as part of their photosynthetic reactions), while animals burn carbohydrates and oxygen. On Earth, we still find a net surplus of plant-made carbohydrates, along with the hydrocarbons (natural gas, petroleum, and coal) that began with decaying plant matter. We have come to understand and appreciate the cycle of life, in which plants had to precede animals, and through which we humans find the food, oxygen, and hydrocarbon fuels we consider essential. What remains unknown, at least in any true detail, is the way in which the tremendous variety of plants and animals began to emerge from a small number of simple molecules.

Questions

1. How would you define "life"? Can you see any exceptions you would like to make to this definition? For example, would robot machines, programmed to make other robots, be "alive" by your definition? Why or why not?

2. Why do we say that the composition of living organisms resembles the composition of the stars? What does this imply for theories of how life might have originated throughout the universe?

3. Why is carbon such an important element in all living creatures on Earth?

4. What are polymers? In what form do smaller molecules, the sort we call monomers, string themselves together: in straight lines, in rings, or in more complicated structures?

5. How do DNA molecules replicate, or reproduce themselves?

6. What other key functions besides governing the reproductive process do DNA molecules regulate?

7. What do protein molecules do in living systems? Carbohydrate molecules? Molecules called "lipids" and "fats"?

8. How do animals use the sunlight energy that plants have stored through photosynthesis? Describe the chemical reactions that first stored the energy, and later, in almost the reverse reaction, released it.

9. Where did oil, coal, and natural gas acquire the energy that they contain, ready to be released upon burning?

10. Would we expect another form of life, in some other planetary system, to be made of about the same elements as life on Earth? Would we expect such life to use the same kinds of amino acid molecules as we do? Should we expect such life to have developed DNA molecules like those in living organisms on Earth? Why?

Further Reading

Asimov, Isaac. 1962. *The genetic code.* New York: Signet Books.

Beadle, George, and Beadle, Muriel. 1966. *The language of life.* New York: Doubleday & Company.

The Biosphere. 1970. [eleven articles.] San Francisco: W. H. Freeman and Company.

Calvin, Melvin. 1969. *Chemical evolution: Molecular evolution towards the origin of life on the earth and elsewhere.* Oxford: Oxford University Press.

Dawkins, Richard. 1976. *The selfish gene.* Oxford: Oxford University Press.

Edsall, John, and Wyman, Jeffries. 1958. *Biophysical chemistry.* New York: Academic Press.

Eiseley, Loren. 1946. *The immense journey.* New York: Vintage Books.

Lanham, Url. 1978. *The sapphire planet.* New York: Columbia University Press.

Mason, Brian. 1966. *Principles of geochemistry.* 3rd ed. New York: John Wiley & Sons.

Watson, James D. 1976. *Molecular biology of the gene.* 3rd ed. Menlo Park, Calif.: W. A. Benjamin.

CHAPTER 9

The Origin of Life

Life on Earth began at least 3.4 billion years ago, and may have appeared within a few hundred million years after the Earth formed, just about 4.6 billion years ago. To trace the way in which life originated, we must study the conditions on the "primitive" Earth, soon after our planet formed. In particular, we must determine the composition of the Earth's primitive atmosphere, since the chemical reactions that are essential to life have always involved the blanket of gas that surrounds the Earth.

Because of erosion and the motions of the Earth's crust, the geological record of the first billion years of this planet has vanished forever, dragged down into the mantle, including evidence of the impact craters that doubtless covered the Earth's surface, as they still cover the surfaces of Mercury, Venus, Mars, and the moon. These motions are driven by convection currents within the Earth's mantle, which derive their energy from heat liberated by radioactive minerals in the Earth's deep interior. Not only has the primitive crust vanished, but even the present configuration of the continents is a relatively recent phenomenon, for just 200 million years ago they were much closer together (Figure 9-1).

We can, however, reconstruct some of the Earth's early history by using our (admittedly limited) knowledge of the way in which the planets formed. The primitive Earth that we then visualize is a rocky spheroid, of the same size and composition as the Earth today but with a greater rate of heat energy released from radioactive minerals. The most obvious differences, and the most important ones for creatures on its surface, between the Earth today and the Earth of 4 billion years ago arise from the changes in the Earth's atmosphere. These changes are more difficult to reconstruct.

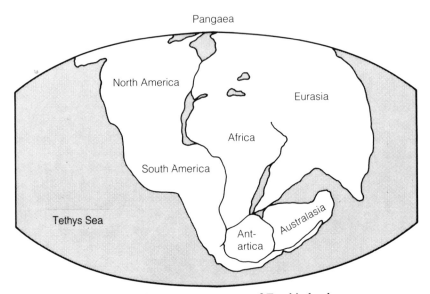

Fig. 9-1   Two hundred million years ago, most of Earth's land areas were concentrated into a single continent, Pangaea, allowing living species to diffuse more easily than is now the case.

How the Earth Got Its Atmosphere

The Earth apparently grew to its present size through the accretion, or sticking together, of smaller particles within the cloud that formed the solar system. Because hydrogen formed the largest fraction of this contracting cloud of gas and dust, many scientists believe that hydrogen must have formed a significant part of the Earth's primitive atmosphere. The presence of large amounts of free hydrogen atoms and molecules ready to combine with other molecules creates what chemists call *reducing* conditions. The traditional picture of the Earth's primitive atmosphere suggests a highly reducing gas mixture, made of methane, ammonia, water vapor, and hydrogen molecules. This is similar to the present atmospheres of the giant planets Jupiter and Saturn, where primitive conditions have persisted down to the present day. The assessment of our original atmosphere as highly reducing has widespread acceptance, but it may require some adjustment in the light of new discoveries.

Through studies of the abundances of the inert gases neon, argon, krypton, and xenon that form a tiny part of our atmosphere, and through careful investigation of the oldest rocks, scientists have recently found evidence that the Earth probably *never* had an atmosphere captured from the cloud of dust and gas that formed the planets. As the process of accretion built the planets, it may have occurred in such a way that the rocky material richest in *volatile* elements joined Earth last (Figure 9-2). This material must have been similar to some of the meteorites and comets that we find in our solar system today (see Chapter 12). The volatile elements, the lightest and most easily vaporized, include the hydrogen, carbon, nitrogen, and oxygen that form our present atmosphere and also life itself.

According to this view of our planet's formation, the volatile matter that permeated the outer layers of Earth (and of the other inner planets) would have included not many individual hydrogen atoms or hydrogen molecules (H_2), but a great number of hydrogen atoms bound into other molecules such as water. Heated by its collision with the forming Earth, the late-accreting material would have its volatiles vaporized, thus producing the Earth's original atmosphere. Ultraviolet light from the young sun would have broken some of these molecules apart, and this process, called photodissociation, along with chemical reactions of the gases with the primitive crust of the planet, would have yielded a mildly reducing atmosphere made of CO, CO_2, N_2, H_2O, and a small amount of H_2. We find support for such a mildly reducing primitive atmosphere, and against a highly reducing, extremely hydrogen-rich one, in the geological record on Earth. Even the oldest rocks, about 3.8 billion years old, show no evidence for an atmosphere as rich in hydrogen atoms and hydrogen molecules as the traditional model requires.

Whether the Earth's primitive atmosphere was highly reducing, as in the traditional view, or mildly reducing, the more modern opinion, we do know some facts about the way in which the atmosphere has changed with time. First, any hydrogen atoms and hydrogen molecules must have escaped from the Earth during its first few hundred million years, because the planet had an insufficient gravitational force to retain them. Hydrogen bound into heavier molecules, such as water vapor, would have remained, but the free hydrogen vanished early in the Earth's history. Second, what-

Fig. 9-2 The material richest in the volatile elements—hydrogen, carbon, nitrogen, and oxygen—joined the Earth last as the proto-Earth grew into the present Earth through the process of accretion. The earlier stages of accretion occurred at higher temperatures than the later stages.

Accreting earth

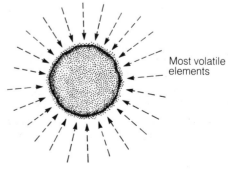

Most volatile elements

Fig. 9-3 Volcanoes, such as this one in Hawaii, release new gases into our atmosphere even today. The venting of gases by molten magma gave the Earth its primitive atmosphere.

ever the primitive atmosphere of the Earth may have been, massive outgassing from the Earth's crust soon overwhelmed it with gases released from rocks heated by the short-lived radioactive elements present at that time. This release of volatile elements, which occurs in a much milder form in volcanoes today, put great amounts of water vapor, carbon dioxide, nitrogen, and carbon monoxide into the atmosphere. Thus, essentially all of the water in our oceans has come out of the rocks that now form the Earth's crust and upper mantle (see Figure 9-3).

Soon after the Earth formed, then, its atmosphere was probably made mostly of water vapor, nitrogen, carbon-oxygen compounds, and a small amount of escaping hydrogen. Once the bulk of the hydrogen had escaped, a significant change occurred: The atmosphere could produce enough ozone to shield the Earth's surface against ultraviolet light from the sun.

Ozone molecules (O_3) each consist of three oxygen atoms, while ordinary oxygen molecules (O_2) are pairs of oxygen atoms. When ultraviolet light strikes molecules of water vapor (H_2O), it will break the molecules apart into hydrogen and oxygen atoms. So long as hydrogen molecules were present in the atmosphere, however, any free oxygen atoms would soon combine with hydrogen. Once the hydrogen had evaporated from Earth, ozone and oxygen molecules *did* form, in small amounts to be sure, but enough to prevent ultraviolet light from reaching the Earth's surface.

Ozone molecules have such a great ability to absorb ultraviolet light that even a small concentration of ozone protects us against the deadly (to us!) ultraviolet light from the sun.

We must therefore add to the water vapor, carbon dioxide, nitrogen, and carbon monoxide in the later atmosphere a small amount of oxygen (perhaps less than 1 percent), and a far smaller, though crucial, amount of ozone. We still seem nowhere near the Earth's present atmosphere, made mostly of nitrogen and oxygen molecules, with small amounts of water vapor, only traces of carbon dioxide, and next to no carbon monoxide. What happened to produce these changes?

The basic answer is: Life appeared on Earth. In particular, life has provided the bulk of the oxygen in our atmosphere. It has helped to remove most of the carbon dioxide, and to retain most of the nitrogen. Let us therefore follow the history of each of the basic components of our atmosphere to see what has happened during the past 4 billion years and how the present atmospheric composition has been established.

First of all, the *water vapor* that was produced in such abundance in a sense still forms part of the atmosphere—the oceans. If the Earth were much warmer, we would not have oceans. A cycle of evaporation and rainfall, followed by water runoff into the seas, keeps a small fraction of the water in the form of atmospheric vapor, but most of it resides as a liquid in the great reservoirs that cover 71 percent of the Earth's surface. These oceans, which set Earth apart from all other planets, have probably been in existence from the earliest period of the Earth's history.

The *carbon dioxide* that once ranked second in abundance among the atmospheric constituents has become locked up as chemical compounds in rocks, principally calcium carbonates such as chalk and limestone. These rocks, which form on the ocean bottoms, use up carbon dioxide dissolved in sea water; as this carbon dioxide disappears into rock deposits, more carbon dioxide dissolves into the water, and so on. Although calcium carbonates will form even if life does not exist, the presence of living marine creatures accelerates the process dramatically. Because marine life has been widespread for billions of years, almost all the carbon dioxide has disappeared from our atmosphere; a small amount remains because the weathering of rocks, respiration, and the decay of organic matter return this gas to the air.

Oxygen, as we have seen, existed only in small amounts before life began to flourish. Green plant photosynthesis has increased the oxygen concentration to 21 percent of our present atmosphere from a much lower abundance. Fossil records show that rocks formed and buried more than about 2.5 billion years ago are suboxidized, formed under oxygen-poor conditions. Fully oxidized rocks come from only the past 2.5 billion years. Confirming evidence for the time of oxygen increase comes from the layered rocks called stromatolites, made from colonies of blue-green bacteria living

Fig. 9-4 Stromatolites formed under conditions that are found only rarely on Earth today. These modern stromatolite beds on the western coast of Australia are exceptional now, but represent the conditions under which stromatolites were made billions of years ago.

at the boundaries between water and rock sediments. Ancient stromatolites reach back with wide dispersion around the Earth to times of 2.0 to 2.3 billion years ago, in agreement with the data from mineral deposits. We may therefore conclude that oxygen-releasing organisms appeared in large numbers shortly after 2.5 billion years ago. Large plants—trees, flowers, and grasses—go back no more than 600 million years, so we have blue-green bacteria to thank for the basic oxygen enrichment (Figure 9-4).

The *nitrogen* that now forms 78 percent of our atmosphere came from outgassing, but its persistence is aided by the existence of life. Every time lightning discharges occur in our atmosphere, some of the atmospheric nitrogen combines with oxygen to form nitric oxides, which rain washes into the soil and oceans (Figure 9-5). If the Earth lacked life, only the slow weathering of the rocks would return this nitrogen to the atmosphere, but denitrifying bacteria, tiny organisms that break apart nitric oxides, live in Earth's soil in immense numbers and play an important role in recycling the nitrogen.

The *carbon monoxide* that was once an important component of the Earth's atmosphere has long ago combined with oxygen to form carbon dioxide. Once photosynthesis began to liberate large amounts of oxygen, this highly reactive element tended to combine with every available chemical compound that was not fully oxidized. As we have seen, the carbon

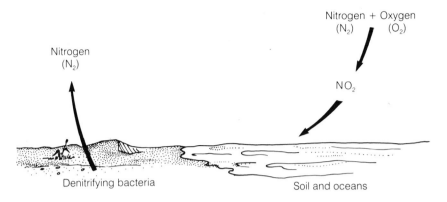

Fig. 9-5 Nitrogen in our atmosphere combines with oxygen to form various oxides, which wash out of the air and into the Earth's soil and oceans. Denitrifying bacteria in the soil return nitrogen to the atmosphere, creating the balance that allows the nitrogen to continue to form 78 percent of the air.

dioxide has mostly been locked up in carbonate rocks, the bulk of which have formed through the activities of living organisms.

Argon, the last important component of the present atmosphere, forms about 1 percent of our air. Here we have one gas that is not affected by life, for argon is an inert element that does not form chemical compounds. Argon arises from the decay of small amounts of radioactive potassium in the Earth's crust, so we should expect to find some argon on any rocky planet in any planetary system, provided that the temperature was not so high at the time the planet formed that potassium could not condense.

We have seen how the Earth's original atmosphere was changed by ultraviolet radiation, by chemical reactions, by outgassing, and by the escape of hydrogen, so the carbon dioxide, nitrogen, hydrogen, and carbon monoxide yielded to the nitrogen and oxygen we now breathe. The water vapor—along with the oceans!—has probably been present throughout the Earth's history. We must now attempt to follow the ways in which life has developed in Earth's changing atmosphere, whose largest changes appear to have been caused by the activities of life itself.

Early Ideas about the Origin of Life

We have stressed the essential *unity* of life at the molecular level, and have suggested that this unity indicates a descent from a set of common ancestors. The next chapter follows this descent in detail, but we can recognize immediately that *the complexity of life has increased with time.* The earliest rocks that indicate the presence of life contain evidence only of *microbial* life, and older rocks show no evidence of life at all. We can easily accept the continuing increase in the complexity of life, but we have more trouble explaining how life began. How did atoms and molecules, on the surface and in the atmosphere of Earth, ever combine into assemblages that are alive?

Human intuition usually invokes outside intervention in this problem of life's origin, a reasonable extrapolation of our observations of how we make things, and how new humans come into existence. Some force or causative agent "makes it happen." The hidden assumption in applying the rule of causation to life is that something "alive" had to exist *before* life could appear on our planet. Similarly, most religions invoke outside intervention, usually in the form of a divine and omnipotent being, to explain the appearance of life.

Some scientists, still imbued with the idea of a causative agent, have suggested that life appears everywhere in the universe. This notion of panspermia, developed in 1903 by Svante Arrhenius, suggests that life floats as spores through the interstellar medium and occasionally comes to rest on a planet, where replication and evolution can occur. As we have learned more about the hazards of space travel, the hypothesis of panspermia has become less and less believable. The combination of high-energy ultraviolet light, X rays, and cosmic rays in interstellar space would probably prove lethal over the length of the trip, a million years or more for random wandering from one planetary system to the next. Furthermore, the panspermia hypothesis does not confront the question we are trying to answer. If the Earth was impregnated by spores from outer space, where did the spores come from? Finally, the relative abundances of rare trace elements in life match those in sea water so closely that they seem to require a terrestrial origin for terrestrial life.

The Chemical Evolution Model

How *did* life arise? A wide variety of laboratory experiments have shown that *in the oxygen-rich atmosphere we now have,* the chemical compounds that form all living systems *cannot arise from spontaneous chemical reactions.* If the transition from inanimate to living matter occurred on Earth, conditions must have been quite different at the time this change occurred. This requirement fits well with what we know, and with what we can deduce, about conditions on the early Earth.

If we adopt the traditional view that Earth's primitive atmosphere consisted of methane, ammonia, and hydrogen molecules, with water able to condense, we can easily demonstrate that in such an environment, the production of organic molecules has a great likelihood. The energy needed to drive the chemical reactions that make organic molecules could have come from many sources, but especially from the sun (Figure 9-6). Solar ultraviolet light would have reached the surface of the primitive Earth because no ozone existed as an ultraviolet filter. Such high-energy photons

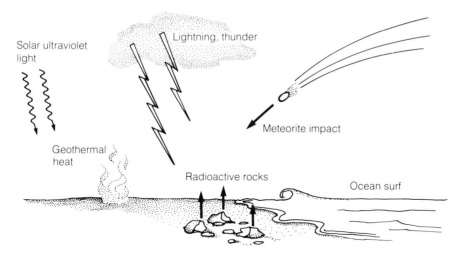

Fig. 9-6 On the primitive Earth, chemical reactions that require energy could have obtained this energy from ultraviolet light from the sun, from lightning discharges, from localized geothermal heat sources, from radioactivity, from meteorite impacts, or from the shock waves produced by ocean surf.

would have been the primary energy source, but additional energy could have come from lightning discharges, from local geothermal sources (hot springs and volcanoes), from radioactive elements in rocks, from impacts by meteorites, and even from shock waves generated by thunder or by ocean surf.

It may seem ironic that solar ultraviolet light *helped* life begin, since today ultraviolet light destroys many molecules essential to life. This apparent paradox can be resolved by noticing that once organic molecules had formed, they could be washed out of the atmosphere into pools and ponds; or they could form at the interfaces between water, land, and air, from which they could enter deeper water. Just a few meters of clear water or a thin layer of other organic compounds can provide adequate screening from solar ultraviolet light (Figure 9-7).

The traditional scientific picture of life's formation on the primitive Earth involves the spontaneous formation of complex molecules from atmospheric gases, and the molecules' later concentration in pools of water. Some of the chemical reactions could have been catalyzed, or speeded up, by soil materials with suitably active surfaces. The runoff of rainwater would have helped to bring the product molecules into ponds, lakes, and tide pools at the edges of oceans. These pools and ponds were rich in organic compounds and dissolved minerals. They probably served as the nurturing, protective environments in which molecules interacted at random, until some of the reactions produced compounds that could serve as guides in the formation of other molecules. Grains of clay at the edges of

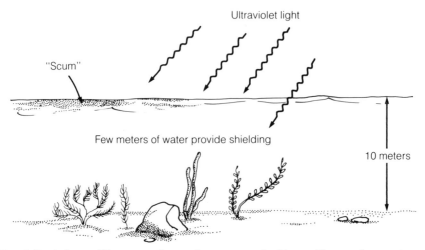

Fig. 9-7   A few millimeters of organic compounds ("scum") or a few meters of water will filter ultraviolet light from sunlight. Today, life on Earth relies on the ozone layer in our atmosphere to provide such a filter.

these ponds could have helped to organize small molecules into larger ones (see page 179). This tide pool mixture of life-producing ingredients had a vague preview in Charles Darwin's conception of "some warm pond" in which life could have begun. Later, J. B. S. Haldane and A. I. Oparin independently developed this model in detail, characterizing the primitive oceans of Earth as a dilute organic broth, a primordial soup in which more complex molecules could assemble.

This model for the origin of complex molecules was tested experimentally by Stanley Miller and Harold Urey in 1953. A schematic representation of their experiment appears in Figure 9-8. The atmosphere of the primitive Earth, a mixture of methane, ammonia, hydrogen, and water vapor, resides in the large flask. Energy comes from an electric discharge, which could be replaced with a source of ultraviolet photons (or with shock waves). The water in the lower flask models some pool of water on the Earth's surface. The heater and condenser provide circulation of water vapor through the system in the same way that water evaporates from pools, moves into the atmosphere, and then condenses as rain, falling back to Earth along with the chemicals formed in the atmosphere. The mechanical aspects of this experiment seem to represent the conditions of the primitive Earth quite well, although we have made no substitute for the mineral-rich surface of our planet.

What happens when we turn on the energy supply and let the experiment run? After a few days of continuous operation, organic molecules build up in the water-containing flask. Chemical analysis of this mixture provides

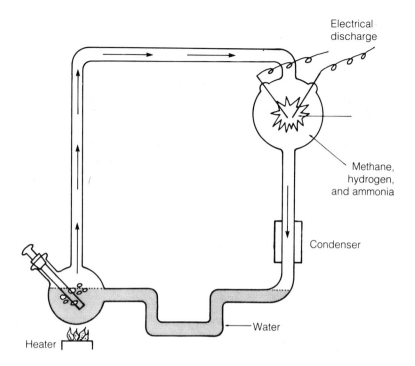

Electrical discharge

Methane, hydrogen, and ammonia

Condenser

Water

Heater

Fig. 9-8 A schematic diagram of the experimental equipment used by Stanley Miller and Harold Urey shows the 5-liter flask that contained water, methane, hydrogen, and ammonia to simulate the Earth's primitive atmosphere and oceans, plus the electrical discharge apparatus that released energy into this mixture. After the first night of operation, a thin layer of hydrocarbon molecules formed on the water, and a few weeks more yielded several types of amino acids.

exciting news: In addition to a large amount of unidentifiable organic "gunk," *amino acids* have formed, including glycine, alanine, and numerous other compounds of interest.

We can illustrate what has occurred with the following chemical formula for the production of glycine:

One ammonia molecule
    +
two methane molecules   one glycine molecule
    +   =   +
two water molecules   five hydrogen molecules
    +
ENERGY

In symbolic language,

$$NH_3 + 2\ CH_4 + 2\ H_2O \xrightarrow{\text{energy}} C_2H_5O_2N + 5\ H_2$$

The formation of other amino acids occurs in a similar manner in the experiment we have described. When we recall that one of the major activities of life as we know it consists of the manufacture of proteins from amino acids, we can see that the production of amino acids represents a large step away from the realm of inanimate matter, and a general confirmation of the idea that the basic substances of which life is made could form under conditions that are likely to have existed on the primitive Earth, and on Earthlike planets in other planetary systems. A further illustration of the ease with which these compounds can form under natural conditions comes from the recent discovery of amino acids in meteorites (see Chapter 12).

In contemporary life, of course, amino acids do not come from meteorites or from laboratories duplicating the Miller-Urey experiment that we have described. Instead, they are made by living systems, as we can easily demonstrate by investigating the *symmetry* of the amino acid molecules.

The amino acids found in the soup that arises in the Miller-Urey experiment include equal amounts of right-handed and left-handed molecules. As we have seen (page 152), in living systems the amino acids in proteins are exclusively of the left-handed variety! Life on Earth apparently selected one of the two possible forms at the time of its origin and has maintained this discrimination ever since. As we pointed out previously, this selection confers an advantage, since it allows more precise programming of the structures of more complex molecules made from these amino acid units.

Thus, life elsewhere, *if* it uses amino acids, should also have chosen one of the two asymmetries. Here we have a good test for distinguishing between biologically produced molecules and those that formed spontaneously. The latter should always occur in equal mixtures of both asymmetries, while the former should show a marked preference for one asymmetry over the other.

Did Life Really Originate in This Manner?

Let us return from chemistry laboratories to the primitive Earth. We have identified a way in which the building blocks of proteins could form. What about the components of lipids, carbohydrates, and the nucleic acids?

We find that some of the simplest fatty acids formed directly in the Miller-Urey experiment: acetic acid, formic acid, and proprionic acid. Fatty acids help to form fats (lipids), so once again we feel some confidence that we are on the right track. But after these triumphs, the picture becomes less clear.

From our discussion of DNA in the last chapter, we know that it has a sugar, deoxyribose, together with four bases, adenine, guanine, thymine, and cytosine, and a phosphate, phosphoric acid. RNA uses uracil instead of thymine. None of these substances arises in the Miller-Urey experiment, but the experiment *can* produce formaldehyde, cyanoacetylene, hydrogen cyanide, and urea. These compounds take us part of the way to our destination. If we heat formaldehyde in alkaline solutions, or in the presence of clays, we produce sugars. We thus form carbohydrates, and in the process develop a pathway to produce one of the key structures in nucleic acids. Further reactions of hydrogen cyanide can produce adenine, while urea and cyanoacetylene can react to form cytosine. Phosphates could presumably come from rock weathering, with subsequent water runoff into collecting areas, and we observe these processes occurring even today.

So far, so good, but no one has yet found a way to make all of the essential building blocks from this mixture of compounds. We may simply be ignorant about the appropriate chemical reaction, or perhaps we merely need some *other* combination of special local environments to complete the picture. On the other hand, we may have misled ourselves somewhere in our discussion, and we might need an altogether different chemical system to bring life into existence.

As we have mentioned, the basic scenario of the Miller-Urey experiment has been challenged during the past decade by scientists who feel that the Earth's atmosphere never had as much hydrogen as the mixture of gases used in the experiment. Philip Abelson and others, however, have shown that complex organic molecules can form even in a *weakly* reducing atmosphere, one with only a small amount of free hydrogen. For example, when we shine ultraviolet photons on a mixture of carbon dioxide, carbon monoxide, and nitrogen together with just a small amount of hydrogen molecules, hydrogen cyanide and water appear. Hydrogen cyanide, combining with itself in alkaline solutions (the early oceans), can produce amino acids, if ultraviolet photons continue to arrive. This reaction also produces cyanamide. Cyanamide can make amino acids link together, the first step in the formation of proteins, when cyanamide mixes with amino acids in a dilute solution irradiated with ultraviolet light. Using the language of chemistry, we write

Three hydrogen-cyanide molecules
+   one glycine molecule
two water molecules   =   +
+   one cyanamide molecule
ENERGY

In symbols,

$$3\ HCN + 2\ H_2O \xrightarrow{\text{energy}} C_2H_5O_2N + CN_2H_2$$

We conclude from these experiments that we do not require the extremely hydrogen-rich conditions of the Miller-Urey experiment for the first steps in the production of important compounds for living systems. Rather, the essential requirement seems to be *the absence of free oxygen in the atmosphere*. Further experiments must be performed with more realistic versions of the primitive Earth. We can test the "realism" of these models, to some extent, by mathematical calculations that follow the formation of our atmosphere, and the chemical reactions that the postulated constituents of the atmosphere undergo with one another and with likely substances on Earth's surface. We must also follow the escape of light molecules from the Earth's gravitational field. These mathematical models, however, even though we can calculate them at amazing speed, can be only as good as our knowledge of the lists of ingredients and of the chemical pathways they follow.[1] In other words, calculations can provide a unique answer only if we know all the conditions, but we simply *don't* know all the facts of life for the primitive Earth. The same holds true for the experimental approach to the origin of life. This has the advantage of allowing the phenomena to speak for themselves, but we cannot design a satisfactory experiment without knowing all the environmental conditions. At best, we can show that *some* mixtures and *some* mechanisms produce plausible results, and, in this limited sense, we have been quite successful.

To illustrate the great variety of possibilities for conditions on the primitive Earth, we can easily postulate a completely different scenario for the formation of life. During the first few hundred million years of their history, all the inner planets must have undergone an intense bombardment by meteoroids of all sizes and presumably by comets as well (see Chapter 12). But we think that life may have emerged on Earth during precisely this time! Could the meteoritic bombardment have played some role in the origin of life? On Mercury and the moon, the absence of a dense atmosphere would have prevented the survival of any complex organic molecules that were present in the infalling objects: The molecules would dissociate immediately upon impact. But even the primitive Earth had an atmosphere thick enough to slow a large fraction of the incoming objects. Thus, organic compounds, including amino acids, could have arrived intact on Earth's surface. This would undoubtedly have given a head start to prebiological chemistry on Earth, on Venus, and, to a lesser extent, on Mars. The same head start would have occurred, we may speculate, on *any* rocky planet, provided the planet had a primitive atmosphere.

We can easily imagine the build-up of organic molecules in ponds, again as the result of water flowing over a pock-marked landscape. The craters formed by the larger impacts could themselves have been the sites for further reactions among these compounds. A continuing source of hydrogen would have come from the breakdown of the organic compounds brought by the meteorites, partially compensating for the hydrogen that

---

[1]Computer scientists express this fact in a pithy phrase: "Garbage in, garbage out."

constantly escapes from the Earth's upper atmosphere. In this way, an environment containing free hydrogen could persist for a longer time than we would otherwise predict.

Polymerization

This cometary or meteoritic scenario does not change in a radical way our approach to the question of how life began. It still envisions spontaneous chemical reactions among likely compounds, and this general scheme should work. But remember where our discussion rests. We have considered only the production of the *components* from which proteins, lipids, and nucleic acids are made. How do we assemble these much larger molecules? DNA and protein molecules are polymers, long chains that repeat a basic pattern over and over. Thus, our problem consists of making molecules join together in a repeating fashion, the process of *polymerization*. In the case of DNA, this process must include the development of the marvelous double helix (see page 155). How did all this occur? We don't know. No one has yet achieved the polymerization of these ingredients under natural conditions. We can put DNA into a suitable bath and it will replicate, but we have not yet been able to make the *initial* DNA spontaneously occur, even though we know its composition and structure.

The initial formation of DNA and other polymers remains one of the greatest puzzles to be solved by experiments to duplicate the chemical reactions that occurred on the primitive Earth. Present theories suggest that *clays* probably played a critical role in promoting the polymerization that must have preceded the development of cells. The current, sketchy model that biologists use suggests that molecules began to organize themselves on the surfaces of grains of clay, at the edges of ponds that periodically froze or dried out (Figure 9-9). This drying or freezing, or both, would

Fig. 9-9   Certain types of clays could have provided excellent sites for molecules to form. The grains of clay would help the compounds that collected organize themselves into long-chain molecules as the water in which they floated dried up.

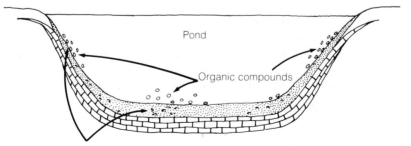

Grains of clay on pond bottoms and sides

tend to *concentrate* the solutions of organic compounds, and help to eliminate the water molecules that appear during polymerization. Clays have importance because they possess about the largest surface areas of any fine material that we know, and the organization of the mineral lattices within the clay grains seems particularly well suited for a wide assortment of organic compounds to stick to the grains' exposed surfaces. The information contained in the lattice structures that atoms assume in mineral configurations might thus have allowed these structures to serve as the first templates for the organization of organic matter.

Experiments have shown that a clay called montmorillonite can indeed line up important substances, such as adenosine and guanine, in ways that would promote polymerization. In addition, sugars, fatty acids, amino acids, and proteins all interact with this mineral. Montmorillonite has a wide distribution on Earth today, while a similar clay, nontronite, forms the best model on Earth for the soil of Mars. Clay minerals may also be among the dust grains observed in the interstellar medium, where they could help to produce the organic molecules recently discovered by radio astronomers (see page 78). On Earth, clays appear predominantly at the edges and on the bottoms of bodies of water, just where we would want them if they were to help with the process of polymerization during the early history of our planet.

Beyond Polymers

Let us summarize our results from trying to understand the origin of life on Earth. We have pursued the following path:

1. A primitive atmosphere containing a few percent hydrogen and no free oxygen;
2. The formation of a widely dispersed "primordial soup" of organic compounds and phosphates, the result of the energy provided by solar ultraviolet light, by lightning discharges, and by other sources;
3. The formation of additional compounds within the soup as the result of continuing chemical reaction, aided by local sources of heat, water runoff, and evaporation and/or freezing;
4. The formation of polymers, probably on the surfaces of clays.

This sequence does not make a unique specification for the origin of life. In fact, we are not even sure of the *composition* of the Earth's primitive atmosphere. Our model also fails in completeness, for we can't see how all of the chemical units of life have formed and we don't know how they joined together to make the polymers that we now see as the essential components of life. Nevertheless, we have at least demonstrated that *some* of the important compounds could form as long as the primitive atmosphere had at least a small amount of free hydrogen, and this gives us hope that further efforts will add to our basic picture and will eventually lead to a solution of the polymerization problem. Our model gains strength from the fact that nature apparently makes organic compounds quite easily, since we have found a whole list of such molecules in the interstellar medium, in comets, and in meteorites.

At the stage where polymers form, we stand at the brink of living systems. But we do not know how nature bridged the gulf between inanimate and animate matter. Quite probably, many different assemblages occurred, and many of them may have been successful at some sort of reproduction. We could generally define "life" at that stage of its development as *a system capable of self-replication.* But this time was already well advanced. Before this, random protein synthesis might have occurred, in response to the existence of polymers vaguely resembling modern RNA. The production of a protein that could produce *more* RNA from the original RNA, a *polymerase,* could have started life on its way to success. Presumably, life's first attempts at replication must have been haphazard and extremely crude by the modern standards of contemporary DNA. The early form of the genetic code probably had fewer restrictions and simply discriminated among different types of amino acids rather than selecting specific amino acids.

All this happened not inside living cells, as life processes do today, but in the ponds and pools where the initial chemical reactions occurred. The key events that led to the evolution of cells, membrane-bounded regions within which organic molecules can interact, remain uncertain. We do know, however, that organic polymers in high concentration are likely to join together and separate from the solution in the form of droplets. Such droplets, called *coacervates,* could serve as the prototypes of cells (Figure 9-10). If the right kind of long-chain molecules exist, they will tend to form a membrane that encloses the droplets.

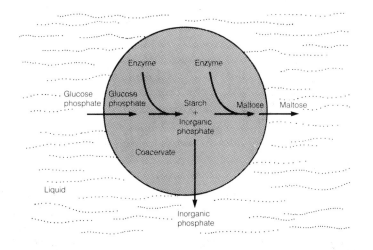

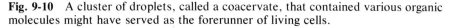

Fig. 9-10 A cluster of droplets, called a coacervate, that contained various organic molecules might have served as the forerunner of living cells.

Laboratory studies of coacervates have shown that under the appropriate conditions, they can form systems within which simple chemical reactions occur. But no one has yet produced a self-maintaining system such as we need for a valid model of the first cells.

In another approach to the problem, scientists have demonstrated that if proteinlike polymers develop from the amino acids produced in simulated "primitive" environments, these polymers will form microspherules that resemble cells in size and appearance, if the polymers are heated under the proper conditions (Figure 9-11). Such experiments at least illustrate the opportunities available for the further organization of matter, even if they do not represent the precise path toward the development of cells. The next step toward modern life would be the incorporation of a self-replicating system within the boundaries of these proto-cells, a system capable of directing the formation, maintenance, and renewal of the entire cell.

This conjecture leads us to a level of organization in matter that we can recognize on the contemporary Earth. The primitive organisms whose development we have suggested resemble a modern virus. This does not mean that an influenza virus is a "living fossil," as a living dinosaur, to be discovered on some lost island (or in the depths of Loch Ness!), would be in the animal kingdom. Modern viruses show careful adaptation to their

Fig. 9-11 If we heat certain polymers that are quite similar to the proteins in animals, some of the long chains will form microspherules that look much like cells.

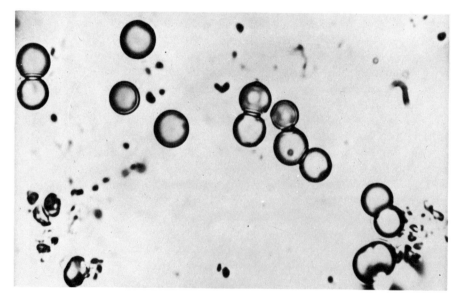

hosts, and the hosts have appeared on Earth quite recently. Indeed, some evidence suggests that viruses are all derived from cells, which then would have preceded viruses in the course of life's development. But in their structure and their functioning, modern viruses have great similarity to our hypothetical primitive organism: Each consists of nucleic acid with a protein covering, and each can survive long periods of dormancy.

Summary

In our attempts to discover how life originated on Earth, we must deal with the fact that life began at least 3.4 billion, perhaps 4 billion, years ago, when conditions on Earth were quite different from conditions now. The geological record of these early eras has vanished, so we must add deductive methods to the actual evidence to estimate the conditions on the primitive Earth.

The chemical composition of living matter suggests that life began when conditions on Earth were far more oxygen-deficient than is now the case. This seems reasonable in view of the fact that most of the oxygen in our atmosphere has come from green plant photosynthesis, starting about 2.5 billion years ago. The Earth's primitive atmosphere may have consisted mainly of hydrogen-rich compounds, such as methane and ammonia, or, as recent evidence suggests, it may have been a mixture of carbon dioxide, water vapor, carbon monoxide, and nitrogen. Experiments have shown that if either of these mixtures of gases are placed above a bath of water (to simulate the primitive oceans) and are exposed to ultraviolet light (to simulate sunlight, unfiltered by the Earth's present ozone layer), then simple organic molecules will form. These organic molecules include amino acids, the basic units of much larger protein molecules. Other compounds, also important in forming the small molecules found in living organisms, are produced in these laboratory simulations.

Some of the small organic molecules formed in this way on the primitive Earth may have collected in pools and ponds, where they assembled themselves into the long chains typical of proteins and nucleic acids. These long chains, or polymers, repeat a basic pattern over and over again. The best way to promote such polymerization seems to be to collect molecular units on grains of clay, which then undergo periodic freezing and thawing, or evaporation and wetting.

As various types of molecules passed through these stages, the first self-replicating molecular structures, capable of duplicating themselves when they had split apart, arose as part of the natural trial-and-error process that saw all sorts of molecules appear, if only briefly. Polymers that could reproduce themselves had immense advantages over polymers that formed at random, so we are not surprised that the basic mechanism for duplication, that in DNA, appears in *every* organism that reproduces itself on Earth.

Questions

1. Why do we think that life on Earth began when the Earth had a smaller amount of oxygen than is now the case?

2. Why can't we determine what the Earth was like, soon after it formed (4.6 billion years ago) by studying the geological record laid down in the rocks as they formed?

3. Where did the oxygen now in the Earth's atmosphere come from?

4. Where did the nitrogen in our atmosphere come from? What processes tend to remove nitrogen from the atmosphere? What processes replenish it?

5. How did life arise on Earth, several billion years ago? What were the essential ingredients needed for this process to occur?

6. What can laboratory simulations of the conditions that may have existed on the primitive Earth show us about the formation of organic molecules? What difficulties stand in the way of making an accurate experiment that duplicates the primitive Earth?

7. Why are *clays* thought to be of prime importance in making complex molecules?

8. Is it possible that organic molecules from outer space played an important role in the formation of life on Earth? How?

9. At what point in the formation of more and more complicated molecules should we say that "life" appeared? Why?

10. What were the first ancestors of modern cells like? What environmental hazards did they overcome?

Further Reading

Dickerson, Richard. 1978. Chemical evolution and the origin of life. *Scientific American* 239:3, 70.

Gould, Stephen. 1977. *Ontogeny and phylogeny.* Cambridge, Mass: Harvard University Press.

Kenyon, Dean, and Steinman, Gary. 1969. *Biochemical predestination.* New York: E. P. Dutton & Co.

Miller, Stanley, 1974. The first laboratory synthesis of organic compounds under primitive earth conditions. In *The heritage of Copernicus,* ed. J. Neyman. Cambridge, Mass.: The M.I.T. Press.

Miller, Stanley, and Orgel, Leslie. 1974. *The origins of life on Earth.* Englewood Cliffs, N.J.: Prentice-Hall.

Ponnamperuma, Cyril. 1972. *The origins of life.* New York: E. P. Dutton & Co.

CHAPTER 10

From Molecules to Minds

We have followed the speculations about the origin of life on Earth that suggest a process of polymerization, the spontaneous formation of long-chain molecules, perhaps on grains of clay at the edges of shallow ponds that periodically evaporated. By trial and error, some of the polymers, including some that could serve as templates for others, found themselves isolated from the primordial soup in droplets bounded by other compounds. This model for the development of life suggests that these polymer-containing droplets were the protocells from which the first true cells must have developed.

Prokaryotes

Because viruses seem to be derived from cells, and because they cannot function alone, biologists usually say that the simplest living organisms, those with the least complexity of structural organization, are the simplest *cells,* bacteria and blue-green algae.[1] These cells are *prokaryotes,* cells without special centers or nuclei. (Their Greek name means "before the nucleus.") Prokaryotes contain long strands of DNA, each with several thousand genes, hundreds of times more than the number of genes in a virus. Both DNA and RNA appear in prokaryotic cells, while viruses have *either* DNA or RNA as their nucleic acids. The amount of DNA in a prokaryotic cell exceeds the amount of DNA or RNA in a virus by 10,000

---

[1] Strictly speaking, these organisms are not algae but a special kind of photosynthetic bacteria. We therefore call them "blue-greens" or "blue-green bacteria."

times, so prokaryotes have far more complex reproductive capacities than viruses. The earliest cells were probably of this type (Figure 10-1).

Prokaryotes are free-living entities: They can reproduce, and make proteins, without a plant or animal "host" in which to live. These organisms represent the simplest biological system capable of independent life. When prokaryotes first evolved, they had to use the nutrients in their natural environments, the organic compounds produced by the interaction of ultraviolet sunlight and other energy sources with the chemical mixture in which they found themselves. As this source of food was depleted by the changes in our planet's atmosphere (see Chapter 9), some microscopic prokaryotes developed the basic process that led to the vast proliferation of advanced forms of life: the conversion of sunlight into stored chemical energy for future use. In other words, prokaryotes "invented" photosynthesis. Although the earliest form of this process did not result in the liberation of oxygen, it represented a tremendous advance for the bacteria that could carry it out. A billion years of evolution may have been required for organisms to develop the type of photosynthesis that provides us with the oxygen we breathe.

We could easily substitute "starlight" for "sunlight" in our description of bacterial photosynthesis; blue-green bacteria would be just as happy using the light from Alpha Centauri as they are with the light from our own sun. We can immediately see the potential for converting a strange planet in some distant solar system into an environment that humans could inhabit simply by seeding its atmosphere and oceans with the appropriate

Fig. 10-1 These ordinary-looking spherules are thought to be the oldest fossils of organisms yet found, 3.4 billion years old. The microspherules are less than a hundredth of a millimeter in diameter.

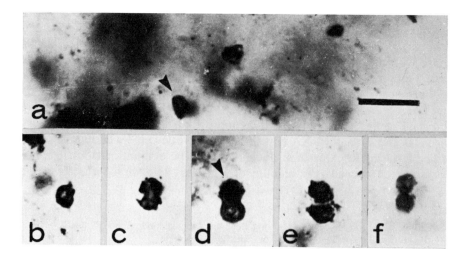

microorganisms. We might, however, have to adjust the trace element composition of the other planetary environment before this "directed panspermia" could succeed.

As we attempt to unravel life's early evolution on Earth, we should note that we can still see evidence of the transition to an oxygen-rich environment in today's prokaryotes. Some prokaryotes live only in the complete absence of oxygen; others can live with it or without it, while still others require oxygen for survival. This diversity suggests that the conversion of the Earth's atmosphere from oxygen poor to oxygen rich, through the liberation of oxygen by photosynthesis, probably occurred during the era when prokaryotes first diversified on Earth. When we reach the next step in biological complexity, cells with nuclei, we find almost without exception that these cells must have gaseous oxygen in order to live.

The transition to an oxygen-rich atmosphere apparently occurred between 2.5 and 2 billion years ago but might have begun even earlier. At first, oxygen in the atmosphere would have been poisonous to all the prokaryotes, but its availability as an easy source of energy for life's metabolism must have created a strong selectional pressure in favor of organisms that could tolerate, and then use, this new gas. Oxygen's first appearance in our atmosphere occurred as the result of the dissociation of water vapor molecules by ultraviolet light rather than through the action of living creatures; we find a tiny quantity of oxygen in the atmosphere of Mars because of this nonbiological process. Thus, organisms could first get a small taste of oxygen and, as they learned to like it and to use it, could tolerate more and more of it.

It is intriguing to find that chlorophyll molecules, which release oxygen in plants, have the same structural unit (called a porphyrin) that we find in some important oxygen-using molecules, such as hemoglobin (the iron-containing pigment in our red blood cells), hemocyanin (the copper-containing equivalent in many marine animals), and cytochromes, chemical complexes found in most advanced cells. This coincidence suggests that life forms *used the structures they had already developed* in their responses to environmental changes. In other words, selectional processes similar to those that affect entire organisms (natural selection) may also operate at the molecular level. In this view, the ability to *use* free oxygen should have evolved at approximately the same time as the ability to *liberate* oxygen.

Eukaryotes

The appearance of free oxygen in the atmosphere of our planet coincided with further evolutionary development. We cannot trace the exact path from the presence of many types of prokaryotes to the appearance of the *eukaryotes,* cells that contain true nuclei (Figure 10-2). Intermediate stages

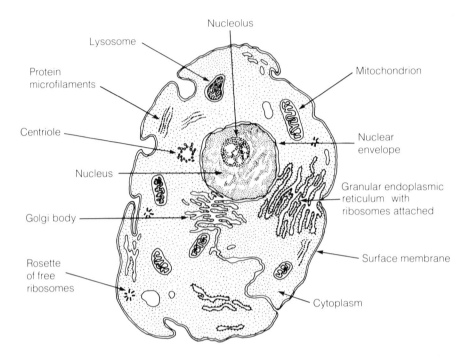

Fig. 10-2 Eukaryotes consist of cells with true nuclei; that is, cells in which the DNA-containing chromosomes are enclosed by a membrane.

of evolution undoubtedly occurred, leading to the development of eukaryotes at a time between 2 billion and 700 million years ago. Eukaryotes represent the highest form of *cellular* complexity that has yet evolved on Earth. These cells keep their chromosomes, the DNA-containing portion of the cell, within the protection of the membrane-enclosed nuclei that give them their name, which means "with nucleus" in Greek.

In eukaryotic cells, the chromosomes typically contain hundreds of thousands of genes, and the percentage of DNA in a cell has increased by 10 to 1000 times above the percentage found in bacteria. Eukaryotes can store much more information in their genetic material than prokaryotes can, and their increased complexity of cell structure and cell function reflects this. Eukaryotic cells have important internal structures: They contain specialized subunits, called organelles, that have their own genes and protein-synthesizing machinery, and carry out specific functions (Figure 10-2). All animal and plant cells, for example, contain organelles called mitochondria for respiration, and all plant cells contain chloroplasts for photosynthesis.

How did eukaryotes evolve from prokaryotes? The traditional view suggests a gradual development, by which prokaryotes slowly acquired a progressively clearer substructure. In contrast to this gradualist view, another hypothesis, recently championed by Lynn Margulis, suggests that this internal specialization arose from a symbiotic, or mutually beneficial, relationship between two or more prokaryotic cells. We can imagine a prokaryote that would benefit from the ability to perform photosynthesis but cannot do so. If the prokaryote incorporated a simple, chlorophyll-laden relative, it would suddenly achieve this capability. This suggested ancestry implies that mitochondria may be derived from bacteria, and chloroplasts from blue-greens. More specific evidence for this evolutionary path exists in the fact that chloroplasts and mitochondria have gene systems that resemble the systems found in bacteria.[2]

From the stage of the first eukaryotes, evolution led to the increasing specialization of cells, in function and structure, and to their incorporation into larger, far more complex units, such as organs and large individuals. The bodies of humans and other large mammals each contain about a thousand trillion cells, of about a thousand different varieties. The amazing complexity of structure in a human being has apparently arisen through the process of natural selection, nothing more or less than differential reproduction among competitors. Our development of self-consciousness, so strikingly important, has come from the same competition.

As organisms continued to reproduce themselves, they eventually developed the technique of *sexual reproduction*. All eukaryotic cells can reproduce asexually, but the advantages of sex had such great impact that the most complex eukaryotic organisms rely on it. Sexual reproduction allows a new offspring to obtain about half its total number of genes from each parent, because each parent contributes part of its genetic makeup (Figure 10-3). This fact probably allows natural selection to proceed more rapidly than it does for asexual reproduction. It is reassuring to learn that scientists have proved that the most efficient sexual reproduction, from an evolutionary point of view, is that with just two sexes, because three or more sexes increase the difficulty of reproduction enormously, even as they add to the possibilities of differences in offspring.

Until now, we have followed the history of life on Earth in stately billions of years. The Earth formed 4.6 billion years ago. The first life of which we have any record (so far) lived 3.4 billion years ago. Geological evidence suggests that the atmosphere became oxygen rich between 2.5 and 2 billion years ago. At the same time, stromatolites formed, fossil evidence of blue-green and other photosynthetic bacteria. The first eukaryotes appeared perhaps a billion years later (give or take half a billion

[2]The suggestion has recently been made that prokaryotes and eukaryotes both evolved from a common ancestor. Biologists are currently studying the implications of this new hypothesis.

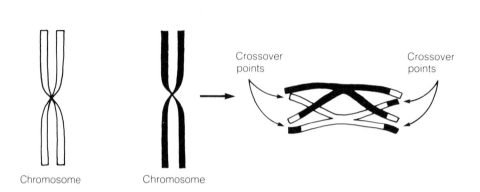

Chromosome   Chromosome

Fig. 10-3 In sexual reproduction, the fertilized egg cells each contain chromosomes with DNA from each parent. When the egg cell divides and redivides, the chromosomes pair up to form connections at "crossover points" where the exchange of genetic material occurs. This shuffles the chromosomal material, so that after a number of such divisions, each new cell has a roughly equal mixture of maternal and paternal DNA.

years), somewhere over a billion years ago. Thus, life has been in continuous existence for most of Earth's 4.6 billion years, but during the bulk of this time, life has consisted of communities of many species of microorganisms too small to be seen as individuals by human eyes.

When we think about the evolution of life, we usually picture monkeys and humans arising from more primitive mammals, or birds and reptiles developing from earlier vertebrates. Such events, important though they are to us, represent the most recent stages of evolution, spanning less than 10 percent of the history of life on Earth. The long, slow changes that produced the first prokaryotes, and eventually the first eukaryotes from prokaryotes, form most of life's development, so we should not overlook their importance to us, or the probability that life on some other planet would spend much of its history in microscopic form.

The Great Leap Forward

From our emphasis on the billions of years when life was "only" microorganisms, we should not conclude that an increase in biological complexity always requires such eons of time. Instead, life on Earth suggests that the early stages of life take far longer than the later ones, and that the initial steps toward complexity are more difficult than the later, fantastic increases in the complexity of living creatures.

Just about 600 million years ago, some combination of circumstances, presumably involving the large amounts of oxygen that blue-green bacteria and plants had liberated, and the widespread distribution of newly habitable environments, led to a sharp increase in the distribution, number, and variety of fossils that marks the beginning of the *Cambrian era*. This era must have followed a long period of the gradual evolution of soft-bodied organisms from single-celled eukaryotes, to creatures much like today's jellyfish, which flourished just 50 or 100 million years before the Cambrian era began (Figure 10-4). We know far less about life before the Cambrian era than we do about later life, because only at the start of the Cambrian era did organisms evolve "hard parts"—shells, carapaces, and exoskeletons—that could be well preserved in sedimentary rocks to form the fossils we find today.

The Cambrian era brought a remarkable speeding up in the rate of diversification of large, complex organisms. Four billion years elapsed from the origin of the Earth to the appearance of trilobites, whose abundant fossils mark the transition to the Cambrian era (Figure 10-5). With two compound eyes that may have provided binocular vision, these animals mark a significant evolutionary step. Yet trilobites remain extremely primitive by human standards; they would hardly be able to construct and to use radio telescopes to communicate with other trilobites on some distant planet. But just 600 million years after the appearance of these distant relatives of the scorpion, some peculiar, apelike hominids moved from the jungle forests of ancient Africa into the open plains. A scant 3 million years after this transition, we humans wonder about contact with other inhabitants of the universe.

Fig. 10-4 These fossils of feather corals from the Ediacaran period, 700 million years old, appeared shortly before the explosive development of species of the Cambrian era. These are among the oldest known multicellular organisms in the Earth's history of life.

Fig. 10-5 The trilobites that flourished in the seas at the start of the Cambrian era had two eyes and a rather complex bodily structure. This photograph shows a trilobite fossil about two centimeters long.

We can best appreciate the acceleration involved in the different steps of evolution with a diagram (Figure 10-6), or with a calendar's analogy. Think of the entire history of Earth as a single calendar year, starting January 1. Then bacteria made the atmosphere oxygen rich as summer began, while the Cambrian era started only on November 13. Our first

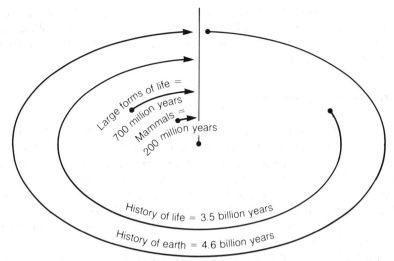

Fig. 10-6 If we represent the Earth's history by one turn of a giant wheel, then life has existed for at least three quarters of a turn, but large forms of life have been around for only one sixth of a cycle. Hominids have existed for less than one thousandth of a full turn of 4.6 billion years.

human ancestors, the hominids, appeared at dinnertime on December 31, and human ability to send New Year's greetings to the other side of the galaxy would occur barely in time, at 11:59:59.9155 P.M., less than one hundred millionth of the Earth's lifetime!

This time scale tells us something useful in our quest for life elsewhere in the galaxy. We assume that the development of life on Earth may be fairly typical of life's development, wherever it occurs. This is the assumption of mediocrity, suggested by Josef Shklovskii and Carl Sagan. Thus, even without knowing the details of how life began and how it developed on our planet, we can assume that wherever life occurs, it will take *about* 4.6 billion years—say, 3 to 6 billion years—for intelligent life to develop, once a planet has formed. But intelligent life, in the sense of a civilization capable of galactic communication, has existed on Earth for only the last tenth of a second out of a year's total. On another planet, evolution might differ only slightly from that of life on Earth, so that life might require 4.7 billion or even 6 billion years to reach the same capability. In this case, our attempts to make contact with other life would be doomed to failure if the other planet formed at the same time as Earth. On the other hand, small changes in the rate of evolution might have brought another planet, formed when Earth formed, to our state of self-consciousness several hundred million years ago. Finally, since other planets themselves could be billions of years older or younger than the sun and its planets, we have another factor that could put us at a very different period of development compared with another kind of life. Until we find another planet with life and compare that planet's evolutionary time scale with our own, we hall have to be content with the rough estimate of 3 to 6 billion years from a planet's formation to the emergence of a civilization capable of interstellar communication.

Suitable Stars for Life

This apparently trivial statement has immediate consequences in directing our decisions about where to look for life in our galaxy. We must obviously limit our search to planetary systems whose central stars have total main-sequence lifetimes that exceed a few billion years. The reason for this requirement is that the luminosity of the star must remain approximately uniform during the period that life evolves. Once the star leaves the main sequence, its increase in luminosity will alter the planetary environments dramatically and irreversibly.

If we impose a lifetime of at least 4 or 5 billion years on main-sequence stars, we restrict our search to stars with spectral types of F5 or cooler; that is, to F5 through F9, G, K, and M stars. This might seem a difficult constraint, since the brightest stars in the sky are mostly O, B, and A stars, but in fact the great majority of stars do satisfy the criterion we have just set. For example, all but two of the 25 stars closest to the sun are cooler than spectral type F5.

The other requirement on stellar lifetimes implicit from our terrestrial perspective must be that we have a star *at least* 4 or 5 billion years old. In other words, if we pick a G2 star, whose spectral type is identical with the sun's, but a star that has been on the main sequence for only 2 or 3 billion years, then the best we could hope to find on its Earthlike planet (if it *has* an Earthlike planet!) could be a lively population of bacteria. This discovery would surely interest us, but we would have to go there to find it and the crew of astronauts who had been on this journey might deservedly feel a sense of letdown.

Even if we have the right kind of star, one that has been on the main sequence for at least 4 or 5 billion years and is shining steadily on an Earthlike planet, what assurance do we have that we would find intelligent life? What does it mean for a planet to be Earthlike? And what are the fundamental conditions we need for intelligent life to develop?

Life on Other Planets

Our quick review of biological evolution on Earth cannot do justice to such an immensely complex field of science. But for our purposes, even a general outline can help to define those aspects of life and its development on Earth which we can try to generalize to other planets.

The first and strongest impression we find from our survey of life is the importance of *liquid water*. Terrestrial life originated in water, and water remains vital to life's continued existence. Large organisms did not evolve the adaptations that would permit them to live on land until 350 million years ago, at least 3 billion and perhaps 4 billion years after life appeared on this planet. We can imagine other liquid environments, such as seas of ammonia, but life seems to require a liquid medium that can transport and concentrate important molecules. The liquid state does not arise easily in nature, since it requires that the local temperature stay within a narrow range. And to form polymers, we require not just liquid in any form but liquid that collects on a solid surface. These two constraints, solid surfaces and a narrow temperature range, are easily satisfied by a planet in a roughly circular orbit at the right distance from its star, but perhaps other such environments can also be found in interstellar space. Meanwhile, we can assert that in our own solar system these conditions were satisfied only on Earth, and that it is not a coincidence that our planet exhibits the most abundant, and perhaps the only, biological activity in our family of planets.

Once life has begun, does it need an exposed land mass for its further development? We cannot answer this question with confidence. On Earth, the most advanced life forms evolved on land. The greater variety inherent in land environments and the greater availability of oxygen seems to increase the rate at which natural selection occurs: About 80 percent of all known species of plants and animals are found on land. Furthermore, the demands on form and function in a marine environment may not favor the natural selection of intelligent organisms so much as the conditions on land do. For example, the need for streamlining tends to prevent the development of complex limbs with the ability to manipulate objects and ultimately to make tools. Indeed, the most intelligent marine creatures, whales and dolphins, are mammals whose ancestors were land animals similar to otters. Other things being equal, we might look for planets on which water remains liquid but does not cover the planet completely: We should have land *and* oceans!

But would other things be equal? Are we making an error by assuming life elsewhere must be like the life on our planet? After all, octopi, starfishes, and lobsters have appendages capable of manipulation and are not especially streamlined. Perhaps an ocean-covered planet could develop intelligent marine life from a line similar to our mollusks rather than to our mammals. The warmth of the seas themselves might provide the equivalent of the mammalian bloodstream. Would such organisms learn astronomy and consider the possiblity of communicating with others like themselves across the depths of space? We must simply leave this possiblity open, with no way to verify it until we find an ocean-covered planet.

Suppose that we have a planet in orbit around a star of the proper age, that the planet's surface has both seas and land, and that life on this planet has undergone the same type of "early" evolution as Earth. How likely would the appearance of an intelligent form of life then be? In other words, how improbable is intelligent life? To try to answer this question, we must first take a closer look at the process of evolution.

Evolution and the Development of Intelligence

At the molecular level of life, we have described the process that causes an evolutionary change, one that could ultimately lead to new adaptive traits in species, in terms of random changes in DNA as it replicates, leading to new DNA molecules that are not exact copies of the old ones (see page 158). No one knows exactly why such mutations occur, but DNA molecules show a remarkable stability against these changes. Mutations only appear in advanced animals about once per gene for every 100,000 cell divisions. If such a mistake occurs in the "germ plasm" of an individual, it changes the characteristics of that individual's offspring. As the physician Lewis Thomas has written, "The capacity to blunder slightly is the real marvel of DNA. Without this special attribute, we would still be anaerobic bacteria and there would be no music."

Some changes may aid an organism in its attempts to survive and to reproduce, while other changes may hinder these attempts and still others will have no effect at all. *Natural selection* describes the testing of these various changes over the course of time. If a certain change does aid an organism, the organism will have proportionately more descendants, all of whom may embody the particular variation. As natural selection proceeds, new species of organisms can appear, as successive generations differ more and more from their ancestors.

When we examine the record of evolution revealed by fossils on Earth and as we fill in the missing record with our imagination, we notice two important facts. First, species do not repeat. After the trilobites disappeared, other arthropods came and went, but trilobites have never re-emerged from the ever-branching tree of biological diversity. Second, during the period when life moved from the first eukaryotes to human beings, some species have remained essentially unchanged. We still have blue-green bacteria, along with many other prokaryotes (anaerobic bacteria in particular) that have survived, with presumably only small changes, since their first appearance on Earth.

Although some species do not change, most do. Natural selection does have an orientation, one that leads toward the development of more complex species of life. While the rest of the universe steadily increases its disorder, life heads in the opposite direction at a pace that seems to quicken as complexity increases. No blue-green bacteria exist on the moon, but humans have walked there.[3] The price of this capability comes high, for humans have developed the ability to exterminate themselves voluntarily, a talent which the blue-greens lack.

Is intelligence inevitable? To determine whether intelligence is likely to be either a widespread or a very unlikely phenomenon in the universe, we must consider the way in which intelligence appeared on Earth and must also try to determine whether intelligence proves beneficial or harmful to a species. Does intelligence carry with it the seeds of catastrophe, leading the species that have it to their destruction? As yet this is a matter of personal opinion, with no way to judge the direction in which we are headed. So far, we can only say that our example on Earth seems to show that natural selection does favor the development of intelligent, self-conscious life. What happens next remains hidden from us.

Let us first define what we mean by intelligence. For our restricted purposes in searching for life, we call an intelligent species one that has developed the ability to communicate—either passively or actively—over interstellar distances. This strictly technological definition has a direct bearing on our search for life, since the only alien species we are likely to discover must have this communication ability. By using this definition, we give our imaginations free rein in thinking about the other characteristics an intelligent species might have—for example, in structure, size, musical ability, and belligerence.

Recent thinking about evolution on Earth tends toward the idea that intelligence has such great survival value to a species that it forms an inevitable step in the development of life. (Here biologists are considering a more general definition of intelligence than our technological one.) Aside from our (potential) ability to leave the Earth, we humans already have the means to preserve ourselves against extinction on our own planet, for we can modify our environment to maintain the conditions best suited to our existence. We can heal our sick, feed our hungry, move to any part of the globe, and pass from generation to generation the knowledge of how to achieve these goals. This ability to collect and to transfer information makes us "intelligent" in the traditional sense and has a tremendous survival value.

The speed with which intelligence has developed on Earth increases our hope that this curious property of matter has also appeared many times over in our galaxy. Let us imagine, for a moment, that all the humans on Earth suddenly disappeared. How much time would pass before the emergence of a new species with intelligence? Two hundred million years ago,

---

[3]But we might temper our human pride in this accomplishment with the realization that over a billion bacteria "walked" with each astronaut!

15 cm

Fig. 10-7 The first mammals to appear on Earth 200 million years ago resembled modern shrews and mice and hunted insects while keeping an eye out for predatory dinosaurs.

Earth's only mammals were small creatures that resembled shrews and mice (Figure 10-7). Three million years ago, the first hominids appeared. These two times roughly define the interval needed for intelligent creatures to evolve, once complex animals appear. Two hundred million years is a long time, but if we destroyed ourselves, this span of time might bring Earth another intelligent species, perhaps wiser than humans. Then the length of time needed for the emergence of intelligent life on Earth would have increased from 4.6 to 4.8 billion years. Would the stars notice this trivial difference?

With the perspective that we obtain by thinking about the totality of the Earth's history, we see that we need not worry much about what would have happened if the Turks had conquered Vienna in 1529, or if Hitler's dreams had been realized. The development of intelligence would surely have proceeded with little change despite these various catastrophes, minor episodes at best on an astronomical time scale. We might have had to wait a thousand years, or even several thousand years, but once again people would have become curious about those lights in the sky and wondered about how thinking occurs, or how humans can live at peace with the universe.

If we humans did destroy ourselves in some way that did not annihilate our fellow creatures, we can even try to identify our cousins most likely to succeed us as intelligent "masters" of Earth. Some might favor the dolphins, who could again venture onto the land they abandoned millions of years ago. Or perhaps another kind of simian could swing out of the trees and begin the hominid pattern of cave dwelling and primitive agriculture. But we can also imagine far different pathways of evolution toward intelligence, such as descendants of a tribe of clever raccoons, or advanced societies of tool-making octopi. The many examples we find of parallel evolution—for example, the independently evolved eyes of an octopus, different in structure (they cannot "see" colors) but quite similar in use to the binocular vision of mammals—suggests that if an environmental opening exists, some species will evolve to fill it.

We can thus rest assured that if humans disappeared, the process of natural selection that led to our development would again produce intelligent creatures from the vast gene pool on our planet. The new lord of Earth might be a species of mammal with binocular vision and tool-manipulating ability. But beyond that, resemblances to ourselves might not be close. The vanished trilobites have left a message: Once a species disappears, it never returns. This tells us, too, that we should not expect to meet *humans* on other worlds. To quote Loren Eiseley, who is describing a finite universe:

> Life, even cellular life, may exist out yonder in the dark. But high or low in nature, it will not wear the shape of man. That shape is the evolutionary product of strange, long wandering through the attics of the forest roof, and so great are the chances of failure, that nothing precisely and identically human is likely ever to come that way again.

Future Evolution on Earth

Has evolution stopped with the emergence of human beings? We have no reason to think so. But humans have introduced a new influence on evolution: attempts at conscious control. We see striking evidence of this influence in the tremendous growth of the human population. One of the factors that originally led Charles Darwin to his idea of natural selection was his observation that succeeding generations of a given species are usually no more numerous than preceding generations. Since most organisms have more offspring than parents, this implies that most of the offspring never survive to reproduce. Such differential success at survival and reproduction causes an increase in the number of "successful" individuals, at the expense of the less successful ones.

Human beings seek to modify this process. We work together to overcome the effects of disease, famine, genetic defects, and natural disasters. As a result of our efforts, the number of human individuals does increase from one generation to the next. One big jump in the human population came with the introduction of agriculture; another came when we learned how to avoid the most common fatal diseases. During the past few generations, the human population has doubled every 35 years. This tremendous rate of increase creates enormous problems that will continue to challenge our political, social, and technological abilities. But if we do manage to bring population growth under control, we can anticipate a long future for human beings on this planet. What can we expect to occur in an evolutionary sense, if we manage to avoid crowding ourselves out of existence?

It is easy to guess where we are headed. As human understanding of natural selection, and the urge to control it, have become more fully developed, we have reached the stage of experiments with genetic engineering, cloning, and the prolongation of human lifetimes. Interest in these efforts seems likely to increase, spurred by the desire to achieve the same goal that humans have pursued since first they became aware of the rhythm of their lives: immortality.

Scientists do not see any fundamental reason, as opposed to a present lack of understanding in detail, why we should expect to remain unable to structure forms of life as we choose, and prolong life for very long periods of time. If our distant descendants achieve this capability, will they choose to reach for immortality? Or will they deliberately shun this path and decide to live and die as other forms of life do? Some people might find eternal life a dreadful bore, or wish to yield their place to new humans. We can only guess at the consequences of having the ability to prolong life indefinitely, but one thing seems clear: If advanced civilizations exist elsewhere in the universe, this question must have been asked and answered many times. As the astronomer Frank Drake has pointed out, the results of this inquiry could have significant effects on our search for intelligent life. Somewhere among the stars, a race of immortals may have emerged, spending their time in the rituals and entertainments that please them, perhaps in contact with other races like themselves, and even (unlikely though it may seem) eager to introduce less advanced societies to their level of consciousness.

The Web of Life

We should not forget that despite our importance to ourselves and to each other, human beings represent only a tiny fraction of the living beings on this planet. The importance of these other species becomes increasingly evident to us as we learn more about the science of *ecology,* the study of the relationships among organisms and between organisms and their environments. We all know examples of the ways in which one organism depends on another: Viruses have their bacteria, ants have their aphids, humans have their cereal grains. But ecologists now emphasize the fact that *our planet has no closed ecological system.* In other words, we cannot surround some environment with an impenetrable wall and expect the organisms within it to survive. We *all* depend on one another in complex ways that we now perceive only dimly.

This may seem surprising, but we can see the fact more clearly from our discussion of the Earth's atmosphere, which offers a good illustration of this complex interaction. What would Earth look like *if life had not developed?* Suppose nature's first experiments in the primordial soup had failed. Then hydrogen would still have escaped, since our planet's gravitational force cannot retain hydrogen atoms. Hence organic compounds, broken apart by solar ultraviolet photons, would gradually become oxidized: They would combine with oxygen molecules, liberated by the ultra-

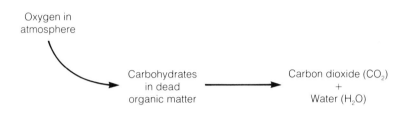

Fig. 10-8 Without life, decaying organic compounds would oxidize, combining with oxygen molecules to put almost all the carbon once in organic molecules into carbon dioxide.

violet destruction of water molecules (Figure 10-8). With no new sources of organic compounds, the carbon on Earth would gradually become part of carbon-dioxide molecules.

We can estimate how much carbon dioxide would appear on a lifeless Earth by adding up all the organic carbon buried as shales, coal, oil, and the carbon in carbonate rocks. Limestones, for instance, are calcium carbonates, primarily made of calcium, carbon, and oxygen, formed through the presence of living creatures. If we calculate how much carbon dioxide would correspond to the carbon in the Earth's organic molecules and in carbonates, the result may surprise us. If all of this carbon dioxide entered our atmosphere, we would have an atmosphere about *70 times* more massive than our present one; in other words, the atmospheric pressure would increase by 70 times. To reach this extreme value, we would have to eliminate not only the effects of life, but also the formation of carbonate rocks that arises from weathering caused by carbon dioxide dissolved in water. In other words, if the Earth had *no life* and *no water* (at least, no liquid water), then our atmosphere would consist of nearly pure carbon dioxide, with many times the thickness of our present atmosphere (Figure 10-9). With water absent, some nitrogen might persist in the carbon dioxide atmosphere and some argon would continue to exist. This description (see Chapter 13) fits perfectly the atmosphere of our sister planet, Venus!

We may conclude that our analysis of what happens to a planet in the inner solar system from life's presence or absence does make sense, for, as we shall see, Venus now seems quite incapable of having either life or liquid water. Furthermore, our analysis has enough generality to apply not only to planets in our *own* solar system, but also to other planets in systems far from the sun, since all planetary systems probably have rocky inner planets.

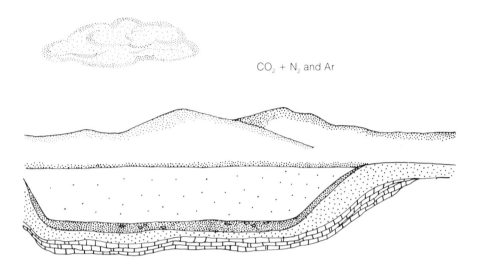

$CO_2 + N_2$ and Ar

Fig. 10-9   A lifeless Earth would have a thick carbon dioxide atmosphere with some nitrogen and argon but almost no oxygen.

Gaia

A close relationship clearly exists between the composition of our atmosphere and life on our planet. Life began, so we think, because the Earth's early atmosphere contained the compounds essential for the genesis of living organisms. Life *persists* because the atmosphere provides the medium through which chemical exchange can occur, along with a thermal blanket and a shield against ultraviolet photons and cosmic ray particles.

The intricacy of this chemical interaction between *life* on Earth and Earth's *atmosphere* has led two scientists, Lynn Margulis and James Lovelock, to suggest that life on Earth acutally regulates the compositon of the lower atmosphere by controlling the amounts of some of the important gases that are present—the gases containing elements that are essential for the continuation of life. The growth of organisms thus has a strong influence on some of the chemical reactions taking place in the environment, including those affecting the gases that regulate the mean planetary temperature. The mean temperature, in turn, will affect the growth of organisms, and this kind of feedback is found repeatedly as one examines the many ways in which life interacts with the inanimate environment (Figure 10-10).

The total system of life on Earth—all the organisms as well as the gases, liquids, and solids they produce and consume—Margulis and Lovelock call *Gaia,* after the Greek goddess of Earth. Gaia is the product of nearly 4 billion years of evolution, during which life has differentiated and expanded to become the multitude of organisms we find today. This hypothesis suggests that *Gaia will strive to maintain that particular equilibrium within which life can survive most successfully.* Many different mechanisms must be involved in environmental regulation—just as bees use many mechanisms to maintain the optimum temperature and humidity in their hives. For example, if the average temperature on Earth should decrease—perhaps as the result of changes in the eccentricity of the Earth's orbit around the sun—Gaia may respond by promoting the differential growth of organisms leading to liberation of more carbon dioxide into our atmosphere. The excess carbon dioxide would trap more of the infrared photons from the Earth and thus would bring the temperature back to its former value.

The concept of Gaia as a living organism remains intriguing and challenging. It finds support from the fact that changes in the Earth's average temperature have been remarkably small over the past billions of years, no more than a few degrees Celsius from the value that we find today. Gaia dramatically emphasizes our relative insignificance as a land-based mammalian species, for even if we regard ourselves as Gaia's "central nervous

Fig. 10-10   The Gaia hypothesis states that life on Earth controls the amounts of nitrogen, oxygen, and other atoms exchanged between the atmosphere and living organisms, and regulates the temperature by controlling the growth of organisms that results from this exchange of gases. Since the growth of organisms feeds back into the control of temperature, the total system of Gaia forms a feedback loop that can be self-regulating.

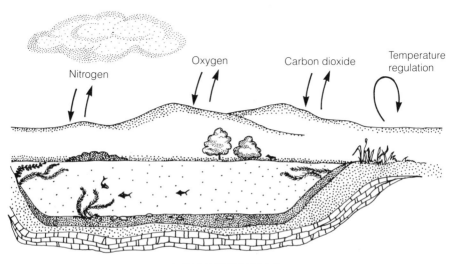

LIFE ON EARTH (Gaia)

system," we see that Gaia might some day replace us with a more effective, less destructive species. In return, the concept of Gaia helps us to see ourselves as part of the web of life. We depend on an intricately intertwined, living system for the food we eat, for the air we breathe, and perhaps for the climate in which we live. Tampering with the Earth's ecological balance therefore risks more than eliminating a few species of organisms or producing esthetically unpleasant effects: We face the potential of seriously disturbing the balance that Gaia has achieved, and the response may occur in ways that are harmful to us.

We must also keep our sense of the interdependence of life firmly in mind as we examine the other planets in our solar system. Isolated species almost certainly do not inhabit otherwise barren worlds. Instead, we expect to find organisms that live in close harmony with their environment, a harmony in which life itself, if it is anything like life on Earth, has caused environmental changes that we may be able to detect.

Summary

Prokaryotes, the simplest organisms, consist of cells lacking nuclei but possessing both DNA and RNA molecules to govern their functioning and reproduction. Photosynthesis began in prokaryotes such as the blue-green bacteria that provided much of Earth's oxygen. As oxygen flooded the Earth's atmosphere, organisms that could use this oxygen as a source of energy developed some 2.5 to 2 billion years ago.

From prokaryotes or from a common ancestor there evolved eukaryotes, cells with well-defined nuclei in which the cell's genetic material is stored. Eukaryotes have subunits called organelles that can perform specialized functions, and they contain a much larger percentage of genetic material than prokaryotes do. The increasing specialization of functions within eukaryotic cells, and of different cells within larger organisms, has led to creatures with many trillions of cells concentrated in particular organs, and with particular tasks that benefit the entire organism.

About 600 million years ago, a billion years or more after the eukaryotes had appeared, a tremendous increase in the variety, size, and distribution of plants and animals occurred. From this time, the start of the Cambrian era, we can date the flourishing of large land plants and of all the varieties of animals more complex than soft-bodied sea creatures. Mammals first appeared about 200 million years ago, at the height of the age of reptiles, while our hominid ancestors seem to go back about 3 million years.

Thus, five sixths of the history of life on Earth involves prokaryotes and eukaryotes no more complex than jellyfish. The planets that might exist in orbit around other stars might well have only these sorts of life, if any life has arisen, or they might be a few billion years ahead of us and have evolved new, perhaps immortal, forms of life. From our use of life on

Earth as a typical example, we may conclude that life begins in water but apparently needs land to evolve into complex forms, capable of intelligence, which we define as the ability to communicate with other forms of life on other planets. The final lesson of life seems to be that all forms of life owe their continued existence to their mutual interconnection, and that the entire surface of our planet may be said to act like a single organism, Gaia, which strives to use its subunits in the best possible way.

Questions

1. Why are cells important to living organisms? Why can we say that the first cell was a small pond, full of organic molecules and water?

2. What is a virus? Why do we have difficulty in deciding whether or not viruses are alive?

3. What is the key difference between prokaryotes and eukaryotes? Why can some prokaryotes live only if oxygen is *absent* from their environment?

4. How long ago did eukaryotes develop into a great number of widely different species? How long after the start of life on Earth was this era of great speciation?

5. Suppose that someday we discovered a planet similar to Earth which had formed only 3.6 billion years ago, rather than the 4.6 billion years we allow for Earth. If both planets follow roughly similar paths for evolution, should we expect to find many sorts of land "animals" on the other planet? Should we expect to find birds? Insects? Sponges?

6. Suppose that life took 2 billion years to appear on the hypothetical planet of Question 5. How far would life then have advanced, if its development proceeds at the same speed as it has on Earth?

7. Why has water been so important to life on Earth? Should we expect that life on any planet would require liquid water?

8. What are mutations? Why are they important in the evolution of life?

9. Why does evolution tend to produce *more* complex organisms from *less* complex ones, rather than the other way around?

10. Do you think that the evolution of what we call "intelligence" should occur naturally on any planet with life? Why?

11. What fraction of the Earth's total lifetime (4.6 billion years) does humanity's recorded history (about 4600 years) represent? How significantly has the human race changed the face of the Earth within that fraction?

12. What are the major changes that would occur in the composition of the Earth's atmosphere and surface layers if life were to disappear from this planet?

Further Reading

Bronowski, Jacob. 1972. *The ascent of man*. New York: Little, Brown and Company.

Gibor, Aharon, ed. 1976. *Conditions for life*. San Francisco: W. H. Freeman and Company.

Jastrow, Robert. 1977. *Until the Sun dies*. New York: W. W. Norton & Company.

Leakey, Richard. 1977. *Origins*. New York: E. P. Dutton & Co.

Lovelock, J. E. 1979. *Gaia: A new look at life on Earth*. Oxford: Oxford University Press.

Mayr, Ernst. 1978. Evolution. *Scientific American* 239:3, 46.

Sagan, Carl. 1977. *The dragons of Eden*. New York: Random House.

Schopf, J. William. 1978. The evolution of the earliest cells. *Scientific American* 239:110.

Thomas, Lewis. 1974. *Lives of a cell*. New York: Viking Press.

Young, J. Z. 1971. *An introduction to the study of man*. Oxford: Oxford University Press.

CHAPTER 11

How Strange Can Life Be?

We have followed the development of life on Earth in order to draw some tentative conclusions for assessing the probabilities of life elsewhere. Although we might think at first that anything could be possible, thus making conclusions based on terrestrial life hopelessly biased, we should remember that the laws of physics and chemistry make some situations far more probable than others. These "laws," the summary of our experience in analyzing the universe around us, appear to be valid as far as we can test them, including the analysis of light from stars and from the most distant galaxies. Unless we choose to abandon the idea that nature follows the same rules in various parts of the universe, our conclusions about the overall probability of life appearing in different environments should have merit, although we must remind ourselves that we are dealing only with probabilities, not certainties. If we *do* abandon this idea, we can make no predictions at all!

The Chemistry of Alien Life

The history of investigating life on Earth includes repeated, unexpected discoveries of strange creatures, tiny and large, flourishing in environments that seem at first to be unlikely havens for life. For example, we have said that sunlight is necessary for photosynthesis to occur, but a species of bacteria, called a thiococcus, can perform photosynthesis in what we could call complete darkness. These bacteria, thriving in the muddy waters of South American jungles, contain a pigment that absorbs infrared sunlight, whose low energy per photon lies outside the sensitivity range of human eyes.

We can say that life must have water. Indeed life must, but that water can be comparable in acidity to a dilute solution of sulfuric acid, or else the water can contain large amounts of dissolved lime and thus be quite basic. Life can assume enormous stationary forms, as in a redwood tree, or be tiny and highly mobile, as in a paramecium. We cannot hope to predict the exact forms of life that we might meet on an alien planet, nor can we predict the kind of adaptation that such living organisms would have made to their environments. But can we make *any* definite statement about the *chemistry* of alien life, about the molecules that form living organisms? On Earth, we can make statements of great breadth. Both the giant redwood and the microscopic paramecium live because DNA and RNA molecules help them make protein. Both these organisms rely on a specific chemistry, which turns out to be the same chemistry for all organisms. This chemistry uses one element, carbon, as the basic structural unit to form complex molecules, and it uses water as a solvent. Water provides a fluid in which nutrient molecules can float; it preserves chemical equilibrium in living cells, it helps regulate the temperature within them, and it forms a large fraction of each organism's body weight (40 percent in dry plants, 70 percent in humans, 95 percent in jellyfish).

To maintain a solvent in the liquid state requires that the temperature stay within a fairly narrow range, 0 to 100° C in the case of water. Thus, the requirements for a common element, carbon, and a common solvent, liquid water, would automatically add a third requirement, the proper temperature range, if life is to arise.

But must life elsewhere use the same chemistry as life on Earth? We hardly expect life on other planets to have molecules exactly like amino acids, proteins, DNA, and RNA, though it might have molecules that perform similar functions. But we must consider this key question: Would other forms of life be based on carbon for structure, and on water as a solvent? Or could we expect a different structural element and a different solvent?

We can attempt to outline the basic requirements for any alien life chemistry. It must be able to form large, complex molecules, for only in this way can it store the amount of information a living organism needs to function properly. It must use a fluid to carry nutrients in, and waste products out, within which chemical reactions can occur, thus allowing life to persist and to reproduce. Though we can conceive of situations in which this fluid is a gas, a liquid will serve better, because it will keep the chemical reactions within a localized region, as in the cells of living creatures on Earth, and at a constant temperature, thus making it easier for the reactions to continue at a nearly constant rate.

Given these two basic requirements—complex molecules and a liquid solvent—what options are open to life? Has chance alone dictated that we have carbon-based life on Earth, or does something favor this element over all others? And why does terrestrial life depend so much on water? Why not ammonia, or alcohol, or another fluid?

The Superiority of Carbon

Let us consider carbon first. To replace it, we would need a relatively *abundant* element that can combine with four hydrogen atoms to form a stable molecule (Figure 11-1). The first requirement stems from our prejudice in favor of the kinds of life that will be abundant in the universe, not the rare and exotic forms that might exist in some unlikely combination, but rather the life that arose under the conditions that have prevailed throughout most of the universe. The second requirement, the ability to combine with four hydrogen atoms, arises from the need for complexity: Since four atoms are as many hydrogen atoms as *any* atom can accept, an atom that competes with carbon as the backbone of life must do at least this well.

We can easily illustrate this point. Think of all the different ways that *oxygen* can combine with hydrogen to form a stable molecule. There are two: water (H_2O) and hydrogen peroxide (H_2O_2) as shown in Figure 11-2. How many stable molecules can *nitrogen* form with hydrogen? Two again: ammonia (NH_3) and hydrazine (N_2H_4). Now ask, how many molecules can *carbon* form with hydrogen? If you don't know the answer, you are in good company—neither does anyone else! But to get an idea, you can consult the *Handbook of Chemistry and Physics*. There you will find the heaviest such molecule listed as enneaphyllin, whose chemical formula is $C_{90}H_{154}$! (See Figure 11-2). Remember that these are just the carbon compounds with hydrogen alone. Colleges teach organic chemistry as a separate subject because carbon has a truly amazing tendency to form all sorts of compounds with every other element.

Does any other element perform as well as carbon? Silicon, the usual candidate to replace carbon, lies directly underneath carbon in the periodic

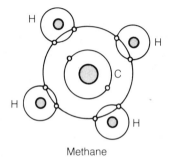

Methane

Fig. 11-1 Carbon atoms can each combine with as many as four other atoms, as in methane (CH_4).

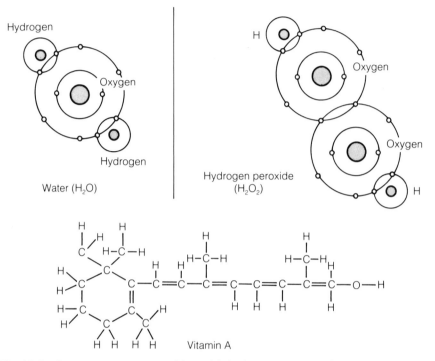

Fig. 11-2 Oxygen atoms can combine with hydrogen atoms to form just two stable molecules: water (top left) and hydrogen peroxide (top right). In contrast, carbon and hydrogen atoms can form many complex molecules, such as vitamin A (bottom).

table of elements. Since silicon also has a considerable abundance in the universe, and since silicon can indeed combine with four hydrogen atoms to form silane (SiH_4), we often meet silicon-based life in science-fiction encounters. But if we examine silicon more closely, we find that the bond between two silicon atoms has only half the strength of the bond between two carbon atoms. Thus, the bonds between silicon atoms are twice as weak, twice as easy to break in chemical reactions. Furthermore, although the strength of carbon-carbon bonds about equals the strength of carbon-hydrogen or carbon-oxygen bonds, we find that silicon-hydrogen or silicon-oxygen bonds have *more* strength than silicon-silicon bonds do. Therefore, long chains or rings of atoms based on Si-Si-Si- structure are unlikely, whereas multiple carbon linkages of just this kind dominate organic chemistry and indeed distinguish it from other sorts of chemical activity.

We could get around this difficulty by making long molecular chains, or polymers, with silicon-oxygen bonds instead of silicon-silicon bonds. Such polymers form silicones, which are both stable and chemically versatile.

Silicones, however, tend to react very little with other molecules, a key fact for their usage in cosmetics and in lubricants for machinery. Nevertheless, we can certainly imagine that special circumstances, or special catalysts (enzymes), could produce more lively interactions among silicone-like polymers.

The basic difficulty with silicon as the backbone of life is its affinity for oxygen. Even if silicon exists in the most "reducing" conditions in an atmosphere loaded with hydrogen, it will form silane, SiH_4, only at temperatures above 1000° K. At lower temperatures, silicon forms silicon dioxide (SiO_2). Proof of this comes from observations of Jupiter's atmosphere, which we know is predominantly hydrogen and which contains NH_3, PH_3, CH_4, and H_2O, the fully "reduced" (hydrogen-laden) forms of the elements nitrogen, phosphorus, carbon and oxygen, *No* SiH_4, however, has been detected on Jupiter. Why? Because the silicon, instead of forming silane, has combined with oxygen to form silicon dioxide. Since oxygen atoms outnumber silicon atoms by a factor of 25 in the universe, silicon atoms are likely to end up in silicon-oxygen compounds (see page 35). And once silicon dioxide has formed, it is very difficult to decompose.

Let us compare carbon dioxide (CO_2), which remains gaseous even at low temperatures (down to −75° C), is soluble in water, and is capable of relatively easy break-up into its constituent carbon and oxygen atoms, with silicon dioxide (SiO_2), which is gaseous only at high temperatures (above 800° C), is extremely insoluble (in almost everything except hydrofluoric acid!), and requires large amounts of energy to be broken apart into its silicon and oxygen atoms. The former substance has the advantage over the latter in every respect in providing a molecule useful to living organisms. Conversely, methane (CH_4), the fully reduced form of carbon, has much greater stability than silane, the fully reduced form of silicon. Methane exists even in our highly oxidizing atmosphere, and although its lifetime there is short (the entire amount must be replaced by methane-generating bacteria on a time scale of about 10 years), it is not nearly so short as that of silane, which bursts spontaneously into flame when exposed to air! This gives us the chance to speculate playfully that if silicon-based creatures did exist and visited the Earth, we would have a natural explanation for the fire-spouting dragons of medieval legends. It seems far more likely, however, that silicon's deficiencies as a structural element rule out the possibility of silicon-based life in almost all situations.

We can test this conclusion by examining silicon and carbon in the universe at large; the results support our statement. Astronomers have found no silicones or silanes in meteorites, in comets, in the interstellar medium, in the atmospheres of planets, or in the outer layers of cool stars. Instead, they find silicates, molecules of oxidized silicon, in combination with various other elements. In fact, we live on an excellent example of silicon chemistry, the planet Earth, which consists largely of silicates. On the other hand, complex carbon-based *(organic)* molecules appear in meteorites, comets, interstellar clouds, planetary atmospheres, and cool stars, along with carbon monoxide and carbon dioxide. Carbon passes easily between its fully oxidized (CO_2) and its fully reduced (CH_4) forms, while silicon does not.

We may reasonably conclude that although life based on silicon may be possible, it will be extremely uncommon at best. Life based on carbon seems favored as the dominant kind of life in the universe. This may sound dogmatic, but we should remember that the abundances of the various elements appear to be roughly the same everywhere we can measure them in the universe. Thus, the elements still rarer than silicon have extremely poor prospects as the basis for the chemistry of life elsewhere.

Solvents

Let us now turn to the question of solvents. What makes water so wonderful? Why does this substance have such critical importance to life as we know it? To be truly useful, any solvent must remain liquid within a large range of temperatures, so that the variations in conditions on a planet or satellite do not make the solvent freeze or boil. This temperature range should cover those temperatures that allow chemical reactions to proceed at a reasonably rapid pace, yet will not cause the destruction of important molecules by collisions. We would also prefer a solvent that helps the organisms that contain it to regulate their temperatures, and (by definition!) the solvent should have the ability to dissolve other chemical compounds, since organisms will use it to transport nutrients and to carry off wastes. We can see how well water fulfills these requirements in comparison with two other possibilities, ammonia (NH_3) and methyl alcohol (CH_3OH).

We should first notice that water does indeed remain a liquid over a large range of temperatures (Table 11-1).

TABLE 11-1

Temperature Ranges at which Solvents Remain Liquid

Solvent	Temperature for Liquid	Range of Temperature
Water	0 to 100° C	100° C
Ammonia	−78 to −33° C	45° C
Methyl alcohol	−94 to +65° C	159° C

In this temperature-based comparison, water outperforms ammonia, but methyl alcohol, by virtue of its greater temperature range as a liquid (despite the lower maximum temperature), seems almost as good as water.

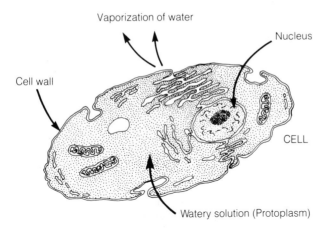

Fig. 11-3 Because water has such a high heat of vaporization, a living cell can respond to a temperature increase by vaporizing a relatively small amount of water, and can maintain about the same temperature.

The next comparison among solvents deals with their ability to help regulate the temperature. Solvents' heat regulation capabilities depend both on the *heat capacity,* which is the amount of energy required to raise 1 gram of the solvent by 1 degree Celsius in temperature, and also on the *heat of vaporization,* the amount of energy needed to change the liquid into a vapor once it has reached its boiling point. We want a solvent to have as large a heat capacity and a heat of vaporization as possible, to minimize the effect of sudden temperature changes on the physical conditions within the solvent. A large heat capacity implies that a big change in the external temperature will affect the organism only slightly. We encounter this principle on a large scale in terrestrial climates: Land near the ocean will have a much milder climate than a region at the same latitude, far from large bodies of water. The heat capacity of water is 1 calorie per gram-degree, intermediate between the heat capacities of ammonia (1.23 calories per gram-degree) and of methyl alcohol (0.6 calorie per gram-degree). In its heat of vaporization, however, water far surpasses both its competitors. To vaporize water requires 595 calories per gram, while ammonia needs only 300 calories and methyl alcohol only 290.

Water's high heat of vaporization means that a small amount of evaporative cooling will deal with the heat released within a cell as life processes occur (Figure 11-3). Indeed, the fact that we consider the normal temperature for human beings to be 37° C (98.6° F), and not 36 to 38° Celsius (96.8 to 100.4° F) gives an excellent indication of how well our bodies can regulate their temperatures. Although we don't yet understand just what conditions are required for intelligence to exist, we can appreciate the fact that mammals need a precise temperature regulation to maintain their rates of chemical reactions at the right level and to keep the structural elements of their bodies in proper condition. Mammals are distinguished from "lower" forms of life by the greater complexity of their brains and by the chemical reactions that go on in these brains; precise temperature regulation allows these reactions to occur properly.

The *surface tension* of water, twice that of ammonia and three times that of alcohol, exceeds the surface tension of any other liquid known. This property undoubtedly had a vital role in making aggregates of organic compounds before cells evolved, for the surface tension would force some compounds together and would preserve the boundaries between watery mixtures of different organic molecules. Surface tension continues to concentrate solutions of solids at the interfaces of different media in an organism; for example, at the boundary of a cell wall.

For its heat capacity, heat of vaporization, and surface tension, water is hard to beat. But in a way these are extra benefits, since life's essential fluid medium must first of all act as a solvent; that is, *dissolve* a wide variety of chemical compounds in order to carry them easily into and out of living systems. Here again, water is outstanding: It has more than twice the ability of either ammonia or methyl alcohol to carry other molecules in solution.

We have compared water to two other fluids and have found that it has several advantages in addition to being the best solvent. Why don't we consider many other possible liquids as well? Other substances, such as hydrogen sulfide (H_2S) and hydrogen chloride (HCl), that could remain liquid under reasonable temperature conditions must be far less abundant than our basic three, since they include relatively rare elements (such as sulfur and chlorine) rather than the hydrogen, nitrogen, oxygen, and carbon that form water, ammonia, and methyl alcohol. Furthermore, these alternative possibilities break up into their constituent atoms more easily. Thus, if water wins over ammonia and methyl alcohol, it probably surpasses all other fluids that are likely to exist in other astronomical situations.

We ought to pay a moment's attention to an extremely special property of water, one it shares with almost no other fluids. Water expands as it freezes, so that solid water (ice) floats on liquid water. If ice sank, ponds that froze in winter would freeze all the way down, killing most of the organisms within them. This sort of freezing would occur in ponds of ammonia (at $-78°$ C) and of methyl alcohol (at $-95°$ C). We may, however, consider our worries about freezing ponds to be an example of "high-latitude chauvinism," since most of the Earth never freezes. Life may in fact have originated in low-latitude areas, surviving and evolving in ponds that never froze on top.

What about another planet, with other conditions on its surface? We can imagine several advantages in using a substance, such as ammonia, that does *not* expand when it freezes. The difficulty in a solvent that expands resides in the stress that freezing produces against the walls that contain the fluid. Frozen water in pipes can burst the steel walls, and the cells of organisms that freeze will rupture, killing the organisms. But if life's chemistry relied on a solvent that did not expand upon freezing, low temperatures might simply produce a dormant state, from which life could easily recover when warm weather reached the place of hibernation.

What an extraordinary advantage for space travel! The immense distances between stars present an insoluble problem for space travel with conventional rockets, if we require relatively brief journeys (see page 372). The need for speed arises from the desire to make the trip within the crew's lifetime. Could we extend that lifetime by a large factor simply by putting the crew on ice? Humans can't survive such freeze-drying, since we consist mostly of water, but imagine a race of astronauts whose life chemistry depended on ammonia as a solvent: They might use the fact that ammonia does not expand upon freezing as the way to visit most of our galaxy, waking up only briefly at each stop!

Before we grow too excited about ammonia, however, we ought to recall a key advantage water has over ammonia. Liquid water is self-shielding against ultraviolet light: Some of the water molecules will be dissociated, releasing oxygen and hydrogen into the atmosphere, and some of the oxygen atoms will link into ozone (O_3) molecules, which absorb ultraviolet light (Figure 11-4). Since most stars produce large amounts of ultraviolet radiation, the self-protection that water provides has great importance, for without such an ozone shield, organic molecules would quickly be destroyed by ultraviolet light, and oceans themselves would slowly be dissociated, molecule by molecule, over billions of years. In contrast to water, the dissociation of ammonia produces *nitrogen,* not oxygen atoms, which can neither form a shield against ultraviolet light nor produce a source of

Fig. 11-4 When ultraviolet light dissociates water molecules, some of the oxygen atoms released will link into ozone molecules, which act as a shield against ultraviolet light.

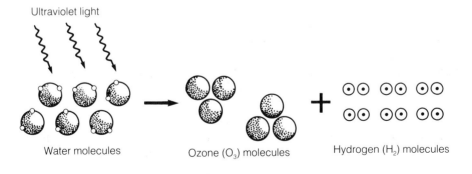

Ultraviolet light

Water molecules   Ozone (O_3) molecules   Hydrogen (H_2) molecules

chemical energy similar to that available from oxidation. Hence, a planet with ammonia oceans would require a separate source of shielding—perhaps a permanent cover of dust-laden clouds—to prevent the destruction of the oceans, and of any life on the planet, by ultraviolet light. Water carries this shielding capacity along with the oxygen in every molecule.

Our considerations of the various possible solvents are not exhaustive, but they serve to make a key point: Life on Earth relies on carbon chemistry, with water as a solvent, not by accident, but because carbon and water have an inherent superiority. The case for carbon may be stronger than that for water, but we should not feel narrow-minded in anticipating that *most* alien life forms should rely on the same element, and the same solvent, for their basic life chemistry.

Nonchemical Life

Until now, we have restricted our discussion of life to the *chemistry* that we know, with possible variations in the temperature, density, and elemental composition taken into account. Confining our considerations to chemistry means discussing the interactions among atoms, which can form many sorts of molecules, as simple as water or as complex as DNA. Atoms interact with one another through electromagnetic forces, which arise from the positive and negative electric charges the atoms contain; thus, when atoms join into molecules, elctromagnetic forces, not gravitational or strong forces, provide the linkage.

When we try to consider strange forms of life, however, we must go beyond the chemical interactions that govern life on Earth. Similarly, we must not restrict ourselves to planetary surfaces in our search for life. Planets do provide the likeliest sites for *chemical* life to develop, because on planets we expect to find the necessary conditions—a relatively high density of matter and temperatures appropriate to chemical reactions—that will allow chemical interactions to proceed at a relatively rapid rate. If we have the right conditions, together with the mixture of elements typical of stars, then the origin of life on planets around stars may well turn out to be a better than even chance.

Even if life based on chemical reactions turns out to be widespread on planetary surfaces, we must not neglect forms of life much stranger to us. Some important examples, which do not exhaust the entire range of possibilities, are life in dense interstellar clouds, on the surfaces of neutron stars, and of entire galaxies.

In these three cases, we shall consider living entities with characteristic sizes that are very much larger or smaller than the forms of life with which we are familiar. Even here, however, we must bear in mind a very important characteristic of life on Earth: the presence of individuals. We humans number in the billions, while most species of mammals, birds, and reptiles have millions, or at least many thousands, of representatives. Only those species close to extinction include just a few thousand, or a few hundred, individual species members. This fact reminds us that life on Earth has evolved through the interactions of enormous numbers of individual animals and plants, precisely because in order for natural selection to make progress—that is, to distinguish those individuals most fit for survival and reproduction—each species had to have a large pool of members who could produce offspring with slightly different characteristics. Most evolutionary "experiments" end in failure, so a species of animals or plants with only a few hundred individuals cannot hope to carry its evolutionary development much further in its native habitat. Instead, such a species fights for simple survival.

Should we expect things to be different on another planet, or in interstellar space? Probably not. We can easily imagine conditions in which certain life forms had no problems of competition for survival and reproduction, and never entered a new environment, but then we would have a greatly reduced pressure for evolutionary changes. As long as competition or changes in environment exist, and as long as mutations occur (see page 158), then we expect natural selection to lead to new types of species, which themselves will evolve over many millions of years. The key to this process remains the interplay of a large number of *individuals* in a species, with a large number of *species* as the result of this interplay.

We thus seem directed toward the conclusion that complex forms of life, forms that have evolved to the point of achieving intelligence, cannot be the product of just a few events that produce a few individuals. In contrast, living organisms by the billions and trillions testify to the natural result of competition for reproductive success on Earth.

Black Clouds

In a marvelous work of fiction, the eminent British astronomer Fred Hoyle has described the arrival near Earth of a giant interstellar cloud, capable of thought, of directed motion, and of an indefinite lifetime, a cloud whose life processes depend on electromagnetic forces (as ours do) and whose thoughts consist of radio messages from one part of the cloud to another (Figure 11-5). Its excursion into our sun's vicinity occurs because the cloud must periodically replenish its supply of stored energy by absorbing large amounts of starlight, in this case from our own star. In the course of its passage, the cloud discovers intelligent beings on one of the sun's planets, since they manage to establish radio communication with the cloud, which reveals to them many mind-bending facts about the universe.

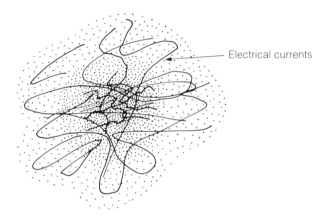

Electrical currents

Fig. 11-5 Fred Hoyle's black cloud is an interstellar mixture of gas and dust in which electrical currents carry thoughts and starlight provides the basic energy source.

The cloud's "brain" consists of complex networks of molecules, which can be increased in number and specialization as the cloud chooses. Electromagnetic currents pass among these molecules, and there is little conceptual difference in the way that the cloud's brain works and the way that our brains do.

Fred Hoyle's black cloud, a living organism as large as the orbit of Venus and almost as massive as the planet Jupiter, reminds us of the difficulties and advantages that our experiences provide when we speculate about other forms of life. On one hand, we can imagine the existence of a living "black cloud" that uses starlight energy to construct greater and greater degrees of ordering, thinks and feels through electromagnetic impulses sent along appropriate pathways. On the other hand, our knowledge of how terrestrial organisms evolved suggests the *necessity of many previous steps through which evolution can proceed.* A black cloud certainly *could* exist, but how its existence would ever arise remains a darker mystery.

If we, like Fred Hoyle, imagine earlier stages in the evolution of black clouds, primitive cloudlets of lesser abilities, then we return to our favoritism for planetary surfaces. Why? Because matter in interstellar space has such low densities (compared with matter on planets) that any interaction among particles will occur far more slowly there. Therefore, the time for life to arise should span not the billion or so years that applies to the Earth but thousands or millions of times longer—far longer than the age of the universe since its expansion began. For the same reason, even if life could originate in matter at the low densities of interstellar clouds, we would expect that all life processes should take longer, because of the greater distances and lower densities of matter.

Billions of years had to pass on Earth before complex organisms appeared, so long, in fact, that the time for "intelligent" life to arise on Earth measures a noticeable fraction of the entire age of the universe (about one quarter). Although some interstellar clouds might indeed possess the "right" temperatures (about −50 to 80 °C) for life to begin in a relatively short time, most are far colder than this (−250 to −200 °C). Furthermore, the denser clouds tend to be the colder ones. But even in those clouds with the proper temperatures and the "right" mixture of elements, the fact that the density of matter falls a million billion (10^{15}) times below the density of our atmosphere would present a serious obstacle to the possibility of organizing matter to a sufficiently great degree for the emergence of life, given the large distances between one "complex" molecule and another in one of these clouds.

Life on Neutron Stars

Until now, we have looked at the consequences for life of our well-known laws of chemistry, the summarization of the effects of electromagnetic forces. Following a stimulating suggestion by Frank Drake, let us consider briefly how another kind of force, the *strong force,* could produce an entirely different sort of life on the surfaces of neutron stars.

As we saw in Chapter 7, neutron stars arise from the collapsing cores of highly evolved stars (Figure 7-4). The temperature on the surface of a neutron star must be about 1 million degrees above absolute zero, more than a thousand times the 300 degrees that characterizes the Earth's surface. Furthermore, the gravitational force on the surface of a neutron star exceeds that on Earth by about 1 *trillion* times. The high temperature and immense gravitational forces mean that no molecule or atom could survive for long on the surface of a neutron star. High-speed collisions among molecules would break them apart, and would knock any atom's electrons loose. As a result, we can hardly expect to find any sort of life based on electromagnetic interactions, which make atoms and molecules, on neutron star surfaces. Still, we should not lose all hope, for life based on strong forces remains conceivable.

Strong forces hold together the nuclei of all atoms more complex than hydrogen. In nature, we find nuclei with anywhere from one particle (the single proton of hydrogen) to 238 particles (the 92 protons and 146 neutrons of uranium-238). Nuclei with more elementary particles than 238 can be made in physics laboratories, but they fall apart (decay) into smaller nuclei after an amazingly short lifetime, *by human standards.* The most massive "artificial" nucleus yet made has 265 particles, 105 protons and 160 neutrons, and decays in less than a billionth of a second.

We do not, however, live at the size level of protons and neutrons. Suppose that we were, say, a proton some 10^{-13} centimeter in size, and traveled at a speed of 1000 kilometers per second, a speed typical of protons at a temperature of 1 million degrees absolute. We would then move a distance equal to our own size in 10^{-21} (one billion trillionth) of a second! Compare this with a human being whose size measures 170 or so centimeters, and who walks a distance equal to this size in about 1 second. We can see that the tiny span of 10^{-21} second would have the same meaning to a proton as a second does to a human being out for a stroll. At the size level of elementary particles, things tend to happen in times that are measured in 10^{-21} second, just as things happen to humans in times measured in seconds. More precisely, the electromagnetic interactions within our body chemistry characteristically occur on time scales of a few *thousandths* of a second (as, for example, when our eyes perceive visible-light photons), a few *tenths* of a second (for example, to send a message all the way through our bodies), or a few *seconds* (say, the time for us to reverse our position completely).

Now imagine the seething surface of a neutron star, where elementary particles travel and collide at speeds of thousands of kilometers per second, which is equivalent to millionths of centimeters in trillionths of a second. Such collisions could produce massive nuclei, with thousands, or even tens of thousands, of elementary particles within them, that would last for perhaps a million billionth (10^{-15}) of a second. After this brief time had elapsed, the massive nuclei would fall apart into smaller nuclei; but from the point of view of an elementary particle, for which 10^{-21} second gives a typical time scale, these massive nuclei do not decay instantaneously, but rather last for millions of times longer than the average time between interactions. In other words, a massive nucleus might have a million different collisions or other interactions before it decayed into some other sorts of nuclei.

We can speculate that evolutionary processes on the surface of a neutron star *might* produce forms of life, individuals that interact with their environment and with other individuals in an organized way. If all this came true, then the development and evolution of such life would happen far more rapidly than our Earthborne experience admits. Since the typical time scale for strong force life, 10^{-21} seconds, falls a billion billion times below the thousandth of a second that characterizes our electromagnetic force life, we might expect evolution to proceed that much more quickly, so that the origin of life would require not about 1 billion years, but about one billionth of a year, or one thirtieth of a second! Short as this time may seem to *us,* it allows for billions upon billions of interactions among ever more complex nuclei on the surface of a neutron star.

If we have the courage to follow this scenario further, we can see that on the surfaces of neutron stars, whole civilizations could rise and fall, and rise and fall a million times over, faster than the human eye can wink. The individual members of these civilizations would have sizes of about

10^{-11} centimeter, would live for 10^{-15} second, and, if they use photons for communication, would favor gamma-ray photons with perhaps 10^{10} times the frequency of visible-light photons. Such gamma-ray photons arise from interactions among elementary particles, just as visible-light photons typically arise from interactions among atoms.

If this sort of neutron star life does exist, then we probably cannot hope that a civilization that arose would use radio waves, either for their own communication or in an attempt to find other civilizations. Nonetheless, these speculations about neutron star creatures, based simply on the facts we know about the strong forces that attract elementary particles, should not be forgotten when we consider how many civilizations may now exist in our own galaxy. But we do have a difficulty in communication that arises from the great differences in perspective, especially of time, between ourselves and a neutron star civilization. We can hardly hope to establish a meaningful interchange with a civilization that lasts only a billionth of a second—or would they still have something to tell us?

Gravitational Life

Since we have taken a mental excursion to consider the possibility of fast-moving, strong force life on neutron stars, we may deem it proper to consider the opposite extreme, the idea of life based primarily on gravitational interaction. Here the typical subunit should be an object large enough for gravity to dominate over electromagnetic and strong forces: a star. If individual stars play the role of individual atoms or molecules in Earth life, or of individual nuclei in neutron star life, then what does it mean to notice that stars cluster together in galaxies? Are galaxies alive?

No, we think not. (But consider the speculations about Gaia on page 202; perhaps the Gaias themselves form individual parts of a super-Gaia, the ultimate organism!) Our definition of life does not simply require a great degree of organization. Galaxies are fairly well ordered, especially spiral galaxies, but not so much so as, for example, paramecia (Figure 11-6). Furthermore, galaxies *seem* to lack any sort of purpose, but here we suffer from human chauvinism. Things *happen* in galaxies, but on time scales that range from seconds (for an individual star to collapse) to hundreds of millions of years (for an entire galaxy to rotate). *Stars typically interact with one another on a time scale of many millions of years.* If life could originate from the repeated effects of such close interactions—as when molecules interacted in the primordial soup on Earth—then scarely any time has passed, measured in terms of the basic gravitational unit of time, the time for a single interaction among stars. Hence, life based on gravitational forces appears to be at much the same stage in its (supposed) development as life on Earth was just a few years after the Earth had

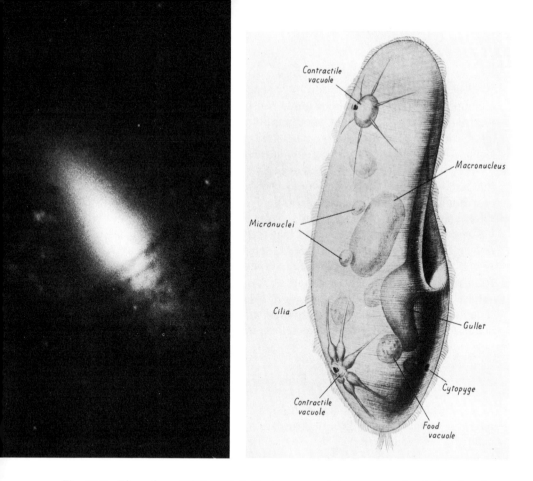

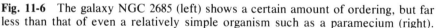

Fig. 11-6 The galaxy NGC 2685 (left) shows a certain amount of ordering, but far less than that of even a relatively simple organism such as a paramecium (right).

formed. Gravitationally based life could yet appear, but we expect this to occur after billions of billions of years, not the mere billions of years needed for life based on electromagnetic forces.

The Advantages of Being Average

We have, so far, considered the possibilities of life based on strong forces, on electromagnetic forces, and on gravitational forces. These are the basic forces we know, but we may be prisoners of our limited experience and limited imaginations. In the following chapters, we shall deal with the possibility of life based on electromagnetic forces, and usually of life that exists on planetary surfaces. We have seen why this appears to be our best bet, but let us take the broad view and say that other kinds of life may exist, which would contribute a bonus to the total number of civilizations with whom we can hope to communicate.

If we count only the possibilities of life on other planets, we may indeed underestimate the number of civilizations with whom we might communicate by leaving out the possibilities of life in interstellar clouds, or on neutron stars, or in entire galaxies, or of life which has left the planet of its origin. Furthermore, we shall be biased not only in favor of life like our own, but towards civilizations similar to ours. But we elect to take this risk because our current view of the universe suggests that such civilizations will be much more numerous than the other forms of life we have considered.

Even if we do restrict ourselves to looking for life on planets with time scales roughly the same as ours, we still have a great variety of possibilities. In our own solar system alone, these chances range from the ammonia clouds of Jupiter to the Sahara desert, from the Grand Canyon of Mars to the frozen methane of Pluto. Other planets in other systems may present a still greater range of environments where life would have a chance to develop. Thus, even with our restriction to planets, which we do not regard as absolute, we have the possibility of finding millions of civilizations in our own galaxy, hence millions of *billions* of civilizations (at least!) in the universe. These numbers will provide a start in our quest to find other civilizations with whom we may have a quiet talk before we pass on.

Summary

Life that is based on chemical reactions—that is, on the interaction of atoms to form complex molecules—appears to require carbon as its key structural element. Only carbon atoms can form chemical bonds with hydrogen, oxygen, nitrogen (and other less abundant elements) in a way that readily promotes the development of a wide variety of information-bearing polymers. Silicon can also form polymers, but these are too stable under ordinary conditions to serve as the basis of life. The chemical affinity of silicon for oxygen implies that at temperatures low enough for complex molecular structures to exist, silicon will be bound up as silicates, the rocks that supply our footing on this planet. If carbon is the crucial element in all chemical life, we are still not much restricted, since carbon has a high abundance everywhere in the universe.

Life also seems to require a solvent, a fluid medium in which atoms and molecules can encounter one another and undergo chemical reactions. The ability of water to dissolve other substances makes it one of the most favored solvents. In addition, water's heat capacity, its heat of vaporization, its ability to remain liquid in an appropriate temperature range, its cosmic abundance, and its chemical stability all single it out as exceptionally well suited for use by all living organisms on Earth.

Ammonia might serve in water's place under certain highly specialized conditions, but we expect the majority of living systems in the universe to rely on the same fluid medium that enables us to survive. The requirement for a liquid solvent imposes a restriction on the temperature range that life can tolerate.

For all types of life, it seems important to have a variety of individuals. Otherwise, the processes of natural selection will not have enough material on which to operate, discriminating among various living creatures on the basis of reproductive success.

We can consider at least two other sorts of life besides chemical life: elementary particle life, made of superlarge nuclei that interact through strong forces and last for far less than a trillionth of a second; and gravitational life, made of huge objects so far apart that gravitational forces predominate over electromagnetic, strong, or weak forces.

Gravitational life can hardly have arisen in the 15 or 20 billion years since the big bang, because this amount of time does not allow gravitational interactions to produce structures more complex than galaxies, galaxy clusters, and stars. Neutron star life might indeed exist, since the high temperatures and large densities on neutron star surfaces allow all sorts of nuclei to form and break apart in tiny fractions of a second. We would, however, have a hard time communicating with any such forms of life, since the natural time scale would be about one billionth of a trillionth of a second, and the most favored wavelengths would be those of extremely high-energy gamma rays.

Even if we restrict ourselves to chemical life, we need not look only at planetary surfaces. Interstellar clouds of gas and dust might acquire extremely complicated structures and might be able to reach the state of self-consciousness even without undergoing any reproductive process. Such black clouds could use internal electric currents in much the same way that our own bodies do, to carry messages to and from the central "brain." But here again, the time scale required to achieve the necessary complexity may be prohibitive because of the relatively large distances among individual molecules.

Questions

1. Why does carbon seem to be the best element for building complex molecules throughout the universe?

2. Would silicon serve as well as carbon in giving structure to large molecules? Why?

3. What does the large abundance of silicate molecules in comets, meteorites, planets, and stars tell us about the willingness of silicon to combine with oxgyen? What are the implications of this result for forming large, complex molecules, using silicon atoms as the key structural element?

4. Why does water's high "heat of vaporization" and large heat capacity give water advantages for use as a solvent in living organisms?

5. What are the advantages of a solvent that expands when it freezes? What are the disadvantages?

6. What physical conditions make life in dense interstellar clouds an unlikely event, compared to life on planetary surfaces?

7. Why would neutron star life have fantastically short lifetimes, compared to living organisms on Earth?

8. Why would gravitationally based life be slow to originate and to evolve, compared with life on Earth?

Further Reading

Bernal, James. 1969. *The world, the flesh, and the devil.* 2nd ed. Bloomington: University of Indiana Press.

Grobstein, Carl. 1974. *The strategy of life.* 2nd ed. San Francisco: W. H. Freeman and Company.

Shklovskii, Josef, and Sagan, Carl. 1965. *Intelligent life in the universe.* San Francisco: Holden-Day.

Sidlovsky, John, and Bateman, Richard. 1965. *Biophysics.* New York John Wiley & Sons.

Science Fiction

Hoyle, Fred. 1957. *The black cloud.* New York: Signet Books.

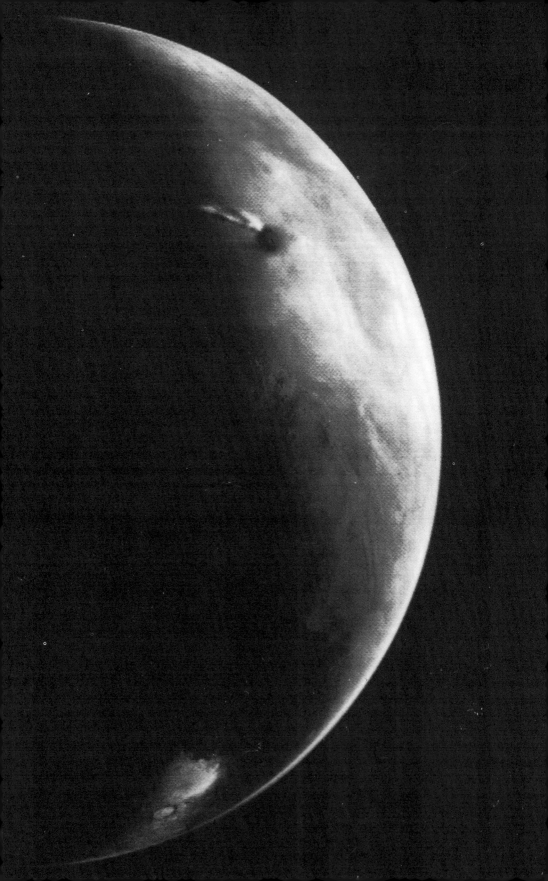

PART FOUR

The Search for Life in the Solar System

Empty space is like a kingdom, and heaven
and earth no more than a single individual
person in that kingdom. . . . How unreasonable
it would be to suppose that
besides the heaven and earth which we can
see there are no other heavens and no
other earths?
  —TENG MU, 13TH-CENTURY PHILOSOPHER

The sun's nine planets, together with the still more primitive material that resides in comets and in meteorites, provide us with numerous sites where life, or at least prebiological complex molecules, may be found. Once we could only speculate about possible sites for life outside the Earth; now we can investigate them in person and with automated spacecraft. Though we have not found any definitive proof of life anywhere besides our own planet, we have already within the past decade acquired vast amounts of useful information that may, in the fullness of time, permit us to understand why life can appear in some environments and not in others. To the extent that our solar system furnishes us with a representative planetary system, our local search for life also brings us information about a multitude of possible habitats for life within our own galaxy and beyond.

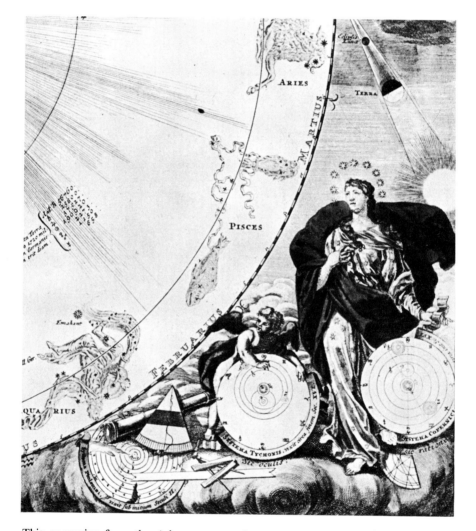

This engraving from the *Atlas novus coelestis (New atlas of the heavens)*, published in 1742, shows various models of the solar system: the Ptolemaic system (left), the Tychonic system (center), and the (correct) Copernican system (right).

CHAPTER 12

The Origin and Early History of the Solar System

During our own lifetimes, the search for life on other worlds has begun in earnest, first with photographic reconnaissance of the five closest neighbors of the Earth, then with actual landings on the moon, Mars, and Venus. We have studied our fellow planets not simply to discover whether or not they were inhabited, but to learn more about their present environments and what they can tell us about the past. But these investigations always return to our central concerns: How did life begin in our solar system, and how widely has it dispersed?

When we move on to search for life in the universe, we can see that a key question is: How many of all the stars that exist—single, double, or multiple—have planets? In other words, how many solar systems like ours exist in our galaxy? We shall return to an explicit discussion of this question in Chapter 17. We now want to concentrate on the planets we know. Our attempts to unravel their origin and history should help us to understand the significance of life on Earth and to predict the characteristics of other, still undiscovered solar systems that may contain planets like our own.

Our attempts to reconstruct the history of how the solar system condensed into one star, nine planets, their satellites, and pieces of smaller debris rest on our examination of the most primitive parts of this assemblage, those that have changed the least during the 4.6 billion years since the solar system formed. The most easily studied representatives of the early days of our planetary system are the comets, frozen lumps of dirty ice that orbit the sun at immense distances from us; the meteoroids and asteroids, rocky or metallic bodies in roughly planetlike orbits around the sun; and Mercury and the moon, the two members of the inner solar system whose surfaces have not been weathered by wind and rain, for they have no atmospheres. These objects have provided us with most of what we know about the early solar system; not enough, unfortunately, to reconstruct the entire scenario of its formation, but enough to suggest a plausible model, one that can explain most of what we have discovered thus far. We expect to find additional records of the conditions in the early solar system during the next decade, as we study the atmospheres of the giant planets and the surfaces of their satellites from space probes.

The Formation of the Solar System

The sun's nine planets orbit the sun in nearly circular trajectories that all lie in roughly the same plane (Figure 12-1). The sun contains 99.9 percent of the mass in the solar system, and the four giant planets, Jupiter most of all, have the bulk of the 0.1 percent residue. The Earth, largest of the four inner planets, has only 1/318 of Jupiter's mass, and 1/329,000 of the sun's mass.

The four giant planets differ most strikingly from the four inner planets (Mercury, Venus, Earth, and Mars) in their *size* and *composition*. The giant

Fig. 12-1 The sun's nine planets orbit the sun in the same direction and in almost the same plane. Between the orbits of Mars and Jupiter are smaller objects called asteroids, whose more elongated orbits sometimes take them outside the orbit of Saturn, or inside the orbit of Mars, and occasionally close by Earth. Pluto, the outermost planet, sometimes passes inside the orbit of Neptune (see Fig. 16-10).

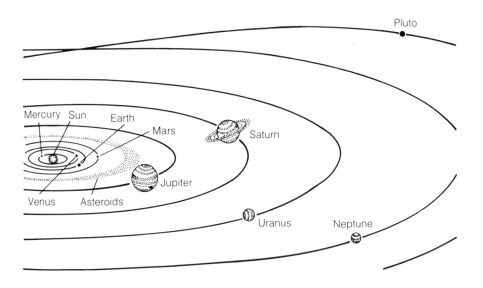

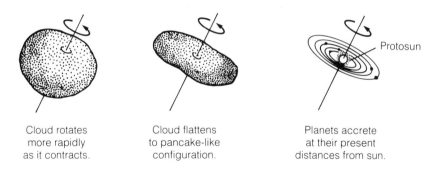

Cloud rotates
more rapidly
as it contracts.

Cloud flattens
to pancake-like
configuration.

Protosun

Planets accrete
at their present
distances from sun.

Fig. 12-2 As the cloud of gas and dust that formed the solar system began to contract, it must have acquired some rotation, which led to more rapid rotation as the cloud grew smaller (left). This rotation tended to support the cloud against contraction in directions perpendicular to the axis of rotation, and thus led to a pancakelike shape for the contracted, rotating cloud (middle). Within the disklike configuration of matter, the individual planets accreted from the matter rotating at that planet's future distance from the sun (right).

planets are large, gaseous, rarefied, and hydrogen rich, while the inner planets are small, rocky, dense, and hydrogen poor. Because the giant planets consist mostly of hydrogen and helium, they resemble the universe at large, while the inner planets are distinctly different: The universe is hydrogen rich, but the Earth is not.

A relatively simple explanation exists for the extreme differences between the four giant planets and the four inner planets. (Pluto, the outermost planet, seems an exception to this scheme; it is apparently most similar to one of the satellites of the outer planets.) The hydrogen-rich planets represent relatively unmodified remnants of the primitive material from which the solar system formed, but the rocky inner planets seem to have lost most of the light gases that form the bulk of this material. Instead of primitive atmospheres, captured from the cloud of gas and dust that formed the solar system, they have secondary atmospheres, which were made after the planets had formed.

Astronomers now think that the solar system began as a subcondensation in some interstellar cloud of gas and dust which somehow grew dense enough to start contracting under its own gravitational forces (Figure 12-2).[1] The cloud must have been rotating, slowly at first, more rapidly as it grew smaller, like a figure skater who pulls in her arms to spin more quickly. The original cloud must have spanned a fair fraction of the average distance between stars, perhaps a light year; the present size of the sun's *planetary* system, 40 times the size of Earth's orbit, equals about 1/2000

[1]One theory, supported by evidence concerning the abundance of a rare isotope of aluminum in the oldest meteorites, suggests that our solar system's formation was triggered by a nearby supernova explosion!

TABLE 12-1

CHARACTERISTICS OF PLANETS AND THEIR ORBITS

Planet	Diameter (Earth = 1)	Mass (Earth = 1)	Average Density (gm per cm³)	Orbital Period (years)	Distance from the Sun (Earth = 1)	Gap between Orbital Distances
Mercury	0.38	0.055	5.44	0.241	0.387	0.336
Venus	0.95	0.82	5.24	0.615	0.723	0.277
Earth	1.00	1.00	5.52	1.0	1.0	0.524
Mars	0.53	0.11	3.95	1.88	1.524	[1.25]
[Asteroids]					avg. = [2.77]	[2.43]
Jupiter	11.18	317.8	1.3	11.86	5.20	4.34
Saturn	9.42	95.1	0.7	29.46	9.54	9.64
Uranus	4.08	14.5	1.3	84.01	19.18	10.88
Neptune	3.85	17.2	1.7	164.79	30.06	9.38
Pluto	0.24	0.002	0.7	247.7	39.44	

of a light year. But some of the *comets* that orbit the sun reach tremendous distances at the far points of their trajectories, a good fraction of a light year and even more. Thus, the comets could be lumps of matter that condensed first as the solar system shrank and whose orbits still reflect the size of the protosolar system when the comets formed. Alternatively, comets may have formed closer to the sun; close encounters with planets could later have scattered the comets outward to their present positions.

As the cloud of gas and dust contracted toward its present size, the cloud's rotation tended to support it in the directions perpendicular to its axis of rotation (Figure 12-2). In other words, gravity could pull matter toward the center of the cloud more easily along the rotation axis than away from this axis. This fact caused the cloud to assume a pancakelike shape during the later stages of its contraction, as particles within the cloud began to collide more often and to assume new orbits around the cloud's center as the result of these collisions.

The combined result of the contraction and rotation was a spinning, disklike cloud, of much greater density than before the contraction began, and densest of all at the center, where the protosun began its final condensation. By the time that the sun grew so dense that nuclear fusion reactions began, the pancake-shaped cloud had begun to form agglomerations at various distances from its center. The regular spacing of the planets' orbits from the sun (Table 12-1) apparently reflects the way in which agglomerations of matter collected within the disklike configuration.

As the sun turned on its nuclear fusion reactions 4.6 billion years ago, the protoplanets, especially the inner ones, experienced a noticeable rise in temperature. The two critical factors that determined what a planet would be like were the *size* of the protoplanet and its *distance from the sun*. Small planets cannot retain hydrogen and helium, the lightest gases, especially if they are so warm that their temperatures rise to the point that the lightest molecules escape. When the planets were forming from loose agglomerations in orbit around the sun, the mass that could form *before* the sun turned on helped to determine how well the planet could retain hydrogen and helium. The other determining factor, the planet's distance from the sun, likewise influenced the escape of hydrogen and helium from the planet's gravity, because closer planets became hotter and had proportionately more difficulty in retaining hydrogen and helium with a given amount of gravitational force (Figure 12-3).

We can see the results of these two factors in the planets today. The four inner planets proved unable to retain hydrogen and helium during the billions of years since they formed; in fact, these two lightest gases must have vanished within a few hundred million years after the sun began to shine. The four giant planets, much farther from the sun, have much lower temperatures, and built up much larger masses during their protoplanet stages than the inner planets did. If we *now* moved Jupiter to the Earth's

OUTER PLANETS   INNER PLANETS

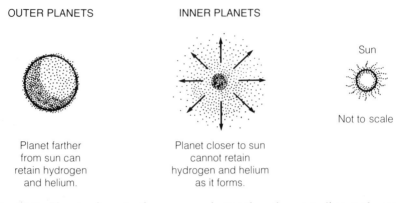

Sun

Not to scale

Planet farther   Planet closer to sun
from sun can   cannot retain
retain hydrogen   hydrogen and helium
and helium.   as it forms.

Fig. 12-3 Planets closer to the sun grew hotter than the more distant planets, and therefore had more difficulty retaining the lightest gases with a given amount of gravitational force. This very difficulty in holding hydrogen and helium kept the planets from growing to large enough masses for hydrogen and helium to be held back against escape.

distance, five times closer to the sun, Jupiter could still retain its hydrogen and helium, because it has 318 times the Earth's mass and exerts a great deal more gravitational force. But Jupiter could grow to this mass only at five times the Earth's distance from the sun; the planet that formed here, our own Earth, was never able to acquire such an enormous mass, and could not retain its hydrogen and helium for long once the sun had turned on.[2] Even among the giant planets, Uranus and Neptune, with "only" 15 and 17 times the Earth's mass, seem to have lower abundances of hydrogen and helium than Jupiter and Saturn, with 318 and 95 times the Earth's mass. This implies that Uranus and Neptune lost some hydrogen and helium as they formed even at the immense distances at which these two planets orbit the sun.

The distribution of planets in *any* planetary system into one group of dense, rocky inner planets and another of rarefied, gaseous outer planets seems reasonable, if our own solar system formed in a "standard" way. We shall find support for this assertion when we study the satellites of Jupiter (Chapter 16), since the inner moons of this giant planet are denser than the outer ones. Jupiter is so large that it warmed the space around it as it formed, producing a separation of minerals with different melting points similar to that which occurred around the sun. We expect the same effect in other planetary systems, although we do not have even one other example yet that could be used as a test (Chapter 17).

[2]The mass of the Earth would approximately equal the mass of Saturn if we added the missing hydrogen and helium to our planet's present store of elements, in the same proportion that we find in the sun and the giant planets.

Comets

The small bodies that are most representative of the primitive solar system are the comets, whose highly elongated orbits show no particular concentration toward the disk of the solar system, thus implying that comets could have formed before the protosolar system had assumed a pancake-like shape. Current theories suggest that a comet basically resembles a huge, dirty snowball with the mass of a large mountain and a diameter of a few kilometers. The "snow" in a comet consists of ordinary water ice, plus carbon dioxide and other unknown frozen gases, perhaps including more complex compounds such as formaldehyde and cyanoacetylene. The "dirt" consists of grains of rocky material of various sizes which have apparently undergone no melting or other transformations, since they are still imbedded in the cometary ices. Astronomers think that this lack of chemical processing ensures that comets do represent pristine samples of the original material from which the solar system formed, and that they may have trapped unaltered grains and molecules from the interstellar medium in their snowy interiors.

For most of its life, each cometary snowball orbits slowly around the sun at immense distances, a thousand times farther than frosty Pluto (Figure 12-4). Far from the sun's warmth, the snowball remains in a cosmic deep freeze only a few degrees above absolute zero. Millions, perhaps billions, of comets orbit the sun, but we know nothing about them until one of them happens to acquire a new orbit, perhaps through the gravitational attraction from nearby stars, that takes it toward the inner solar system. In this case, nothing much happens to the comet until it reaches a distance from the sun comparable to Jupiter's, when some of the ices start to vaporize, and the gas and dust released from the snowball nucleus spread out to form a fuzzy envelope called the coma.

Fig. 12-4 Most comets have tremendously elongated orbits that keep them hundreds or thousands of times farther from the sun than Pluto, the most distant planet.

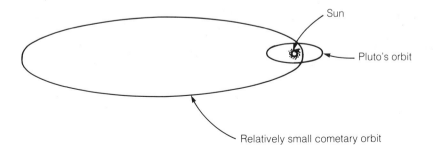

Fig. 12-5 Comet Mrkos, named after its discoverer, appeared in 1957. Streaming out behind the comet's nucleus are a straight tail of gas and a curved tail of dust particles.

As the comet approaches still closer to the sun, more and more gas and dust are released, trailing behind the nucleus as the spectacular tail, often many millions of kilometers in length, that gives a comet its beauty (Figure 12-5). Some comets have two tails, one made of dust and the other of gas, and occasionally a comet can have several tails of each kind. The gravitational attraction of the sun and of the giant planets can wreak havoc in comets as they traverse the planetary part of the solar system. Astronomers have seen some comets break apart as they pass close by the sun, with two or more distinct pieces following the original comet's orbit. Sometimes the gravitational attraction from one of the giant planets can deflect a comet into an orbit much smaller than its original one. Such "short-period" comets, of which many dozens are known, take only a few years to orbit the sun once, as compared with hundreds of thousands, or millions of years, for the original "long-period" comets. Halley's comet, most famous of all for its regular returns at 76-year intervals, has an intermediate orbit that carries it past Neptune but nowhere near the immense distances of the long-period comets.

Comets offer us a chance to investigate primitive solar system material, formed directly from the original gas and dust that made the solar system and little changed during the past 4.6 billion years. They even do us the favor of bringing that well-preserved ancient material near Earth at the time when a long-period comet, perhaps on its first trip close by the sun, passes within a few million kilometers of our planet. Even a tough, well-preserved comet that has made millions of orbits should contain material that has been protected from high temperatures and thus is more primitive than any matter in the inner solar system. For this reason, many scientists are eager to send a space probe to Halley's comet on its next close encounter with the sun in 1986.

By making spectrographic analyses of the light that comets reflect, astronomers have found that a surprising number of relatively complex molecules exist in these primitive snowballs. Since comets are thought to have condensed in the earliest stage of solar system formation, we might expect to find within them some of the molecules found in interstellar clouds (Table 4-1). Indeed, the discovery of water, carbon monoxide, hydrogen cyanide (HCN), and methyl cyanide (CH_3CN) in comets strengthens the assumed connection between these icy objects and the interstellar clouds from which they may have condensed (Table 12-2).

To strengthen this connection still further, we should look for other organic molecules in comets, in addition to those found within the past five years. What is the "parent" molecule of the fragment we observe as C_3? Could it be HC_5N? And what about the simplest amino acids that are produced in Miller-Urey experiments (see Chapter 9)? Could some of them exist in cometary nuclei? These possibilities could be investigated by a series of spacecraft missions that would first sample the gases close to

TABLE 12-2

MOLECULES DETECTED IN COMETS

Coma			Tail	
H_2O^a	CN	NH_2	H_2O^+	N_2^+
HCN^a	CH	C_3	CO_2^+	CO^+
CH_3CN^a	OH			CH^+
CO_2^b	CO			OH^+
NH_3^b	NH			
	C_2			
	CS			

[a]These identifications require further verification.
[b]Suggested parent molecules of molecules such as CO and NH, but not yet directly detected.

the comet's icy nucleus, and would eventually bring back pieces of comets for study in our Earthbound laboratories. We would then know what kinds of compounds these icy messengers might have brought to the surface of the early Earth, perhaps to help the formation of the atmosphere, perhaps even to give a head start to some of the chemical reactions that led to the origin of life.

Asteroids, Meteoroids, and Meteorites

We have seen that comets, which spend most of their lives in the frozen depths of space at the boundaries of the solar system, represent the best-preserved primitive material that is relatively accessible to us. Another source of primitive matter resides in the lumps of debris called asteroids and meteoroids, which, like comets, orbit the sun in rather elongated trajectories, but which remain fairly close to the sun as they orbit around it (Figure 12-1). The asteroids, ranging in size from a few objects hundreds of kilometers across down to thousands of objects less than 1 kilometer across, circle the sun between the orbits of Mars and Jupiter. Asteroids are thought to represent the pieces of a would-be planet that never could form, because Jupiter's gravitational forces kept the material from collecting into a single large object. Thus, the asteroids could provide us with a valuable set of fossils from the era of 4.6 billion years ago, when planets began to form but had not yet grown to their present sizes.

Smaller interplanetary wanderers, called meteoroids, vary in size from a few hundred meters across to a tiny fraction of a millimeter. Meteoroids are basically small asteroids, and astronomers make a distinction between the two only because meteoroids have orbits around the sun that cross the Earth's orbit, so that they can collide with the Earth. When a small meteoroid encounters the Earth's atmosphere, its large velocity relative to the Earth, which arises from the difference between its orbit and the Earth's, subjects it to great frictional heating in the upper atmosphere. This heating consumes any small meteoroid as a shooting star or meteor. If the meteoroid begins with a sufficiently large mass, it can survive its frictional passage through the atmosphere, and its remnant will reach the Earth's surface as a meteorite.

The surfaces of Mercury, Venus, Mars, and the moon are dotted with many thousands of meteorite craters, most of which were made within a few hundred million years after the solar system formed. At this epoch, meteoroids by the millions must have rained down on all the planets of the inner solar system as the last chunks of matter to reach the planets encountered the new planetary surfaces. Our own planet shows only a few large meteorite craters (Figure 12-6), proof that although the Earth did not escape this intense early bombardment, the first few hundred million years of our geological record have vanished because of erosion and the movement of the crustal plates.

Fig. 12-6 The Great Meteor Crater in Arizona, more than a kilometer in diameter, was made about 20,000 years ago by a meteorite with a mass of many thousand tons.

Fig. 12-7 The meteoroid that nearly collided with Earth in August, 1972, had a speed of 15 kilometers per second, relative to Earth, as it passed through the atmosphere above the Grand Teton mountains in Wyoming. It is seen here as a streak of light above the Grand Tetons.

On August 10, 1972, a reminder of what must have occurred far more often in bygone eras flashed through the skies over Wyoming as a meteoroid burned its way through our atmosphere, missing the surface of the Earth by just 58 kilometers (Figure 12-7). This object was only 4 meters across, though it weighed a thousand tons and would have been devastating upon local impact. A related object, apparently a comet, did hit Earth in Siberia in 1908, felling trees for many kilometers around. The impact of the meteoroid of 1972, small as it was, would have released about the same energy as one of humanity's nuclear weapons.

From examination of all the meteorites found on Earth, scientists have developed a general classification scheme based on the chemical and mineralogical composition of these objects. Most of the meteorites that have been found are "stony," basically lumps of rock; a minority are "stony-iron," with metal-rich inclusions; and a few are made mostly of iron, nickel, and other metals. Radioactive dating of meteorites gives maximum ages of 4.6 billion years, the age of the solar system itself. Most interesting of all is the subclass of stony meteorites called chondrites. These contain rounded inclusions, noticeably distinct from the rest of the material, which are called chondrules. And of the chondrites, by far the most significant for us are the carbonaceous chondrites, in which as much as 5 percent of the mass may consist of various types of carbon compounds. Since these objects show the least amount of modification by heating they are the most primitive of meteorites.

Within the class of carbonaceous chondrites, the most primitive examples, known as Type I, contain the highest percentage of carbon, nitrogen, and water of all the meteorites. Some scientists believe that the Type I carbonaceous chondrites may actually be fragments of old comets rather than debris from the asteroid belt. Whether or not this is true, we do not need to wait for samples of a comet to be returned by some future mission to study the products of prebiological organic chemistry in space. Instead, we can begin this study with the compounds found in the various types of carbonaceous chondrites, and the results of such investigations have already given significant results.

Amino Acids in Meteorites

Testing meteorites for compounds made outside the Earth has always suffered from the long wait between the time when a meteorite falls to Earth and the time that it is found. Many different paths of contamination can then add Earthmade organic molecules to what may have been primitive organic molecules. Luckily, however, scientists on at least two occasions have been able to recover carbonaceous chondrites soon after they fell to Earth. The first of these, the Murchison meteorite that fell in Australia in 1972, was recovered, at least in part, on the following morning (Figure 12-8).

Detailed analysis of the Murchison meteorite revealed the presence of 16 amino acids, the basic building blocks of proteins. Five of these amino acids appear in living organisms on Earth, though the other 11 are rare. The fact that only five of the 16 amino acids ordinarily appear in terrestrial life suggests that contamination of the meteorite had not occurred. Two further tests verified this conclusion. First, the Murchison amino acids showed equal amounts of left-handed and right-handed molecules, while Earthmade amino acids are almost all left-handed within a living organism and become equal mixtures of left- and right-handed molecules only if allowed to sit for hundreds of thousands of years. Second, the ratio of the carbon isotopes in the meteorite, carbon-12 and carbon-13, showed only 88.5 times as much carbon-12 as carbon-13. Living creatures on Earth

Fig. 12-8 The pieces of the Murchison meteorite, recovered soon after they fell in 1972, contain 16 different kinds of amino acids.

exhibit a ratio of these two isotopes' abundances which varies between 90 and 92, a small deviation from 88.5 but enough to verify the nonterrestrial origin of the carbon in the amino acids of the Murchison meteorite.

The discovery of amino acids in this meteorite, and later in the similar Murray meteorite, shows that nature has made amino acids within a relatively hostile environment—the frigid wastes of interplanetary space—from the basic ingredients in the inner solar system during its early history. This formation of amino acids tends to confirm our basic scenario for the origin of life on Earth, in which amino acids form naturally from common ingredients on this planet. Further confirmation comes from the discovery in meteorites of guanine and adenine, two of the cross-linking bases found in DNA and RNA molecules. Fatty acids, and other "life-related" molecules, have also been found in carbonaceous chondrites. Therefore, the chemical processes that produce important compounds for the origin of life seem to occur in various natural environments, and we can expect these processes to be widespread in the universe. Comets, too, may contain life-related molecules, and this possibility has nourished a very controversial theory that comets may even have seeded the Earth with plagues as they passed by (see page 82). A more reasonable speculation in this vein suggests that meteoritic and cometary material, which was surely falling on the early Earth, may have provided ready-made organic molecules, and thus a shortcut to the formation of more complex, prelife molecules on Earth.

Mercury and the Moon

In our search for the most primitive, least altered members of the solar system, Mercury and the moon rank first among the inner objects. Since neither of them has an atmosphere, their surfaces have undergone no significant weathering during the 4.6 billion years since they formed.

Furthermore, the rocks on the moon and Mercury are likely to be more primitive than those on Earth because the two smaller objects have no significant plate-tectonic activity: no continental drift, no formation of volcanoes along fracture zones, and no spreading of the sea floor. To produce such plate motions requires a big heat engine, and this in turn requires a relatively large planet which can release enough heat from radioactive rocks to make these processes work. Even Mars, larger than Mercury or the moon, does not have much plate-tectonic activity, because the heat released by radioactivity is conducted to the surface and escapes into space too rapidly to set in motion processes like the sea floor spreading on Earth. Hence, these small planets still preserve their primitive rocks, for no massive internal convection has dragged what was once the crust down into the mantle of the planet. From studying Mercury and the moon, however, we have learned that as far as life goes, they are hopeless. The closeness of Mercury to the sun raises the planet's temperature well above 300° C on the day side, and bathes the surface more intensely than any other in the sun's ultraviolet light. The moon has similar problems, as we shall see.

Mercury's surface resembles the heavily cratered areas of our natural satellite (Figure 12-9), and reflects light in much the same way as lunar rocks, suggesting that Mercury's surface and the moon's consist of much the same sorts of material. The rotation of Mercury is locked into its 88-day orbital period, but in a delicate sort of resonance in which Mercury rotates exactly *three* times for every *two* orbits around the sun. Therefore, each part of the planet eventually faces the sun's blistering heat before reentering an 88-day night, during which the temperature falls as low as

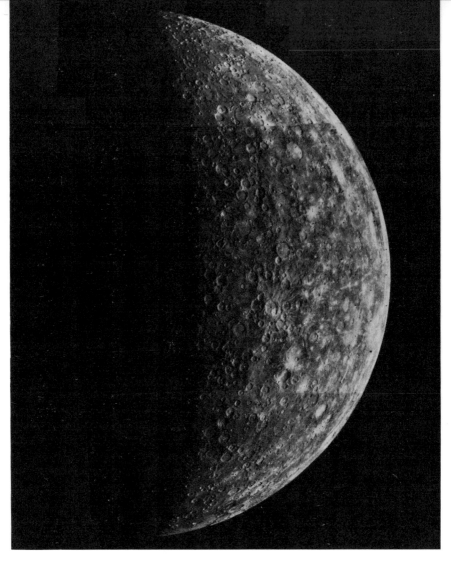

Fig. 12-9 The surface of Mercury, photographed by the Mariner 10 spacecraft in 1974, shows a rugged terrain much like the moon's.

−150° C! Explorers who might land on Mercury to examine its primitive rocks would have to protect themselves against both outlandish heat and sub-Siberian cold, if they planned to remain for several months on end.

The Composition of Mercury and the Moon

By studying Mercury from a distance, we can determine some interesting aspects of our solar system's formation. First, the density of matter in Mercury, 5.44 grams per cubic centimeter, is just a little less than the

Earth's mean density, 5.52 grams per cubic centimeter. Now the large density of the Earth can be explained by the gravitational compression of the matter in our planet's iron-rich core, which must have a density of about 13 grams per cubic centimeter. For comparison, uncompressed iron has a density of 8 grams per cubic centimeter, while the ordinary rocks in the Earth's crust average about 2.8 grams per cubic centimeter. But Mercury is too small to have a compressed core like the Earth's. Mercury's 4900-kilometer diameter puts it closer to the moon (3475 kilometers across) than to Earth's diameter of 12,750 kilometers. But the moon has an average density of 3.34 grams per cubic centimeter, just about what we would expect for an agglomeration of ordinary rocks of this size, and far below the 5.44 grams per cubic centimeter of Mercury.

How can we explain the large density for so small a planet as Mercury? The only good explanation relies on a difference in chemical composition: Mercury must be richer in dense materials than the moon or the other planets. This enrichment probably arose because Mercury formed so close to the sun. As Mercury collected itself, the local temperature in that part of the protosolar system was so high that the lighter elements could not condense as well as the heavier elements. These lighter elements included not only hydrogen and helium but also carbon, oxygen, and silicon, and the result of their relative inability to collect together was that Mercury lost its "fair share" of the silicate rocks, made mostly of silicon and oxygen, that form much of the Earth's crust and mantle, and much of the moon as well. The most abundant heavy element, which should have a considerable enrichment in Mercury, would be iron. Iron also has a high melting temperature, which would keep it from boiling and evaporating.

The Early History of the Earth and the Moon

Three hundred and seventy years ago, the great Italian astronomer Galileo first discovered that our satellite has thousands of craters on its surface. Ever since, astronomers and geologists have debated the craters' origin. Are they the result of moonwide volcanoes? Are they produced by gas bubbles that rose through molten rock and burst on its surface? Or are they the scars left by the impact of rocky projectiles? Galileo saw that the moon's brighter areas, called the uplands, have more craters than the dark areas, called *maria* (the Latin word for seas) because they reminded early astronomers, incorrectly, of terrestrial oceans (Figure 12-10).

The clue to resolving the puzzle of the moon's craters came from the recognition that the difference in the density of craters between the uplands and the maria might be a result of the different ages of these regions. If craters came from impact, which has now been shown to be true for the vast majority, then we can explain the difference in cratering *if* the more heavily cratered uplands are older than the maria, which might have been

Fig. 12-10 The bright lunar uplands have far more craters than the dark lunar "maria," which we know to be old lava flows.

covered with new lava some time after the moon formed. Such immense lava flows, stretching for hundreds of kilometers, could have arisen from collisions with huge meteoroids that eradicated the cratered terrain and replaced it with a newly made "sea" of molten rock.

Like Mercury, the moon is too small to sustain an atmosphere, or to produce plate motions on its surface of a scale that would remove the early crust. Thus, we can again read the record of early events in the solar system by studying the lunar landscape and its rocks. We find the same scenery as on Mercury, countless craters that are silent relics of bombardment by countless objects, ranging in diameter from a few microns up to tens of kilometers. The heavily cratered surface of the moon bears witness to the fact that its crust must have formed early in the history of the solar system. But did the moon form close to the Earth?

Was the moon captured into its present orbit after it had formed? Or did the moon divide from the Earth? If either of these possibilities did in fact occur, catastrophic effects on the Earth's surface must have resulted, which might have had dramatic consequences for the possible origin of life. Many scientists thought that the arguments about where our moon came from would be answered with certainty after detailed exploration of the moon's surface made during the 1970s.

Human Exploration of the Moon

Part of the cultural heritage of the human race seems to be a desire to explore the moon and to speculate about what might be found there. During the 1960s, as instruments circled the unseen side of the moon, landed on its surface, and made tentative chemical analyses, arguments raged not only about the origin of the moon, but also about the possibility, admittedly remote, that astronauts returning from the moon might bring back some strange lunar microbes, capable of infecting Earth with unknown diseases.

Most scientists felt it impossible that life could exist on the moon, but a vocal minority, led by Carl Sagan, pointed out that the moon might have shared some of the Earth's early history, which could have included a primitive atmosphere and abundant water, and that nature seems remarkably adept at making at least the first steps toward life throughout the universe. (In view of the later discovery of complex molecules in interstellar clouds, and of amino acids in the Murchison and Murray meteorites, Sagan seems to have had a good argument.) The opposing viewpoint held that the moon's low gravity could not have held volatile compounds, such as water, long enough for life to develop, and that even if somehow life had evolved and had survived as spores in the lunar subsoil, such life would not be able to interact with organisms on Earth because of differences in chemical structure between the two types of life. In a memorable public exchange, Edward Anders, an expert on meteorites, offered to eat the first dust brought back from the moon as a demonstration of his confidence in the sterility of the lunar surface!

Still, nobody *knew* whether life existed on the moon, and no one could prove what the effects of bringing such life back to Earth would be, so elaborate precautions were taken. This episode has more than passing significance, since we now face precisely the same arguments about samples from Mars. And in some distant time when we first make physical contact with an alien civilization, we must wonder whether their air and soil contain diseases or poisons for us, or ours for them.[2]

[3]Interestingly, UFO reports of alien visitors almost never place them inside protective shields, as we would expect if they were worried about this contamination problem.

Fig. 12-11 Manned exploration of the moon during the early 1970s allowed the examination of our satellite in detail undreamt of only a generation before.

During the past 15 years, more than 50 spacecraft have approached the moon or have landed on its surface. Twelve men from our planet have walked there, gathering samples of lunar rocks and dust, and have set up experimental equipment that has continued to gather data long after the astronauts have left (Figure 12-11). One of the first discoveries to come from inspection of rock samples from the moon has proven to be one of the most important: The rocks in the dark lunar maria are quite similar to terrestrial basaltic rocks. This fact shows that the moon has not always been cold; rather, it must once have been hot enough to produce magma, molten rock that later crystallized into basalts. Studies of lunar rocks in terrestrial laboratories showed differences in composition between lunar rocks from the various maria and from the lunar uplands, and revealed that *all* of these rocks are distinctly different from terrestrial rocks. In particular, the lunar rocks show abundances of volatile elements (those that boil at low temperatures) that are hundreds of times smaller than those in terrestrial basalts.

This difference has great significance. The Earth already shows a notable deficiency in volatile elements, as compared to the abundances of elements in the giant planets and in stars, and the deficient elements on Earth include hydrogen, carbon, and nitrogen, which are vital to life as we know it. A much lower abundance of volatile elements on the moon tends to rule out the possibility of life based on these elements. Furthermore, the evidence suggests that the moon's extreme deficiency in volatile elements has existed ever since the moon formed, so we have no possibility of an early, volatile-rich environment in which prelife chemistry could have flourished. Evidently it *is* safe to eat moon dust, at least in small quantities!

Another key conclusion that emerges from the comparison of lunar rocks with terrestrial rocks is that the moon can never have been part of the Earth; the differences in composition, small though they may be, rule this out from the geologists' point of view. Even though the hypothesis of the moon as ejected from Earth no longer holds water, however, no one knows for sure just where the moon came from. At present, an emerging consensus of lunar specialists favors the idea that the moon formed at about its present location, as part of the Earth's own formation and thus as a sort of double planet, since the moon has $1/81$ of the Earth's mass, far greater than the ratio of any other satellite to its planet.[4] But the way in which the two members of this double-planet system acquired their different compositions remains unclear, and prominent advocates of alternative models of the moon's origin continue to debate the point.

One big reason for the difficulty in resolving the question of where the moon came from is the absence of the first few hundred million years of the lunar geological record. The lunar samples turned out to include *no* primeval, unaltered rocks, 4.6 billion years old, survivors from the primitive solar system material. Instead, most of the lunar rocks are about 4.2 billion years old, older than the 3.8-billion-year record for Earth but still not as old as the solar system. Some scientists feel that still older rocks may be lying on the far side of the moon, where fewer large meteoroid impacts have occurred, such as those that formed the lunar maria (Figure 12-12). But for the time being, we seem forced to conclude that the early meteoroid bombardment was so intense that it destroyed most of the rocks from the first 400 million years of the moon's history.

Thus, the moon has failed to provide us with primitive solar system material and stands aloof in our skies, lifeless, devoid of any atmosphere, baking to 125° C in the two-week lunar daytime and freezing to −125° C during the two weeks of night. What use, then, is the moon to life, aside from inspiring some of Earth's living creatures to added activity during times of full moon?

[4]Pluto may be a possible exception to this statement.

Fig. 12-12 The "back" side of the moon, never visible from the Earth, reveals an enormous crater named after the Russian space scientist Tsiolkovsky. But no vast "maria" appear on the far side of the moon, as they do on the near side.

The answer, surprisingly enough, is that the moon may have played a critical role in the development of life on Earth! The moon has made two key contributions. First, its presence in orbit with the Earth helps to stabilize the orientation of the Earth's rotation axis in space. Were it not for

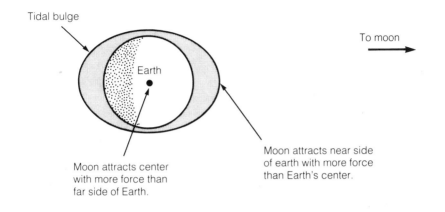

Tidal bulge

Earth

To moon

Moon attracts center
with more force than
far side of Earth.

Moon attracts near side
of earth with more force
than Earth's center.

Fig. 12-13 The *differences* in the gravitational attraction of the moon on various parts of the Earth cause the tides. As a result of these differences, the Earth tends to bulge in directions toward the moon and away from the moon. The Earth's oceans respond more readily than the land does, so the ocean tides slide up and down on the much smaller land movements.

the moon, this axis might wander far more than it does from century to century and over millions of years, producing long-term climate changes that could kill any life that had developed under different conditions. Mars, which has no large satellite, has changed the inclination of its rotation axis in ways that have dramatically affected surface conditions on the planet.

Second, the moon produces large *tides* on Earth. The moon exerts more gravitational force on the near side of the Earth than it does on the Earth's center, and more on the center than it does on the far side (Figure 12-13). The Earth's seas can respond more readily than the land to these differences in gravitational attraction, and so the seas flow, relative to the land, in a twice-daily tidal cycle. In the early years of the Earth, tide pools may have proven to be crucial to the formation and interaction of complex molecules; the tide pools may have provided the first primitive biological cells. Four billion years ago, the moon orbited the Earth at a smaller distance than it does today, and raised higher tides than it does now, so the impact of tidal motion would have been proportionately greater. Tide pools that filled, emptied, and refilled twice a day could have been far more widespread than they are now, if the average tidal variation equaled, say, 2 meters instead of half a meter.

If the moon formed a necessary part of the conditions for life to begin on Earth, would a similar satellite be necessary for life on another planet? Here we know no answer that will satisfy us until we find another planet with life and check to see whether its inhabitants sing the praises of a lifeless, yet perhaps life-giving, fellow traveler through space.

Summary

Comets, meteoroids, and asteroids represent lumps of solar system matter that have changed very little since the time when the solar system formed 4.6 billion years ago. In particular, comets might have formed first of all, and may contain samples of the most primitive solar system material, because most comets orbit the sun in highly elongated trajectories that carry them much farther from the sun than any of the planets. At these huge distances from their central star, no heating has disturbed the original state of the condensed material.

Comets each have far less mass than planets or satellites, and consist of frozen "snowballs" of water, carbon dioxide, carbon monoxide, ammonia, and more complex molecules. Close passage by the sun vaporizes some of the cometary ices, producing a gauzy coma around the nucleus and a long, rarefied tail that streams behind for millions of kilometers. Meteoroids, which seem to be small asteroids, orbit the sun in trajectories that let some of them intersect the Earth's orbit. Frictional heating in our atmosphere then vaporizes most of the object to produce a meteor or shooting star. Larger meteoroids can survive this heating, in part, to fall to Earth as meteorites. An important though rare class of meteorites, the carbonaceous chondrites, have part of their mass in the form of carbon compounds, quite unlike most meteorites, which are iron, iron-nickel, rocky material, or a mixture of these.

Some carbonaceous chondrites have revealed the presence of 16 different amino acids, the basic units of proteins. Since these amino acids appear to have formed in nonbiological processes, their existence suggests that amino acids tend to form rather easily as part of the process of forming a solar system.

The planet Mercury, not subject to any atmospheric weathering, shows a heavily cratered surface, the result of great meteoroid bombardment during its first few hundred million years. Mercury's high density in spite of its small size suggests that the planet must be richer in dense elements, especially iron, than the moon is. Such a conclusion seems reasonable in view of the high temperature that would have prevailed where Mercury formed, which would make the retention of lighter elements more difficult.

Old lava flows on the moon, the lunar maria, are the result of tremendous meteoroid impacts that covered up the craters there. But even the lunar uplands, older than the lava flows, have yielded few rocks older than 4 billion years; the first few hundred million years of lunar history seem to be missing. The moon rocks do show, however, that the moon was never part of the Earth.

Because the moon stabilizes the wanderings of the Earth's rotation axis, and because the moon causes tides, we may owe the existence of life on our planet at least in part to our relatively massive satellite. In this view, the moon has saved us from death-dealing changes of climate and has provided the twice-daily tides that filled and emptied nutrient-rich tide pools, thereby concentrating the compounds within them and promoting the development of further complexity through polymerization on surrounding clays.

Questions

1. Why do we think that comets are the most primitive members of the solar system?
2. What makes shooting stars? Why do most shooting stars never reach the Earth's surface?
3. Why does a comet produce a spectacular tail when it nears the sun? How big is the comet's nucleus, which has most of the comet's mass?
4. What do comets consist of? What does this imply about the primitive solar system's composition?
5. Why does the discovery of amino acids in the carbonaceous chondrite meteorites seem important for theories about the origin of life on Earth?
6. In what ways do Mercury and our moon resemble one another?
7. Why is the average density of matter in Mercury considerably greater than that in the moon? What does this tell us about the elements of which Mercury and the moon are made?
8. Why do the surfaces of Mercury and the moon show so many craters? How has this affected the geological record of the moon's first few hundred million years?
9. Why did some scientists speculate that moon dust might contain microorganisms? How could the airless and waterless moon ever have acquired life?
10. Is the composition of moon rocks identical to that of similar rocks on Earth? What does this tell us about the origin of the moon and the Earth?
11. In what ways has the presence of the moon favored the origin and development of life on Earth?
12. Why were the tides important in producing life on Earth, according to most theories of how life began?

Further Reading

Cameron, Alastair G. W. 1976. The origin and evolution of the solar system. In *The solar system*. San Francisco: W. H. Freeman and Company.

Dole, Stephen. 1974. *Habitable planets for man.* 2nd ed. New York: Elsevier Scientific Publishing Co.

Grossman, Lawrence. 1975. The most primitive objects in the solar system. *Scientific American* 234:30.

Hartmann, William. 1976. The smaller bodies of the solar system. In *The solar system*. San Francisco: W. H. Freeman and Company.

Murray, Bruce. 1976. Mercury. In *The solar system*. San Francisco: W. H. Freeman and Company.

Murray, Bruce, and Burgess, Eric. 1976. *Flight to Mercury*. New York: Columbia University Press.

Mutch, Thomas. 1970. *Geology of the moon: A stratigraphic view*. Princeton: Princeton University Press.

Sagan, Carl. 1973. *The cosmic connection*. New York: Dell Publishing Co.

Schramm, David N., and Clayton, Robert N. 1978. Did a supernova trigger the formation of the solar system? *Scientific American* 239:4, 124.

Whipple, Fred. 1975. The nature of comets. In *New frontiers in astronomy*, ed. by Owen Gingerich. San Francisco: W. H. Freeman and Company.

Wood, John. 1976. The moon. In *The solar system*. San Francisco: W. H. Freeman and Company.

CHAPTER 13

Venus

Venus, near-twin of the Earth in size and mass and the planet that comes closest to the Earth, has been associated with love and beauty by the ancient cultures that admired her brilliance in their morning and evening skies. Because the nearest planet hides her surface beneath a blanket of perpetual clouds, early speculation about life on other worlds suggested mysterious creatures on Venus, forever hidden from our view. Modern astronomy has laid this fancy to rest, however, as we have learned that Venus stifles under a dense smothering blanket of atmosphere, with its surface more than 350° C hotter than boiling water! As a result of this discovery, the key question that this planet poses in our search for life is: Why did Venus turn out so different from Earth? How did two planets, nearly alike to begin with, diverge so widely in their later development?

To obtain some perspective on this problem, we must analyze Venus as best we can by sending radar waves and spacecraft through its atmosphere and by making spectroscopic analyses of the gases that surround it. These modern approaches to Venus have allowed us, within the past decade, to reach a rather complete understanding of our suffocating sister planet.

The Temperature of Venus

The clouds of Venus, far thicker than those on Earth, continuously obscure the planet's surface (Figures 13-1 and 13-2). Thus, when we measure the temperature of the parts of Venus that reflect sunlight, as we can by spectroscopic techniques, we obtain the temperature at the level of the cloud tops, some 55 kilometers above the surface. This temperature, a relatively

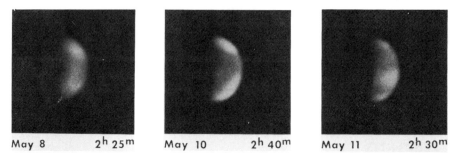

| May 8 | 2ʰ 25ᵐ | May 10 | 2ʰ 40ᵐ | May 11 | 2ʰ 30ᵐ |

Fig. 13-1 Even the best telescopic observations of Venus reveal a nearly feature-less expanse of unbroken clouds, though in ultraviolet light we can see some patchy shadings on the cloudtops.

Earthlike $-33°$ C, does not seem particularly threatening. Radio waves, however, unlike visible light, can penetrate all the way through the clouds. Radio emission from the surface of Venus indicates a surface temperature of 475° C, hot enough to melt lead!

We can also beam radio waves at Venus in such a way that they will pass through the clouds, reflect from the surface, and return to antennas on Earth, carrying information about the surface that reflected them. To study Venus in this way employs a sophisticated radar system, similar in principle to the devices that traffic police use to detect speeding motorists

Fig. 13-2 These pictures of Venus were taken in early 1979 when the Pioneer Venus spacecraft was about 65,000 kilometers above the planet's surface.

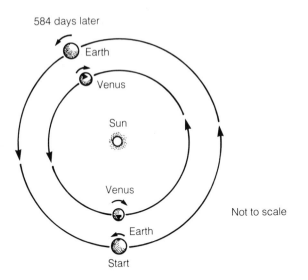

Fig. 13-3 Venus rotates once every 243 days, in the opposite direction to its motion around the sun. Each time Venus and Earth are close to each other, at intervals of 584 days, the *same* side of Venus faces the Earth.

(but with enormously powerful transmitters for Venus reflections!) These techniques showed that Venus *rotates* more slowly than any other planet, taking 243 days, and that this rotation occurs in the opposite direction to the planet's orbit around the sun (Figure 13-3). Radar reflection also demonstrated that the surface of Venus is covered with craters and is as rough as the moon's surface.

The Atmosphere of Venus

How does Venus maintain such a high temperature at its surface? The answer must surely lie with the planet's atmosphere, since we know from spacecraft landings that the surface contains no particularly large amounts of radioactive rocks to give off heat. The key qualities of the atmosphere that lead to the planet's enormous surface temperatures are the atmospheric *composition* and the amount of the atmosphere, which can be expressed in terms of the surface *pressure*.

As verified by a series of instrumented spacecraft sent into the atmosphere of Venus by the Soviet Union and the United States, the atmosphere of Venus consists largely of carbon dioxide, which provides 96 percent of the total. Most of the remainder of Venus's atmosphere is nitrogen molecules (about 3.5 percent), with a small amount of argon and traces of oxygen, water vapor, hydrochloric acid (HCl), hydrofluoric acid (HF), and

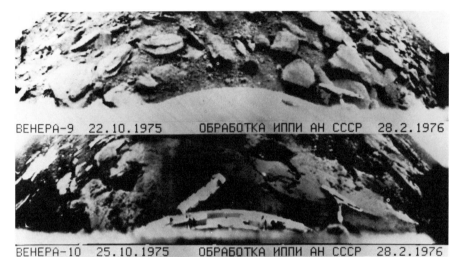

BEHEPA-9 22.10.1975   ОБРАБОТКА ИППИ АН СССР 28.2.1976

BEHEPA-10 25.10.1975   ОБРАБОТКА ИППИ АН СССР 28.2.1976

Fig. 13-4 The Soviet spacecraft Venera 10 sent these pictures back from Venus's surface. The rather familiar-looking boulders show that the surface temperature does not melt rocks.

carbon monoxide. This composition differs noticeably from that of the Earth's atmosphere, which is mostly nitrogen, oxygen, water vapor, and argon, with only traces of carbon dioxide. But still more important, the total atmospheric pressure at Venus's surface turned out to be *90 times* that on Earth. In other words, not only is Venus's atmosphere mostly carbon dioxide, but there is 100 times more carbon dioxide in the atmosphere of Venus than there is nitrogen (80 percent of the total) in the Earth's atmosphere.

With an atmosphere almost 100 times thicker than Earth's and a temperature high enough to melt some minerals at its surface, Venus possesses the hottest surface of any planet, hotter even than Mercury's, and an excellent setting for a movie about the delights of hell. The planet seems as hostile to life as any we could imagine. Pictures radioed back by the two Soviet spacecraft that landed on Venus in 1975 show that some sunlight does reach the surface, though heavily filtered by the continuous clouds (Figure 13-4). The intensity of the lighting on the surface of Venus roughly equals that on Earth when the sky has a complete cover of thick clouds. In view of the fact that the surface temperature is high enough to melt lead and zinc, it is reassuring to find rather ordinary rocks on the surface and not some molten goo. To balance this reassurance (such as it is), we have learned that the clouds of Venus that make this planet so brilliant in our skies are composed of droplets of *sulfuric acid* (H_2SO_4), not of water or of ice crystals, and seem to be the source of intense lightning discharges.

In brief, our sister planet seems no place for a summer vacation, or even a tourist stop.

The Greenhouse Effect

How did Venus get to be like this? Since our sister planet orbits the sun at only 72 percent of Earth's distance, we would expect Venus to be hotter than Earth, but not 400 degrees hotter! Mars, which orbits at one and a half times the Earth's distance from the sun, has an average surface temperature just 50 degrees Celsius below the Earth's average. But if we consider what the thick atmosphere of Venus does to the light that arrives from the sun, we can see how this planet can achieve a surface temperature of nearly 500° C.

The visible light that manages to penetrate the clouds of Venus reaches the planet's surface and tends to warm it up. As a result of this heating, the surface radiates infrared light, as any object at a temperature above absolute zero will. (Objects at thousands of degrees, such as the surfaces of stars, radiate mostly visible-light photons, but even these high-temperature radiators also emit infrared photons.) Infrared light has more difficulty penetrating the atmosphere than visible light does, however (Figure 13-5). The molecules in the atmosphere, especially those of carbon dioxide, the dominant constituent, can absorb great amounts of infrared light. As they do so, the molecules themselves become warmer and also radiate infrared light in all directions. The infrared light radiated downward contributes to the warming of the planet's surface, while the upward radiation

Fig. 13-5 Sunlight that filters through the clouds of Venus will heat the surface, which then radiates infrared light. Since the carbon dioxide molecules in the atmosphere of Venus absorb infrared light extremely well, much of the infrared light will be trapped and the surface and lower atmosphere will grow much hotter than they would be in the absence of an atmosphere.

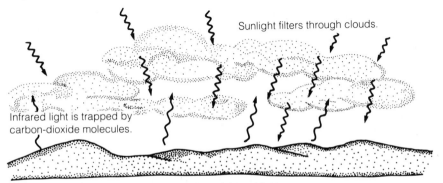

Sunlight filters through clouds.

Infrared light is trapped by carbon-dioxide molecules.

Surface radiates infrared light.

interacts with the atmosphere above in the same way as the infrared light from the ground, heating still higher layers of the atmosphere (Figure 13-5). Eventually, the radiation from the uppermost layers of the atmosphere escapes into space. Since part of this reradiated energy is directed downwards, the temperature of the upper layers will be lower than that of the layers closer to the ground.

These processes of absorption of visible light, radiation of infrared light, and the partial trapping of the infrared light by Venus's atmosphere produce a tremendous heating of the planet's surface and lower atmosphere. Astronomers call the trapping of infrared radiation by a planet's atmosphere the *greenhouse effect,* because a gardener's greenhouse performs a similar trapping: Visible light streams through the windows, heating the plants and soil, which then radiate infrared light that is blocked by the glass panes. Hence, the entire greenhouse becomes warmer than it would be if the glass, which plays the same role as a planet's atmosphere, were removed. A still more familiar example of the greenhouse effect occurs inside an automobile on a hot day, when visible sunlight warms the interior, which radiates infrared light that cannot escape easily through the glass windows; again, the trapping of infrared light produces a noticeable warming.[1]

On Earth, the greenhouse effect plays a crucial role in allowing our kind of life to continue. Our own atmosphere traps some of the infrared radiation from the ground and keeps the Earth's surface and lower atmosphere warmer than they would be if no trapping occurred. Denuded of an atmosphere, the Earth's surface temperature would average about $-20°$ C. In fact, the average temperature is about 35 degrees warmer than this, because water vapor and carbon dioxide molecules in our atmosphere trap some of the infrared radiation. These two molecular types are particularly efficient at absorbing infrared light, while nitrogen and oxygen, the major constituents of our atmosphere, are not. Human beings have already significantly increased the amount of carbon dioxide (released in fuel combustion) in the atmosphere, and have thus tended to make our planet even warmer. This may prove a serious problem during the next few decades, if the present rate of fossil fuel consumption continues.

The greenhouse effect operates full blast on Venus, whose atmosphere, almost 100 times thicker than our own, consists mainly of carbon dioxide. Even though only a small fraction of the arriving sunlight filters down through the clouds, the planet's massive atmosphere, made of efficient absorbers of infrared light, raises the surface temperature by *400 degrees,* in comparison with the temperature the planet would assume if Venus had no atmosphere.

---

[1]The automobile analogy would be exact in a car made entirely of glass, since we are ignoring the heat that enters or leaves the car through the metal roof and doors.

Why Is Venus So Different from Earth?

Now that we understand *how* the atmosphere of Venus keeps its surface so hot, we need to know *why* Venus has such a massive atmosphere, almost completely carbon dioxide, while the Earth does not. As we have seen in Chapter 9, the differences between the atmospheres of Venus and Earth seem to be the result of life on Earth and the absence of life and liquid water on Venus.

The Earth's atmosphere does not contain much carbon dioxide, because limestone rocks, mostly calcium carbonate ($CaCO_3$), have locked up most of the carbon dioxide. A typical calcium carbonate is an agglomeration of millions of tiny sea shells, formed by living sea creatures from the carbon dioxide dissolved in sea water. More carbon dioxide enters the water as the tiny animals use up the dissolved gas, so the amount of carbon dioxide in the atmosphere decreases. If we ground up and heated all the carbonate rocks in the Earth's crust and released the products into the atmosphere, our Earth would find itself with an atmosphere about 70 times thicker than the canopy of air we now enjoy, made mostly of carbon dioxide—just like the atmosphere of Venus! The special conditions that have prevented this atmosphere from developing on Earth—the formation of carbonate rocks by tiny sea animals and by the action of carbon dioxide dissolved in sea water—have been in effect for hundreds of millions of years on our planet, but perhaps only in the earliest history of Venus, if then.

Suppose that the Earth's atmosphere grew much richer in carbon dioxide—say, 10 times as rich as it is now. Even this relatively small change in the atmospheric composition would produce great changes on our planet, because carbon dioxide gas absorbs infrared light so well. We can take this speculation a step further by calculating what would happen if we suddenly moved Earth as close to the sun as Venus. The higher temperature caused by this greater proximity to the sun would make the oceans warmer, speeding up the rate of evaporation and thereby increasing the amount of infrared-absorbing water vapor in the atmosphere. The enhanced absorption of radiation would cause an additional increase in surface temperature, leading to more evaporation, more infrared absorption, and more evaporation until finally the oceans would be entirely in the atmosphere! This process which feeds so successfully on itself has been called a runaway greenhouse effect, since it proceeds until there is no water left on the planet's surface. At this point, the Earth would be extremely hot, and the water vapor molecules in the atmosphere would rise high enough to be broken apart by high-energy ultraviolet radiation from the sun, as shown schematically by the following equations:

$$H_2O + \text{ultraviolet sunlight} \longrightarrow H + OH \qquad (1)$$

$$OH + \text{ultraviolet sunlight} \longrightarrow H + O \qquad (2)$$

The hydrogen atoms produced in this way would escape from the planet's gravitational field, since they have so little mass, while the heavier oxygen atoms would remain behind to combine with other elements.

Thus by the simple experiment of asking ourselves what would happen to the Earth if we moved it as close to the sun as Venus we have found a good explanation of why our sister planet seems so strange to us. Any planet this close to a star like our sun will not be able to maintain liquid water on its surface. This seems a devastating conclusion, since liquid water is essential to the continued existence of life as we know it (see Chapter 11). Liquid water removes carbon dioxide from the atmosphere by dissolving the gas and forming carbonate rocks from it. Life enormously accelerates this process by producing carbonate structures such as shells. Thus without liquid water and without life, the carbon dioxide released into the atmosphere of our hypothetical planet will remain there to produce precisely the hellish conditions that we observe today on Venus. Life on Earth has made our planet quite different from Venus; it would be ironic indeed if human life made Earth more like Venus by releasing ever larger amounts of carbon dioxide into the atmosphere as we burn fossil fuels for our civilization's energy.

While the overall composition of the atmosphere of Venus can be easily understood by comparison with our own atmosphere, one subtle difference invites further consideration. The 1978 Soviet and U.S. Venus probes made the startling discovery that this planet's atmosphere contains 10 to 100 times as much *primordial* argon as the Earth's. Primordial argon is the argon left from the original material of the solar system, as opposed to the argon later released by the decay of radioactive potassium in rocks. How can we explain the surplus of primordial argon on Venus? Does this anomaly mean that our efforts to find a common explanation for the origin and evolution of the atmospheres of the inner planets are doomed?

The difference in abundances of primordial argon on Venus and Earth is at least an indication that we should be prepared to accept the possibility that the different planets somehow began with rather different proportions of the elements—carbon, nitrogen, oxygen, and the noble gases—that we now find in their atmospheres. But an explanation may also be found in the form of some local, specific phenomenon, which has enhanced just the abundances of the noble gases, such as helium and argon, on Venus, perhaps because of this planet's greater proximity to the sun. But should we then expect a steady decrease in noble gas abundances from Venus to Earth to Mars? This is the kind of question that astronomers must explore as we continue our investigations of our planetary neighbors.

When scientists first realized that the surface and lower atmosphere of Venus are extremely hot, they wondered if they could not find *some* place on Venus where life might exist. What about the polar regions? Couldn't they be cool enough, perhaps, for liquid water to be present? Unfortunately, the same massive atmosphere that raises the surface temperature prevents this hypothesis from being true. Like a gigantic oven, the atmosphere circulates heat all over the planet, keeping the entire surface, dayside, nightside, pole to pole—roasting at 475° C, with temperature differences probably no greater than 10 or 20 degrees over the entire planet. This conclusion, predicted by calculations of how the atmosphere behaves, has been verified by observations of the radio waves emitted by the planet.

Thus, the surface of Venus, sweltering under an atmosphere as thick as half a mile of ocean, offers no safe haven for life. But what about life in Venus's *atmosphere?* Could this be an ecological niche we have overlooked? As an example, we may consider a layer at an altitude of 35 kilometers above the surface, where the pressure equals 800 millibars, about the same as that in Denver (Figure 13-6). The temperature there is a balmy 27° C. Sunlight filtered through the haze will be more intense than at the planet's surface, since we are now located above some of the cloud deck. Since Venus has no ozone layer, this sunlight may include deadly (to us!)

Fig. 13-6 At an altitude of 35 kilometers above the surface of Venus, the temperature has fallen by 450 degrees to a pleasant 27° C and the pressure has decreased to less than 1 percent of its surface value.

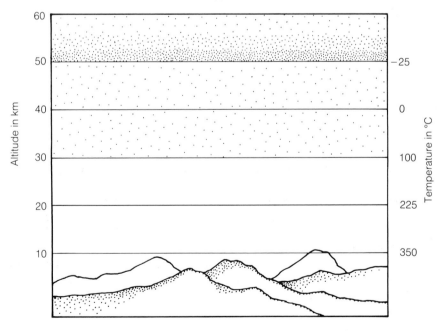

Surface temperature 475° C

ultraviolet light, but we can imagine organisms with simple outer shells that protect them against ultraviolet. Thus, at first glance, the upper atmosphere of Venus seems a pleasant environment; a crew of astronauts could float beneath a balloon, with simple oxygen masks and goggles, searching for the local equivalents of birds and butterflies.

But this survey overlooks two important facts: the atmosphere of Venus is exceedingly dry (the relative humidity never exceeds 0.01 percent) and the haze amidst which we are floating probably consists of sulfuric acid! Some types of life on Earth could, in fact, survive in such environments, but only for a limited period of time. Life as we know it needs water to grow and to reproduce, and the atmosphere of Venus does not seem to provide this basic requirement for our kind of life.

Could a different kind of life have begun and evolved on Venus? Or were conditions on Venus different in past eras, so that our sort of life could have developed and then adapted to the changed conditions? Could human beings alter the present conditions on Venus, perhaps by bringing water-laden asteroids to the planet, thus making it habitable to an oversupply of space-starved Earth dwellers? These questions cannot now be answered, but all the evidence at our disposal points to a lifeless Venus, smothered by its carbon dioxide atmosphere. As for altering the planet, we may surely develop the ability someday, and perhaps the will, but that time will not occur until long after we have either resolved our growing population problems or perished in the attempt.

Meanwhile, human beings continue to send spacecraft to investigate conditions on this brilliant and enigmatic planet. Both the Soviet and U.S. missions that reached Venus in December 1978 included probes that landed on the planet's surface, and the United States also put a spacecraft in orbit around Venus, where it has continued to study the planet for more than one Venerean year of 225 days. Radar studies by the Pioneer Venus spacecraft in 1979 revealed the presence on Venus of a rift valley 5 kilometers deep, 300 kilometers wide, and at least 1500 kilometers long, the largest canyon in the solar system. In addition, the radar also found a mountain on Venus higher than Mount Everest, and a vast plateau region larger than the Tibetan Plateau on Earth.

Within the next five years, the Soviet Union and France plan to float a package of instruments high in the atmosphere of Venus by using a specially designed balloon. The United States plans to continue its exploration with the help of an orbiter equipped with a powerful radar that will provide images of the surface of Venus comparable in detail to those that television cameras sent from cloud-free Mars. Though these spacecraft are unlikely to change our current view of Venus as an inhospitable inferno, we may yet be surprised to learn how tenaciously life can maintain a grip on existence, if it can only get started.

Summary

Venus, the planet closest to the Earth's size and mass, has an entirely different atmosphere and surface than our own planet. Wrapped in a continuous blanket of sulfuric acid clouds, Venus never shows its surface to outside observers, a surface that bakes at a fantastic 475° C. This high temperature arises from the tremendous greenhouse effect that the thick atmosphere produces: The short-wavelength, visible-light rays from the sun can penetrate the atmosphere to reach the surface of Venus more easily than the long-wavelength, infrared photons that the surface radiates can penetrate outward. A similar greenhouse effect on Earth keeps our planet's surface about 35 degrees warmer than it would be in the absence of an atmosphere. But the atmosphere of our sister planet, 90 times thicker than Earth's and made primarily of carbon dioxide, an efficient absorber of infrared light, has a far greater effect.

Radar waves penetrating the clouds of Venus have revealed the planet's slow rotation period (243 days) and the fact that the surface of Venus is geologically diverse. The runaway greenhouse effect—absorption of infrared light by water vapor that raises the temperature, thereby producing more water vapor and still higher temperatures, and preventing the formation of limestone rocks that would use up carbon dioxide—has made Venus an entirely different kind of planet than Earth. We can conclude that relatively subtle differences between one planet and another, such as the Earth's greater distance from the sun, can lead to great variations between planets during the billions of years after their formation.

Questions

1. How can we tell that the surface of Venus is covered with craters, since we can never see the planet's surface?

2. In the Earth's oceans, the pressure increases by 1 "atmosphere" for every 10 meters of depth. How deep into the oceans would we have to descend for the total pressure to equal that at Venus's surface, about 90 atmospheres?

3. Why does the atmosphere of Venus produce a much greater greenhouse effect than the Earth's atmosphere does?

4. Why is carbon dioxide particularly important in producing the greenhouse effect on Venus and on Earth? How does the release of more carbon dioxide into the Earth's atmosphere change the temperature at the Earth's surface?

5. Why does the temperature on the dark side of Venus almost equal the temperature on the day side, even though the dark side remains dark for months on end?

6. What has happened to most of the carbon dioxide that was once in the Earth's atmosphere to prevent this planet from having a runaway greenhouse effect like that on Venus?

7. The atmospheric pressure on Venus decreases by a factor of 2 for every 5 kilometers we rise above the surface of Venus. Starting from a surface pressure of 90 atmospheres, how high must we ascend into Venus's atmosphere, until the atmospheric pressure falls to 1 atmosphere, equal to the pressure at sea level on Earth?

8. Why does the level in Venus's atmosphere where the pressure equals 1 atmosphere fail to present a good environment in which life could develop, despite the fact that the temperature there is about 35° C?

Further Reading

Chapman, Clark. 1977. *The inner planets.* New York: Charles Scribner's Sons.

Weaver, Kenneth. 1975. Mariner unveils Mercury and Venus. *National Geographic* 147:858.

Young, Andrew, and Young, Louise. 1976. Venus. In *The solar system.* San Francisco: W. H. Freeman and Company.

CHAPTER 14

Mars

Our ancestors in many cultures came to associate Mars, probably because of its red color, with spilled blood, and named this planet after their gods of war. Later, the fascination with the red planet we described in Chapter 1 led to the discovery of clouds, dust storms, and seasonal changes on the surface of Mars, and to the controversies over the "canals." The strong feeling among many scientists that we might find life on the next planet out from Earth culminated in the two Viking landings on Mars in 1976, and we can anticipate additional missions in the future.

Table 14-1 summarizes the physical characteristics of Mars and Earth. Despite the planet's small size and its greater distance from the sun, we can see a striking similarity between Earth and Mars in length of day, in apparent seasonal changes (the result of similar inclinations of the two planets' rotation axes), and above all in the visibility of a solid surface on which these changes occur. Because 71 percent of Earth's surface lies beneath the seas, the total *land* areas of Mars and Earth are almost equal.

The best observational evidence for Martian life found during the early part of this century consisted of seasonal changes in the planet's surface contrast. As part of these changes, a yearly *wave of darkening* was observed to begin in late spring at the polar cap and then spread toward the equator as summer advanced (Figure 14-1). This just reversed the direction that spring travels on Earth, but astronomers conjectured that the wave of darkening on Mars might also be the result of growing vegetation. If the availability of water on Mars had more importance than the local surface temperature, and if water came only from the melting of the polar cap, then plants might flourish as windborne water vapor, or running water on the ground, spread toward the equator from the pole. The growing plants would

TABLE 14-1

COMPARISON OF THE PHYSICAL CHARACTERISTICS OF THE EARTH AND MARS

Planet	Radius (km)	Mass (gm)	Surface Gravity (Earth = 1)	Length of Day (hours)	Length of Year (years)	Distance from Sun (A.U.)
Earth	6378	5.98×10^{27}	1	24	1.0	1.0
Mars	3395	6.42×10^{26}	0.38	24.5	1.88	1.52

Mars has a diameter 53% of Earth's. The most striking apparent difference between Earth and Mars consists of the huge amounts of water on Earth. Because the earth's surface is three-quarters oceans, Mars and Earth have almost the same surface area.

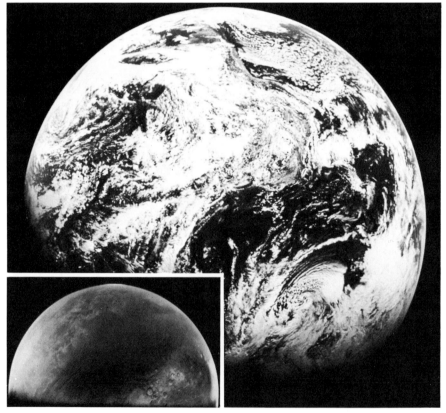

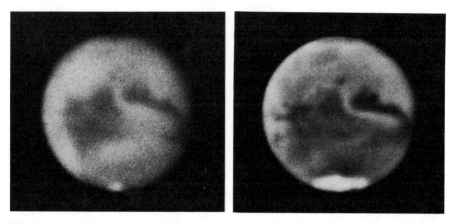

Fig. 14-1 Telescopic observations of Mars show darker features that stand out against the brighter and lighter-colored background. In each hemisphere, spring brings a change in contrast between the dark and light areas. As spring turns to summer, the polar cap shrinks to its minimum size (see left panel).

reflect less light than bare ground, and thus the areas with more vegetation would be darker.

Modern Observations of Mars

As telescopes improved and the equipment used with them grew more sophisticated, this image of Mars began to change. The first challenges came from a series of temperature measurements. Mars is *cold*. The temperature of the warmest spot on Mars, at the warmest moment of the Martian day, during the time when Mars comes closest to the sun, rises to 27° C, about equal to summer in Los Angeles. But the *average* temperature on Mars falls far below the freezing point of water. Even the spot where the noontime temperature reaches 27° C has a nighttime temperature near − 90° C! Under these conditions, any channels of liquid water, natural or Martian made, appear highly unlikely.

Spectroscopic observations of Mars confirmed the suspicion that no running water was to be expected on the red planet. By the early 1960s, astronomers knew that Martian ''air'' consists mostly of carbon dioxide, with only a trace of water vapor. The total amount of gas in the Martian atmosphere, and the pressure that the atmosphere exerts at the Martian surface, must be less than 1 percent of the values for our own atmosphere. The surface pressure plays an important role in the speculation about the possible existence of life on Mars, because if the Martian atmosphere does not exert at least 0.6 percent of our atmosphere's pressure, then *water cannot exist as a liquid, no matter what the surface temperature may be.*

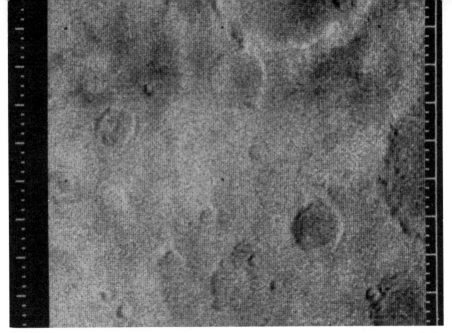

Fig. 14-2 Photographs from Mariner 4 showed that craters cover most of the surface of Mars, giving it a certain resemblance to the surface of the moon.

If the atmospheric pressure falls below this threshold value (that is, below 6 millibars of pressure), then water *vapor* will form directly as ice melts *(sublimes)*, just as dry ice (frozen carbon dioxide) sublimes at the Earth's surface. Without liquid water, life as we know it cannot persist, so the total atmospheric pressure at the surface of Mars seems to be a key to the possibility of life there.

Results from Space Probes

On July 15, 1965, a spacecraft called Mariner 4 sailed past Mars, carrying a small complement of instruments that gave our first close-up look at another planet. Mariner 4 showed that at least some parts of Mars look much like the moon (Figure 14-2). Two follow-up spacecraft went past Mars in 1969, and finally humans sent a spacecraft into orbit around the planet in 1971. The instruments aboard these vehicles had become increasingly advanced, so that the ultraviolet-light spectrometers on Mariners 6 and 7 could discover the presence of ozone on Mars in a proportion of less than one part per million, while the infrared-light spectrometer on Mariner 9 showed that the atmosphere of Mars contains less than 25 parts per billion of methane.[1] The Mariner spacecraft showed that the average surface pressure on Mars falls close to the critical level of 6 millibars that would forbid

[1]The Earth's atmosphere contains 50 parts per billion of ozone and 2000 parts per billion of methane.

liquid water from existing. Careful study of the thousands of pictures sent to Earth by Mariner 9 failed to show any evidence of the notorious "canals." Evidently these linear features were only an optical illusion resulting from chance alignments of dark spots on the surface (see Figure 1-2).

Spectrometers on Mariners 6 and 7 showed that the Martian polar caps are primarily frozen carbon dioxide, not frozen water. This emphasizes how cold Mars must be, since in order for carbon dioxide to freeze, the ground temperature must fall to $-125°$ C. Measurements of infrared radiation from the planet indicated that even at its equator, the surface temperature of Mars falls to $-90°$ C just before dawn on an average day. The temperature changes by more than 100° C each day, far more than even the most extreme temperature changes on the deserts of Earth. These impressive variations testify to the thinness of the Martian atmosphere, which is 150 times less dense than Earth's, and provides only a 5-degree greenhouse.

The small amount of ozone in Mars's atmosphere does not provide an effective shield against ultraviolet light from the sun. When some scientists put this fact together with the probable absence of liquid water and the extreme cold, they concluded that the prospects for life on Mars seemed to be vanishingly small. Other scientists, however, stressed the fact that some terrestrial microorganisms could exist under the harsh Martian conditions, if they obtained suitable shielding from ultraviolet light, perhaps by living under rocks, and had access to some liquid water, just enough to coat grains of soil for some small fraction of a day. Furthermore, these scientists pointed to the existence of sinuous dry valleys with branching tributaries (Figure 14-3) as evidence that liquid water once flowed on the

Fig. 14-3 Winding valleys, often complete with branching tributaries, resemble dry river beds on Earth and strongly suggest that liquid water once flowed on Mars's surface.

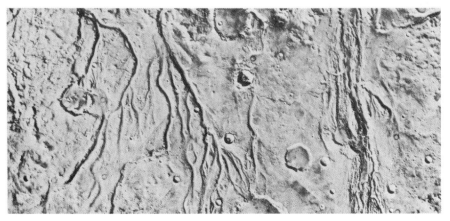

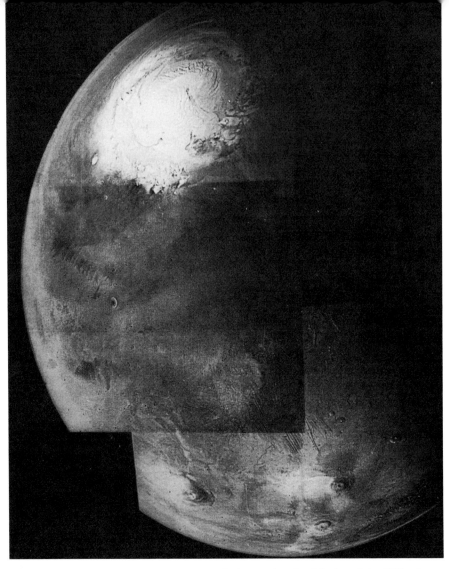

Fig. 14-4 This mosaic of photographs of Mars, taken by Mariner 9 in 1973, shows the north polar cap (top) and the great volcanoes of Mars (bottom).

surface of Mars. The immense Martian volcanoes, such as the mighty Olympus Mons, whose unbattered summit towers 25 kilometers above the surrounding plain (Figures 14-4 and 14-5), show that the planet's crust has undergone activity within relatively recent eras. Could the cosmic luck of the moment have brought us to the red planet during an extreme ice age? Might conditions have been more favorable to life on Mars in the past? And could they become favorable again?

We do know that in past eras, the angle by which Mars's rotation axis is tilted with respect to the planet's orbit around the sun changed its value over millions of years' time. In fact, this angle varies periodically through

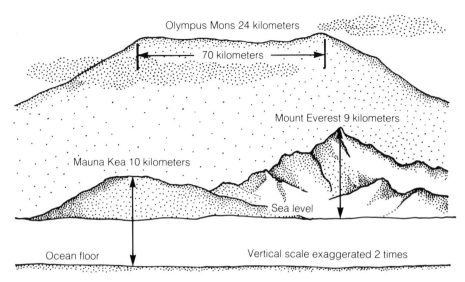

Fig. 14-5 Olympus Mons, the largest known mountain in the solar system (so far!), rears its enormous bulk 25 kilometers above the plains of Mars. This volcano has a diameter of 300 kilometers, more than five times the width of Mauna Loa, the largest volcano on Earth.

greater and lesser values (Figure 14-6). These changes of tilt arise from the combined gravitational forces of Jupiter and the sun on the planet Mars, and they have an important effect on the Martian climate.

Consider what happened when the angle of inclination of the rotation axis exceeded its present value. In those years, the change between summer and winter on Mars must have been more pronounced than the change now, and we can imagine that the polar caps might have disappeared completely in summer for a given hemisphere. When this effect was first discovered, astronomers thought that this sublimation could release sufficient

Fig. 14-6 The angle of inclination between Mars's rotation axis and the perpendicular to the planet's orbit around the sun changes periodically by several degrees. When this angle has a greater value than it does now, the difference between Martian summers and winters must be more pronounced than it is now.

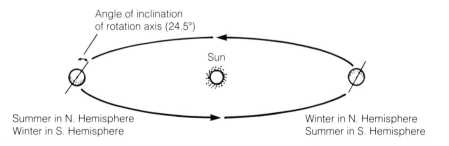

gas (mostly carbon dioxide) into the Martian atmosphere to *raise the at-mospheric pressure enough for liquid water to exist*. This would provide an explanation for the braided channels, closely resembling dried-up river beds, that Mariner 9 had photographed. With such a cyclic appearance and disappearance of liquid water, life on Mars might have developed the ability to survive the hundreds of thousands of years when no liquid water exists by means of some sort of superhibernation that would await the years when the polar caps would sublime once again, allowing liquid water to revive the desiccated organisms. Unfortunately, most of the water-carved channels appear to be not a million or so years old, but a billion or so years. This means that these imagined cycles of liquid water may have occurred only during much earlier eras of Martian history.

The Viking Project

After a hundred years of controversial observations made from Earth and eight years of equally controversial studies from spacecraft, the preparations for the first landing on Mars unfolded in a highly charged atmosphere. The Viking mission plan called for two spacecraft, each with an orbiter and a lander, to arrive at Mars in the summer of 1976. The orbiters could study the suitability of the landing site before the landers made the actual descent and would serve as relay stations to send back the data collected on the surface as well as platforms for remote investigations of the entire planet.

Table 14-2 shows the scientific payload of the Viking spacecraft, the experimental equipment designed to answer the most pressing questions about Mars. What is the composition of the Martian soil? The inorganic analysis would tell us. Is the interior still active enough to produce "Mars-quakes"? A seismometer would give the answer. What does the Martian atmosphere consist of? Data would come from mass spectrometers on the aeroshell that protected the lander during its descent, and on the lander itself, as well as from a spectrometer on the orbiter. What is the temperature on Mars? The orbiter's infrared radiometer could measure both the ground and the atmospheric temperatures, while the lander could sense local variations precisely. What is the weather like? The Viking lander could measure wind speed, wind direction, and atmospheric pressure. What does Mars look like? Images from the orbiter could show surface features slightly larger than the Rose Bowl; later, once the orbit had been adjusted, we could distinguish the grandstand from the playing field. Meanwhile, the landers would take pictures of their surroundings with resolutions that varied from a few millimeters near the lander to a few hundred meters at the horizon. Two cameras on each lander gave the ability to take stereoscopic pictures, and filters allowed for color pictures from Mars. What does Mars smell and taste like? The molecular analysis instrument

TABLE 14-2

<small>Scientific Payload for the Viking Missions to Mars</small>

Investigation	Instruments
Viking Orbiter	
Visual imaging	Two television cameras
Water vapor mapping	Infrared spectrometer
Thermal mapping	Infrared radiometer
Viking Lander	
Imaging	Two facsimile cameras
Biology	Three analyses for metabolism, growth, or photosynthesis
Molecular analysis	Gas chromatograph-mass spectrometer (GCMS)
Inorganic analysis	X-ray fluorescence spectrometer
Meteorology	Pressure, temperature, and wind velocity gauges
Seismology	Three-axis seismometer
Magnetic properties	Magnet on sampler observed by cameras
Physical properties	Various engineering sensors
Radio Propagation	
Orbiter/lander location Atmospheric and planetary data Interplanetary medium General relativity	Orbiter and lander radio and radar systems

would tell us, by examining the composition of the atmosphere and the soil. And does life exist on Mars? A good question! Many instruments could help answer it, but three were designed specifically for this purpose alone. We shall discuss the performance of these three experiments in the next chapter.

With the choice of experiments and of the scientists to build and operate them, the Viking project began in earnest in 1969, seven years before the actual landing. As the pictures of Mars came back from Mariner 9 during 1972, the planet assumed an entirely new identity: In place of the drab, moonlike landscape that previous spacecraft seemed to have dimly glimpsed, there emerged huge volcanoes, deep canyons, and those intriguing, sinuous channels. The search for a suitable landing site on Mars had to follow three constraints. First, the surface pressure had to be at least 4 millibars, or the atmosphere would not provide sufficient friction to slow the lander safely. This eliminated all the higher-elevation parts of Mars. Second, the landing had to occur relatively close to the Martian

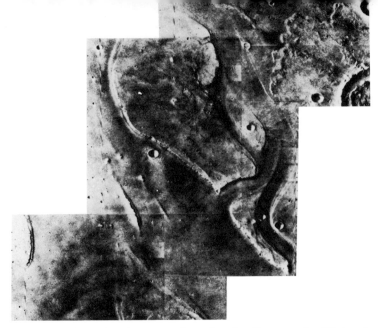

Fig. 14-7   The original site chosen as the prime landing area for Viking 1 turned out to have impact craters in the dried-up stream beds, evidence that billions of years had passed since water carved these channels.

Fig. 14-8   The final site chosen for the landing of Viking 1 lay at the western edge of the Chryse basin, and was favored for its overall smoothness. The cross marks the projected landing site, as photographed from the Viking 1 orbiter. The landing in fact occurred about 15 kilometers to the left of the center of the cross, which itself spans 6 by 8 kilometers.

equator, and not near the poles, so that the lander could establish good radio links with Earth. Third and most important, the site had to be "safe," meaning that the lander could not survive if it settled down on a rock larger than a basketball. Yet the smallest surface features visible from the orbiter were larger than a football stadium!

Fortunately, radar studies made from Earth could tell us the general roughness of various areas on Mars, and were used to eliminate some otherwise attractive landing sites from final consideration. In view of the difficulties that the scientists and engineers faced in finding good landing sites for spacecraft 400 million kilometers from Earth, they deserved their celebrations for landing not once but twice at this distance. Furthermore, the landing sites turned out to be soft enough for the Viking sampler to dig trenches in the soil, rather than hard slabs of rock from which no samples could be taken aboard the spacecraft.

The scientific criteria for a good landing site leaned heavily on the search for life on Mars. Simply put, the Viking team wanted to land in places that were warm, wet, and inhabited. In view of what was already known about Mars, quite liberal definitions were needed if any sites were to make the grade. The final decision before Viking reached Mars was to land once in a region where several of the sinuous channels observed by Mariner 9 seemed to form deltas and a second time at a higher latitude, in the hopes of finding more water near the surface, since the landing would occur in summer for that hemisphere.

Once the first Viking spacecraft went into orbit around Mars in June, 1976, however, the scientists saw that the high-resolution photographs of the prime landing area showed impact craters right in the sinuous channels (Figure 14-7). These craters implied a span of billions of years since the time that water last flowed there. Worse yet, the site seemed to present too many hazards to risk a landing, so the prime landing area was shifted, after many discussions, to the western edge of the Chryse basin, downstream from what appeared to be well-defined water erosion (Figure 14-8).

On July 20, 1976, seven years after the first human walked on the moon, the first Viking lander settled down comfortably on the surface of Mars. The first experiment made there was to take and transmit a picture of one of the lander's footpads, and subsequent images filled in the landscape (Figure 14-9). The choice of a landing site had been good, for only one rock in the immediate vicinity was large enough to have overturned or punctured the lander, if the landing had occurred upon it. Color pictures confirmed the fact that Mars's surface *is* red, the result of iron oxides in the soil. What surprised some people is that the Martian sky also has a reddish cast, unlike the blue skies of Earth. The reason for this red tint is visible in the dirt around the spacecraft: Some of the particles are small enough to remain suspended in the atmosphere, where they give the Martian sky its unusual

Fig. 14-9 Rocks a few centimeters or even a few dozen centimeters in diameter dominate this view of the immediate surroundings of the Viking 1 lander. The housing at lower left contains the sampler to scoop up soil.

color. A similar effect occurs during violent sandstorms on Earth, when somewhat larger particles seem to turn the sky yellow. Astronomers now feel certain that windblown dust is responsible for the wave of darkening that once suggested the presence of vegetation on Mars.

The second landing on Mars, on September 3, 1976, put a spacecraft at 40° latitude but revealed a landscape much like that at the first site (Figure 14-10). Even though two sites hardly provide much of a sample of the entire planet, comparisons of their detailed similarities and differences revealed a huge amount of information.

The New Mars

To the experienced eyes of the Viking scientists, the Martian landscapes seemed hauntingly familiar. The sharpness and variety of the rocks on Mars, together with the undulations of the surface and the dune fields visible near the horizon, all resemble scenes on Earth, which also has regions with reddish soils. Both Earth and Mars contrast sharply with the

Fig. 14-10 The panoramas seen from the Viking 1 landing site (top) and the Viking 2 site (bottom) show an overall resemblance, despite their separation of 3000 kilometers. Rocks ranging in size from a few meters to a few centimeters in diameter dominate both views; no signs of life are visible.

TABLE 14-3

Elemental Composition of Soil at Two Viking Landing Sites

Element	Percentage of Total Composition of Soil	
	Site 1	Site 2
Silicon	20.9 ± 2.5	20.0 ± 2.5
Iron	12.7 ± 2.0	14.2 ± 2.0
Magnesium	5.0 ± 2.5	
Calcium	4.0 ± 0.8	3.6 ± 0.8
Sulfur	3.1 ± 0.5	2.6 ± 0.5
Aluminum	3.0 ± 0.9	
Chlorine	0.7 ± 0.3	0.6 ± 0.3
Titanium	0.5 ± 0.2	0.6 ± 0.2
All others (thought to be mostly oxygen)	50.1 ± 4.3	

moon, where the total lack of an atmosphere has exposed the soil to relentless bombardment by micrometeorites and to the bleaching effects of the sun's full glare of ultraviolet light. The examination of Martian soil revealed other similarities to the Earth, and, still more important, showed that *the chemical composition of the soil at the two landing sites is virtually identical.*

Table 14-3 shows the composition of the soil at the two Viking landing sites in terms of the chemical *elements;* the inorganic soil analyzer could not distinguish among various compounds and minerals formed from individual elements. The abundance of sulfur on Mars seems unusually high, but otherwise the Martian ratio of elements is similar to that in nontronite, an iron-rich clay found on Earth. Fine particles of this material may form the dunes in Figure 14-9 and may also cause the distinctive reddish tint of the Martian sky. But the key conclusion drawn from Table 14-3 is that the surface of Mars consists—at least at the two landing sites sampled—of elements *in the abundance ratios familiar to us on Earth.*

The same holds true for Mars's atmosphere. The predominance of carbon dioxide, first discovered from Earth, was confirmed by the Viking mass spectrometers, which sorted atoms and molecules by mass and thus allowed scientists to deduce which compounds must form the atmosphere. Ninety-five percent of the Martian atmosphere is carbon dioxide, while nitrogen molecules provide 2.7 percent, argon 1.6 percent, oxygen molecules about 0.1 percent, and water vapor and carbon monoxide still smaller traces of the Martian "air." The overwhelming dominance by carbon dioxide reminds us immediately of Venus. Once again, we meet a planet whose atmosphere seems to have evolved in the absence of abundant life and of liquid water.

For Venus, water seems simply not to exist, because the planet orbits so close to the sun that either it grew too hot to form with water, or else the water escaped soon after Venus formed. On Mars, we seem to meet the opposite problem: The bulk of Martian water probably has remained on the planet, but most of it has frozen in the subsurface soil as permafrost, similar to the wintertime freezing of water in Arctic soils on Earth. We seem to have the classic situation of "Goldilocks and the three planets": Venus is too hot, Mars is too cold, and Earth is just right!

From measurements of the abundances of the various gases in Mars's atmosphere, and from our knowledge of how gases escape from a planet such as Mars, we can reconstruct the planet's early atmosphere. We find that Mars may once have had a much denser atmosphere than it does today. Despite the fact that carbon dioxide forms 95 percent of the present Martian atmosphere, *most* of the carbon dioxide near the planet's surface seems to be locked up in the polar caps and in carbonate rocks. We can calculate that if *all* the carbon dioxide that now seems to be missing were present in gaseous form, then the atmospheric pressure could rise to a value between 50 and 100 percent of its value at the Earth's surface.

This interesting fact allows us to explain the mysterious channels, apparently carved by running water. With a surface pressure of 500 millibars, liquid water could easily exist, though this atmosphere might have been unable to keep the planet warm enough to allow a fully developed rain-river-pond-rain cycle to persist. In fact, we have found no signs of any dry lakes, shorelines, or any other large bodies of standing water on Mars. The branching channel systems show no evidence of the meanders or "oxbows" that typify mature, slow-flowing river systems on Earth. Instead, we see what looks like the scars of water moving in haste, as in the flash floods of our own desert regions, but on a vastly larger scale (Figure 14-11). Sometimes the source of water on Mars appears to be "collapsed" terrain, and we can wonder whether the subsurface permafrost might have suddenly melted, as the result of geothermal activity or of meteoritic impact, and whether the resulting flood of water broke through ice dams to spread·across the Martian plains.

This was an active, unstable phase of the planet's history that apparently occurred billions of years ago, and one of the major questions still to be answered is *how long* this phase lasted on Mars. The presence of impact craters in the floors of the large channels suggests that the last massive floods on Mars took place at least 3 billion years ago, perhaps more. But at least *some* period on Mars included the existence of liquid water, so we might find that life began on Mars, only to die out during the past few billion years. Certainly all the major elements that we have identified as essential to life exist on Mars. The famous polar caps, mostly

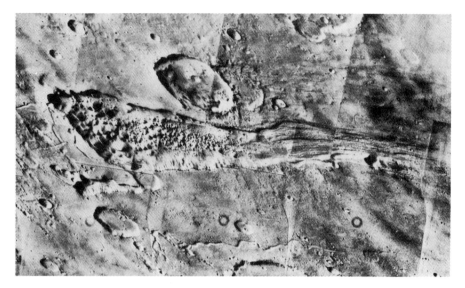

Fig. 14-11   This large triangular valley, several dozen kilometers across, may have arisen from subsidence of the Martian surface, perhaps the result of ice melting below the surface. This subsidence apparently caused a gigantic flash flood that carved the exit channel to the right of the photograph. Similar sources of water may have carved the branching channels seen on the planet's surface.

made of dry ice, also include a significant amount of water ice. We can be sure of this from the fact that the summer sunshine raises the temperature of the north polar cap to the point where all the dry ice sublimes, leaving a small cap of water ice (Figure 14-12). Before the water ice can sublime completely, the next fall's carbon dioxide frost covers it over, preserving some ice for season after season. The edges of the polar cap may become prime candidate areas in future searches for life on Mars.

Meanwhile, the Martian prospect seems grim. A hardy scout on Mars could never build a fire, since no fuel exists, and the atmosphere has too little oxygen for anything to burn. If the scout brought an electric stove, chipped ice from the polar cap, and tried to produce water, she would merely sublime the ice into vapor, never into water. Only with a pressure cooker could anything as simple as boiling an egg take place, and even then, a simple doubling of the pressure, such as occurs in Earth kitchens, would result in boiling water at 5° C! To make liquid water, and keep it liquid at, say, 60° C, the scout would need a pressure cooker that raised the pressure 100 times above the outside pressure. It is thus evident that scouting on Mars will be no picnic, yet a fair amount of scouting seems called for if we are ever to find life on the fourth planet of the sun.

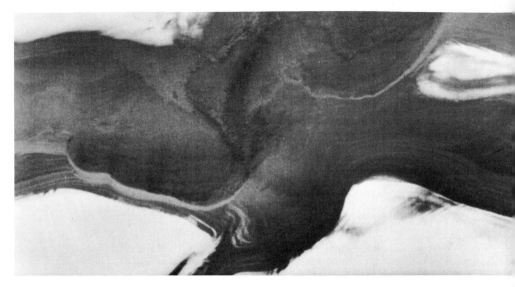

Fig. 14-12 During summer in the northern hemisphere of Mars, the carbon dioxide in the polar cap sublimes into vapor completely, leaving a small cap of water ice, which requires a higher temperature to sublime.

Phobos and Deimos

Mars has two tiny moons, named after the chariot horses of the god of war, that were discovered in 1874 by Asaph Hall, though Johannes Kepler, Voltaire, and Jonathan Swift had all speculated that Mars *ought* to have two moons.[3] In contrast to the Earth's own moon and to the four large satellites of Jupiter, the two Martian moons have diameters of less than 20 kilometers, making them far smaller than the larger asteroids, small enough that a tourist on either of them could launch rocks into space with a good hard toss.

The Viking orbiters took some marvelous photographs of Phobos and Deimos, laying to rest the supposition that such small planetary satellites might be the artificial creations of an advanced civilization. As Figure 14-13 shows, the two moons are too small for their own gravitational forces to deform them into nearly spherical shapes, as occurs for all large objects, such as the Earth and the moon. Battered by impacts from countless meteoroids, Phobos and Deimos probably provide good pictures of what an

---

[3] At the time when Kepler, Voltaire, and Swift wrote, the Earth was known to have one satellite and Jupiter four, so to make a proper geometrical progression, Mars was imagined to have two moons. This simple explanation has not prevented some fantastic speculation that Kepler, Voltaire, and Swift must have been in touch with advanced civilizations who told them of the moons of Mars.

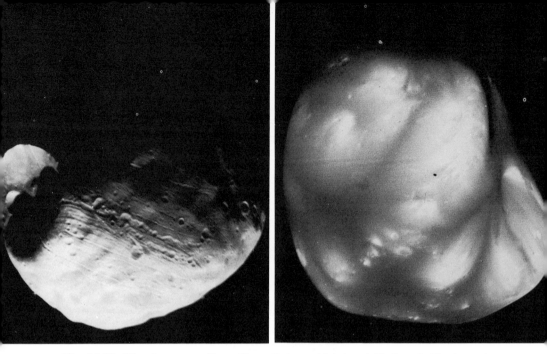

Fig. 14-13 Mars's two small satellites, Deimos (right) and Phobos (left), show the effect of cratering. The impacts on Phobos have also apparently produced the series of parallel grooves visible on the surface of the satellite. We can see much more detail on the surface of Phobos because it is illuminated obliquely. Deimos is seen in full phase, with no shadows to enhance the contrast.

average asteroid looks like. In fact, these satellites may *be* captured asteroids, rather than objects that accreted in their present orbits when Mars formed. Spectroscopic studies suggest that the composition of Phobos and Deimos closely resembles that of the carbonaceous chondrite meteoroids (see page 241).

New photographs of these satellites still reach the Earth more than three years after the Viking spacecraft first reached Mars. The landers are still working, too, taking new pictures at different seasons, finding frost in the winter where only shadows appeared in summer. But some of the Viking experiments have been completed, and among these are the ones directly devoted to the search for life on Mars. This story spreads into a chapter, for the Viking biology experiments represent the best direct search for alien life that we have made so far.

Summary

Mars, with just half Earth's diameter and one tenth Earth's mass, does resemble Earth in its rotation period, in its seasonal changes in appearance, and in possessing an atmosphere. Its greater distance from the sun, however, and—most importantly—an atmospheric pressure less than 1 percent of that on Earth, all combine to preclude the existence of liquid water on Mars. The bulk of the water has either frozen into the subsurface soil as permafrost or into the polar caps, which are mostly frozen carbon dioxide; a tiny fraction exists in Mars's atmosphere, which is also mostly carbon dioxide.

Even though we can show that no liquid water can exist on Mars now except under very special conditions (like carbon dioxide on Earth, which passes from dry ice to vapor directly, water on Mars must be either solid or gaseous), photographs of Mars taken by the Mariner and Viking spacecraft show evidence for liquid water on Mars in the past. This evidence consists of the channels that geologists say must have been carved by flowing water.

The surface of Mars shows a heavily cratered terrain, proof that little erosion and weathering has occurred during the 4 billion years since most of the craters formed. Gigantic volcanoes, some three times as high as any mountain on Earth, show that Mars has geological activity that may even extend to the present day.

The Martian "canals," once highly touted by some astronomers, clearly do not exist at all, but instead must have been optical illusions of a particularly exciting variety. Mars's seasonal changes in color over wide areas of its surface apparently arise from windblown dust, and certainly not from the growth or decay of vegetation. The two moons of Mars, Deimos and Phobos, are so small that they maintain irregular shapes. Their battered surfaces reveal a history of early bombardment. These satellites probably resemble small asteroids in appearance and even in composition.

Questions

1. What sort of observations made before spacecraft visited Mars encouraged people to believe that Mars might have vegetation, and even intelligent beings, living on its surface?

2. What are the "canals" of Mars?

3. Why is it impossible for any liquid water to exist on the surface of Mars?

4. On Earth, the atmospheric pressure decreases by a factor of 2 every time we gain 6 kilometers of altitude. About how high would we have to rise before the atmospheric pressure would fall to the value at Mars's surface, about $1/160$ of the pressure at sea level on Earth?

5. Why does the Martian atmosphere fail to shield the planet's surface against ultraviolet light from the sun? What does this fact imply for the possibility of life on Mars?

6. What evidence suggests that Mars *once* had liquid water on its surface? If this were so, what conditions must have been different for liquid water to exist?

7. Discuss the difficulties that scientists faced in choosing a landing site for the Viking expedition to Mars.

8. What do the Martian polar caps consist of?

9. Where do scientists think that most of the carbon dioxide that was formerly in Mars's atmosphere may now be found?

10. How could you boil an egg on Mars?

Further Reading

Arvidson, R. E., Binder, A. B., and Jones, K. L. 1978. The surface of Mars. *Scientific American* 238:3, 76.

Carr, Michael. 1976. The volcanoes of Mars. *Scientific American* 234:1, 32.

Hartmann, W., and Raper, O. 1974. *The new Mars: The discoveries of Mariner-9.* Washington, D.C.: NASA, U.S. Government Printing Office.

Leovy, Conway. 1977. The atmosphere of Mars. *Scientific American* 237:1, 34.

Moore, Patrick. 1977. *A field guide to Mars.* Boston: Houghton Mifflin Company.

Murray, Bruce, ed. 1973. *Mars and the mind of man.* New York: Harper & Row.

Mutch, Timothy. 1979. *The martian landscape.* Washington, D.C.: NASA, U.S. Government Printing Office.

Pollack, James. 1976. Mars. In *The solar system.* San Francisco: W. H. Freeman and Company.

Veverka, Joseph. 1977. Phobos and Deimos. *Scientific American* 236:2, 30.

Science Fiction
Bradbury, Ray. 1956. *The Martian Chronicles.* New York: Doubleday.

CHAPTER 15

Is There Life on Mars?

How can we determine whether life exists on Mars? The analysis provoked by scientists' interest in this question during the last century has led to the suggestion of four chief ways to detect life on another planet. In order of increasing difficulty for us on Earth, they run as follows: (1) We can *look* from Earth for any major changes on Mars which we might ascribe to another civilization. (2) We can *listen* at various radio wavelengths for signals produced by another civilization. (3) We can *analyze the composition* of the planet's atmosphere, using advanced spectroscopic techniques, to spot subtle changes caused by the presence of life. (4) We can *go* to another planet, to perform complicated experiments that test for the presence of both large and microscopic living organisms.

Notice that the first three of these techniques can be applied directly from Earth, provided we have sensitive telescopes, radio antennas, and spectroscopes. Notice also that the fourth and most direct means of search costs considerably more than any of the first three, though not so much that humans have refused to pay it. Before we look at the results from this fourth test, let us pause to consider what an intelligent Martian might find out about life on *Earth,* using the first three techniques.

Until quite recently, the first two techniques—looking and listening— would have failed to signal the presence of life on Earth to a Martian with the same technology that we possess now. The distorting effects of the Earth's atmosphere would obscure our planet's surface so much that humanmade changes in the landscape could not be seen, and the nighttime glow of our cities would not have reached the level of easy detection. Similarly, no radio transmissions would have been detectable, since humans began to generate large amounts of radio power only in the 1920s.

The third test, however, would have yielded positive results, *even if no civilizations had arisen on Earth.*

Spectroscopic observations made from Mars would reveal the presence of a great number of oxygen molecules in the Earth's atmosphere, along with a small amount of methane (CH_4). This situation would provoke a Martian's curiosity, because methane burns in the presence of oxygen, and even if no fires existed on Earth, ultraviolet light from the sun would rapidly "oxidize" all the methane into carbon dioxide and water. The continued existence of methane on Earth implies the presence of a *source* that replaces the methane as it oxidizes. Terrestrial methane arises primarily from bacteria that live in the intestines of grass-eating animals and in swampy marshes.

In an equally striking way, the large amount of oxygen in the Earth's atmosphere would puzzle an intelligent Martian, since oxgyen is a highly reactive gas that rapidly combines with rocks that are continually being brought up from the interior to the Earth's crust. This process removes oxygen from our planet's atmosphere, so the continued existence of oxygen also requires a source—in this case, the presence of plants that release oxygen through photosynthesis.

Thus, the third test would show Martians that life exists on Earth long before they landed any spacecraft to sample the immediate environment. Likewise, if we had discovered large amounts of oxygen in the Martian atmosphere, we would have considered life to be likely there. But we now know that oxygen forms only 0.13 percent of the thin Martian atmosphere, which has a total pressure only 0.7 percent of our own. This tiny amount of oxygen can be easily explained as the result of photochemical reactions that occur when sunlight strikes the small amount of water vapor in the Martian atmosphere. Methane and other hydrocarbon gases remain undetected on Mars, so the planet shows no signs of the unexpected conditions that mark Earth as biologically active.

In short, the first three tests have failed to show any mark of life on Mars: no "canals," no cities, no radio broadcasts, no unexpected atmospheric gases. As a result, humans put nearly a billion dollars worth of effort into the fourth test, and built the Viking spacecraft, capable of landing on Mars and performing delicate experiments to test for the presence of Martian microbes.

How to Find Martian Microorganisms

Although the greatest excitement we might expect from our search for life would be the discovery of large, advanced creatures capable of communicating with us, the history of life on Earth reminds us that microscopic organisms far outnumber large ones and are far hardier. The fossil record shows that microbial life was the *only* kind of life on Earth for billions of years, far longer than larger creatures have existed. Nor have microorganisms decreased in numbers or in adaptive ability. The dirt in your backyard contains more organisms than the number of stars in our galaxy. We have no reason to expect another planet with life to differ from Earth in the overall development of living systems, so we greatly increase our chances of finding life when we include a search for microorganisms. This realization underlay the design of experiments to find life on Mars.

But how do we find them? We cannot use an ordinary camera to see tiny organisms, and it is difficult (as design studies showed) to send a powerful microscope to Mars and have it function properly on the planet's surface. Instead, Viking had to rely on experiments that would detect microorganisms in slightly more complex ways. Furthermore, the Viking experiments had to deal with the possibility that the carefully planned and marvelously miniaturized laboratory that landed on the surface of Mars might detect Earth microbes that had been carried millions of kilometers to Mars on the lander itself!

The Viking scientists managed to overcome this last problem through immensely careful sterilization of the entire spacecraft. They also wrestled with the best way to obtain samples of Martian soil that might contain microbes, unaffected by the descent of the spacecraft. A simple extendable boom with a scoop at the end was finally judged best, because tests showed that the landing would disturb the soil only slightly, and only directly beneath the spacecraft (Figure 15-1). Suppose an uncontaminated soil sample could in fact be tested for microorganisms. What would they like to eat and drink? Here the Viking scientists fell back on the "fundamental" principles of biology and chemistry, principles that must be at least somewhat biased from the fact that we derive them from one single example of biology, life on Earth.

With this bias firmly in mind, we can offer the following principles: Life on Mars, if it exists, ought to be based on carbon chemistry and should involve a fluid solvent, some common substance that can exist as a liquid under Martian conditions. For Mars, which has no ammonia or alcohol, this means a carbon chemistry with water; that is, a system of life essentially identical in its chemical outlines with life on Earth. This may seem a restrictive conclusion, but the imposition of a chemical similarity at such a basic level permits enormous flexibility in the type of life that might actually exist (see Chapter 11). Then how can we detect tiny Martian microbes of unknown form and behavior?

A logical first step is to study the Martian soil *in detail*. All organisms, even microbes, develop an intimate connection with the places they inhabit. If microorganisms are sufficiently abundant, they will modify the chemistry of their surroundings, thus producing evidence of life that should be detectable through analysis of either the atmosphere or the soil of Mars.

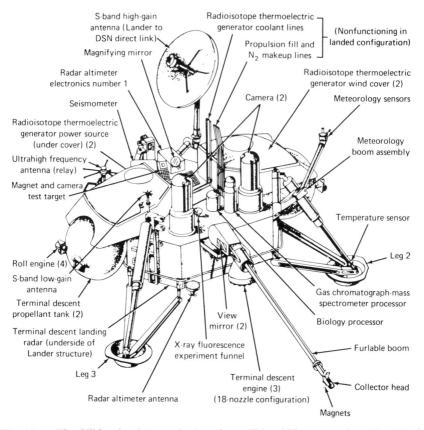

Fig. 15-1   The Viking landers packed an incredible ability to analyze the Martian atmosphere and surface into a small volume, which included an extendable boom with a scoop that could bring soil samples into the laboratory on board.

The Viking Results: Soil Analysis

Once the first Viking lander descended on the surface of Mars and the first soil sample rode the scoop into the test chambers, the results came quickly back to Earth. *No* evidence of organic compounds appeared in the Martian soil, and the atmosphere of Mars showed no compounds unexpected if life did not exist. The Viking lander made its soil analyses with a highly sophisticated instrument called a gas chromatograph-mass spectrometer, or GCMS. The GCMS baked soil in an oven to drive off volatile gases. These gases adhere to the gas chromatograph part of the instrument with different degrees of stickiness, and heating them makes the volatiles leave in sequence. The mass spectrometer analyzes the departing stream of gas to determine which compounds are present.

Figure 15-2 shows an analysis of Antarctic soil on Earth with a laboratory version of the Viking GCMS. Although this soil contains barely enough living organisms to give weakly positive results in the three life-detection experiments, the GCMS shows lots of organic compounds. In other words, the fingerprint of life can be seen more easily than the living creatures themselves. For instance, the Antarctic soil sampled in Figure 15-2 contains 10,000 times more carbon in *organic molecules* than in microorganisms.

We must realize, however, that not all organic compounds require living creatures to produce them. Figure 15-2 shows a GCMS analysis of a meteorite that contains amino acids, the building blocks of proteins. Again we find carbon-laden molecules, but these organic molecules were not made by life. Thus, the detection of organic molecules in general would not show that life exists on Mars, for the simpler kinds of organic compounds could have been brought to Mars by meteorites, or even have arisen from photochemical reactions in the Martian atmosphere.

In practice, these ambiguities did not plague the GCMS experiment, because it found *no* organic compounds in the soil of Mars. On the scale of Figure 15-2, the Martian soil results would be indistinguishable from the zero horizontal line of the graph. These results apply to both landing sites, and to two different samples at each site, including one from underneath a rock that might (so it was thought) have sheltered organisms from deadly ultraviolet light. The only compounds found in the soil were water and carbon dioxide, which was hardly surprising since these gases exist in the planet's atmosphere. The upper limits on all likely organic compounds in the soil, such as hydrocarbons, acetone, furan, and acetonitrile, fall at a few parts per *billion*.

Fig. 15-2   A test model of the Viking GCMS analyzed Antarctic soil (left) and a piece of the Murchison meteorite, which contains amino acids (right). The graphs indicate that a rich variety of organic substances is contained in each of the two samples. Each peak in the graphs represents one or more organic substances. On this scale, the GCMS analysis of Martian soil would show a straight line at zero, with no peaks at all.

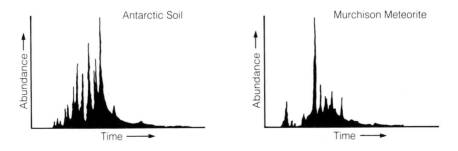

These negative results set powerful constraints on any models of Martian biology: How could life exist on Mars without leaving any trace of its presence? Could Martian organisms be such efficient scavengers that all traces of their wastes, their food, or their corpses could not be found, even with the great degree of sensitivity that the Viking landers brought to Mars? The biologists involved in the Viking project considered this an extremely unlikely possibility.[1]

The Viking Results: Atmospheric Analysis

Study of the composition of the Martian atmosphere also supported the view that no life exists on Mars. The Viking spacecraft did find that nitrogen forms between 2 and 3 percent of the atmosphere of Mars. Since nitrogen exchange between organisms and the atmosphere of Earth forms an essential part of life on this planet, the discovery of nitrogen in the Martian atmosphere seemed a hopeful sign to those who searched for life on Mars. But no evidence of any gases unexpected without the presence of life appeared; the upper limit on the methane abundance at the two landing sites fell below a few parts per million. The infrared spectrometers on the Mariner 9 spacecraft had already set a planetwide upper limit of 25 parts per billion of methane in the entire atmosphere. Thus, there seems little chance that extensive herds of cattle exist on Mars! The spectrometers also found no traces of the silicon analogue of methane, silane. (Silane gas would in fact be unstable in the Martian environment, so silicon-based life would bring us back to the possibility of fire-breathing dragons (see page 211).)

In view of the negative findings of the soil analysis and the equally negative returns from the atmospheric analysis, the prospects for finding life on Mars with the experiments designed for precisely that purpose seemed poor. So the Viking scientists received quite a shock when all three of the biology experiments showed positive results!

The Viking Biology Experiments

After considering many possibilities, the Viking project selected three experiments to search for life on Mars. These experiments, called the gas exchange (GEX), the labeled release (LR), and the pyrolitic release (PR), all reflect scientific experience with life on Earth. Thus, for example, all the organisms that we know derive their energy from two basic processes, *oxidation* (removal of hydrogen, combination with oxygen) and *reduction* (removal of oxygen, combination with hydrogen). Both of these processes

---

[1] An interesting parallel appears in the quests for the *yeti* ("abominable snowman") and the *sasquatch* ("Bigfoot"): Why have no skeletons been found?

deserve investigation on Mars. But how? What would the Maritans "eat" for energy, and under what conditions would they eat?

We knew *something* about the conditions on Mars, and the scientists assumed that life on Mars should be carbon based and should rely on water as a solvent. Thus, they judged it reasonable to offer the Martian microbes some suitable mixture of organic nutrients dissolved in water, to see what—if any—reactions would occur in this diet. The GEX and LR experiments followed just this procedure.

The GEX experiment put a soil sample in contact with several dozen "likely" nutrients (Figure 15-3). These nutrients have such broad appeal for Earth-based life that the GEX quickly became known as the "chicken soup" experiment. Once the soil sample came in contact with the chicken soup, the question was: Would the gas chromatographs detect any *changes* in the composition of the gas above the soil? Such changes could arise from the life processes of the Martian microbes. On Earth, the chicken soup approach would reveal the presence of life through changes in the amount of oxygen, carbon dioxide, or hydrogen in the air above the soil caused by the metabolic activity of organisms in the soil.

The labeled release (LR) experiment aimed at checking on biological activity more directly. This experiment used a set of compounds that were *labeled* by substituting radioactive carbon atoms for some of the ordinary carbon atoms in the compounds (Figure 15-3). This labeled mixture dripped onto the soil sample, and the gases above the sample were tested to see whether any radioactive compounds, such as carbon dioxide or methane, had been released by the life processes of Martian organisms. On Earth, we could call the LR a respiration experiment, to see whether organisms are releasing gases into the atmosphere.

The GEX and LR experiments have two basic drawbacks. First, liquid water cannot exist on Mars now, so Martian microbes may be completely unused to a watery medium. Second, organic nutrients found delicious by terrestrial organisms may be poison to Martian microbes. The third biology experiment, the pyrolitic release (PR), dealt with these drawbacks. In this experiment, Martian soil enjoyed an environment virtually identical to that on the planet's surface, but the atmospheric gases in the test chamber were labeled by adding carbon dioxide and carbon monoxide that had been tagged with radioactive carbon (Figure 15-3). After allowing any organisms to live for a while, the soil was heated to 750° C, and the volatile gases released by this heating passed into a vapor trap and then into a counting device.

Only carbon dioxide and carbon monoxide could pass all the way through the vapor trap. Once these two gases had left the system, the vapor trap was heated to drive off any organic vapors that might have been produced. These vapors could be recognized because they would contain some of the radioactive atoms from the added carbon dioxide or carbon monoxide. In other words, the PR experiment aimed at roasting the corpses of Martian microbes to release carbon atoms that the microbes had incorporated through biological activity. The most important terrestrial activity of this kind consists of plant photosynthesis, in which carbon dioxide in the Earth's atmosphere is converted (fixed) into organic compounds by green plants. Baking the plants vaporizes these organic compounds. In the PR experiment these vapors would be radioactive.

Fig. 15-3 A schematic representation of the GEX, the LR, and the PR experiments shows that each of them analyzes the Martian soil in a different way to test for the presence of life. The GEX experiment exposed a broth of nutrients onto a few grams of soil and then looked for changes in the gas above the soil-and-nutrient mixture. The LR experiment tagged carbon-rich compounds with radioactive carbon-14 atoms in place of some of the usual carbon-12 atoms. These labeled compounds then dripped over the soil sample. Any biological processes should have caused some tagged compounds to appear in the gas above the sample. The PR experiment replaced the normal Mars atmosphere with an equivalent set of gases labeled with radioactive carbon atoms. Any organism that ingested some of these labeled molecules would produce a radioactive signal when the soil in which they lived had been roasted.

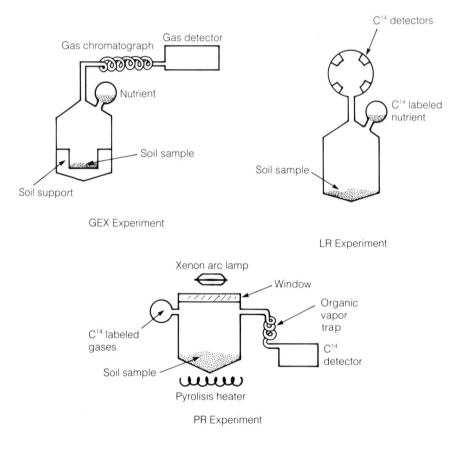

Tests of the PR experiment on Earth showed, however, just how difficult it might be to separate biological activity from chemical reactions. When biologists exposed a sterilized sample of simulated Martian soil to short-wavelength ultraviolet light in the presence of a mixture of gases that simulated the Martian atmosphere, they found that some simple organic molecules, chiefly glycol and formaldehyde, were formed. If the same chemical reactions occurred on the real Mars, they would exactly mimic the results from a simple photosynthetic organism that was turning carbon dioxide into more complex carbon compounds. Hence, the experimenters decided to filter out the short-wavelength ultraviolet light from the light source that simulated the sun inside the experimental chambers, in order to exclude the kinds of photochemical reactions that they had discovered in their laboratory tests.

Results of the Viking Biology Experiments

On the eighth day after the first landing on Mars, the lander scoop dug a trench in the soil and distributed samples to the various experiments. The GEX placed about a gram of soil into a tiny, porous container positioned above the nutrient medium. Two days later, the first analysis of the gas in the container showed an exciting result: A large quantity of oxygen had appeared in the chamber, 15 times the proportion in the planet's atmosphere! The simple exposure of Martian soil to the humidity in the test chamber (caused by the nutrient-laden fluid) had apparently been sufficient to liberate oxygen from the soil. Was this an indication of life on Mars? After months of testing, the biologists concluded that they were observing not biological activity but merely the chemical interaction of Martian soil with a higher pressure of water vapor than had been present for millions of years. In other words, not the "chicken soup" itself, but the humidity it produced, had led to oxygen-releasing chemical reactions in the Martian soil.

The day after the first GEX data, the LR experiment reported: again a positive result! After checking to be sure that the background radioactivity level was low, the LR added about two drops of the radioactive nutrient material to the soil. A sudden rise in the radioactivity of the gases above the soil sample appeared, a more dramatic reaction than biologists find with many life-bearing soils on Earth. Unfortunately for those who hoped for proof of life, the Viking scientists soon realized that the radioactive gas, almost certainly carbon dioxide, could arise from simple chemical reactions that involve peroxides. If, for example, hydrogen peroxide (H_2O_2) exists in Martian soil, it could easily react with an organic compound in the nutrient medium, such as formic acid (HCOOH), to form water and carbon dioxide. A second wetting of the soil showed *no* increase in the amount of radioactivity in the gas in the test chamber. In fact, the additional nutrient apparently absorbed some of the radioactive carbon dioxide that was orginally released. Hence, the scientists concluded not that life exists on Mars, but rather that the Martian soil may contain chemicals such as peroxides which release carbon dioxide when exposed to simple organic compounds.

Since the first two experiments yielded information that seemed ambiguous, the Viking team eagerly awaited the results from the PR experiment, which did not use a water-based nutrient and thus avoided one of the primary agents (water) that the biologists suspected of causing purely chemical reactions in the Martian soil. The PR experiment required an incubation time of five days, during which radioactive carbon monoxide and carbon dioxide stayed in the sample chamber. Analysis of the initial experiment revealed that radioactive carbon had indeed become part of compounds in the soil (Figure 15-4). Weak as this signal was, it seemed clearly positive: To the PR experiment, Martian soil behaved much like an Antarctic soil on Earth, nearly sterile but not entirely so. Thus, this experiment, apparently the most difficult to fool by nonbiological reactions, yielded an undeniably positive result. Yet even in this case the Viking scientists were skeptical that they had found life on Mars.

Fig. 15-4   The PR experiment showed a rise in the amount of radioactivity in the gases above the soil being tested for living organisms once C^{14}-labeled gases were introduced. This suggested that living creatures had incorporated some of the radioactive carbon atoms from the simulated Martian atmosphere.

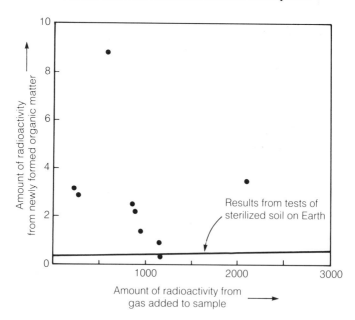

The reason for this skepticism comes from the fact that when the scientists arranged to heat the Martian soil to 175° C and to 90° C for several hours before the radioactive gases were injected, they still found positive results from the PR experiment. The higher temperature reduced the reaction by 90 percent, but a positive signature still emerged; the lower temperature had no effect. Since Mars's surface temperature never reaches even 30° C, the Viking scientists did not believe that any Martian life could have adapted to survive 3 hours at 175° C, which no terrestrial organisms can do. Furthermore, because the GCMS showed that the soil lacks any organic material down to a level of a few parts per billion, the likelihood that a tiny amount of the (supposed) heat-resistant microorganisms could provide the positive results of the PR experiment seemed minuscule.[2]

Nonetheless, all three biology experiments yielded positive results. If these results arise from chemical reactions that mimic the effects of microorganisms, as the Viking scientists have concluded, this simply shows the difficulties we face in trying to distinguish an alien form of life from a vast array of possible chemical reactions. It is a problem remarkably similar to that faced by the scientists on board *H.M.S. Challenger* a century earlier (see page 3). The consensus among the Viking biologists is that Martian soil contains loosely bound oxygen in various compounds, such as peroxides. Because the temperature on Mars is low, and because water is entirely absent from the planet's surface, these compounds can remain far from chemical equilibrium for long periods of time, but if we add even a small amount of water, we produce chemical reactions that mimic the effects of biological activity.

The Viking missions have revealed that Martian soil is chemically active, but probably contains no carbon-based living organisms, at least at the two landing sites. But could we have killed any Martian microbes by landing nearby or by bringing them inside the spacecraft? This seems unlikely, since any such microbes must have adapted to daily temperature variations of almost 100° C, and the temperature inside the spacecraft (27° C) did not exceed the maximum temperature on the planet's surface. The PR experiment, in fact, was designed to simulate the actual Martian environment as closely as possible. Could life on Mars be based on silicon? The GEX experiment, which showed that oxygen appears upon exposing Martian soil to moisture, might be a characteristic reaction of silicon compounds and water. The fact, however, that the reaction was found to die out rapidly following the initial exposure to moisture speaks against a life-based interpretation rather than a chemical one. (See Chapter 11 for other arguments against silicon-based life.)

---

[2]Further testing has suggested that a small amount of ammonia in the first PR soil sample—contributed by leakage from the descent engines on the Viking lander—may have been responsible for the soil chemistry that produced the first, weakly positive result.

Fig. 15-5 The presence of sand dunes on Mars shows that winds blow sand from place to place, but nothing resembling an animal's track or footprint has been seen.

We do have one device to detect life on Mars that remains completely *independent* of chemistry; namely, the cameras on the two Viking landers. The pictures from Mars have been searched with extreme care for any evidence of movement, any burrow, footprint, trail, or artifact of any kind, and nothing has been seen, except the effect of the Martian winds on the dust in the empty landscape (Figure 15-5). Since Mars's atmospheric composition, as well as the soil analysis, argues against the presence of life—especially of life in large quantities—the prospects for any type of life at the two landing sites must be judged poor, despite the tantalizingly positive results from the three biology experiments.

Did the Vikings Land in the Wrong Places?

But what about life elsewhere on Mars? The similarity between the soil analyses at the two landing sites, separated by more than 2000 kilometers, argues against great variations from one place to another. This argument gains strength from current models of the planet's surface, which suggest that the soil is turned over (gardened) by meteorite impacts and by wind down to a depth of several tens of meters, on time scales of the order of tens of thousands of years, and that the soil particles are blown over the whole surface of the planet. We can see this process in action during the intense Martian dust storms, which can obscure the entire planet. Thus, every part of Mars's surface seems likely to come in contact with every other part, so that sampling the dusty material at one place should be equivalent to sampling at all locations, as was indeed the case for the soil analyses made at the two Viking landing sites (see page 280).

A possible exception to this rule arises in the polar regions (Fig. 14-12.) Mars's north pole has a permanent cap of water ice, and we could imagine specialized environments at the edges of this polar cap where conditions could be more favorable to life during the peak of summer. A relatively abundant source of water, one that we know enters the atmosphere and then freezes again, lies in the polar cap. The ground temperatures there, as measured by the Viking orbiters, reach −33° C, about as warm as the highest temperatures measured near the second landing site, but not subject to extreme changes as Mars rotates. In fact, simply because the polar regions *stay cold,* organic material might concentrate there, since it could never heat up and blow away. The south polar region of Mars, not as well observed as the north polar cap, seems to be still colder than the north pole at this epoch, and keeps a small permanent cap of frozen carbon dioxide that may cover a layer of water ice. Thus, the north pole of Mars seems to offer the best high-latitude environment at the present time.

But "best" may not be good enough. Color pictures of Mars show that even near the pole, the bare ground shows the same reddish color as the rest of the planet. Here, too, the soil may be gardened, blown about by the winds, and certainly irradiated by ultraviolet light from the sun. Thus, this soil should be as thoroughly oxidized as the soil tested by the biology experiments made at lower latitudes. Models for the wind circulation on Mars show that the dust that becomes airborne in middle-latitude regions will be carried to the poles, where we can see layers of material that have (presumably) built up as the result of repeated cycles of erosion, airborne transport, and deposition of dust (Fig. 15-6.) Therefore, even here no living organisms should exist in the soil, which should resemble that already tested. Perhaps *under* the ice, at the pole of cold, we might expect

Fig. 15-6 Layered terrain on Mars apparently testifies to the build-up of material through cycles of dust deposition, erosion, and further airborne transport of dust particles.

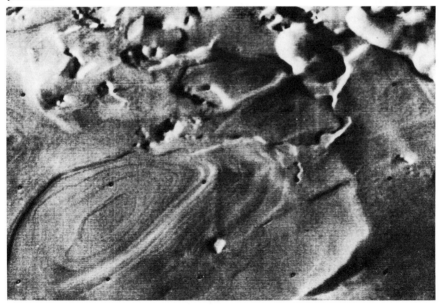

to find some protected carbon compounds, but here the continual low temperatures seem to rule out the existence of active life.

We might imagine isolated "oases" that could harbor life on Mars, but we must remember that every oasis requires a well-populated *external* community for its continued existence. As the sands shift to cover an oasis or dry spells exhaust it, seeds and microorganisms must arrive from somewhere else to repopulate it; otherwise, without this outside reservoir of life, any oasis must soon become lifeless.

Dogmatic statements have fared poorly in the history of science, and we must be reluctant to state categorically that no life exists on Mars. The evidence we have accumulated so far points toward the absence of life, though some new discovery may someday appear to convince biologists that Mars does harbor some living organisms. We *can* say that no known terrestrial organisms, including the toughest of all microbes, could grow in the present Martian environment.

Were conditions always this hostile on Mars? Perhaps no life exists on Mars *now,* but some future mission might still find traces of life's early beginnings there. As we have seen in Chapter 9, we have good reason to believe that primitive conditions on Mars resembled conditions on the primitive Earth. We also have a record of a more hospitable climate long ago on Mars. So life may have begun, only to die out as the Martian climate changed and the planet's atmosphere approached its present low density. We should search for evidence of these early beginnings in the oldest sedimentary rocks on Mars and by probing the walls and floors of the

Fig. 15-7 Future Mars missions might use automated rovers that could roam through dried-up stream beds, taking samples, analyzing them, and sending pictures and data back to Earth by radio.

sinuous channels, just as we look for evidence of microbial fossils in the oldest rocks on Earth. To do this, we need "rovers" that can move through the Martian canyons, beaming pictures back to Earth and taking samples as directed (Figure 15-7). In the best case, we should bring back samples to our laboratories, where we can study them in detail. Missions that would involve this sort of sample return from Mars are now under consideration by NASA for possible development during the late 1980s.

Why Is Mars So Different from Earth?

Meanwhile, we can try to understand why the Mars we see today differs so much from our own planet. The combined effects of this planet's smaller size and its greater distance from the sun have apparently kept Mars from producing an environment suitable for biological development. If Mars were larger, it might have accumulated a greater supply of the volatile elements, such as carbon and nitrogen, that are essential for the existence of a dense, stable atmosphere within the inner solar system. A larger planet would have released (outgassed) more volatile elements from a greater amount of tectonic and volcanic activity, driven by its greater internal heat. This hypothetical Mars would no longer show a surface pockmarked by craters, for the primitive crust would have disappeared, just as it has on Earth. But would the thicker atmosphere have managed to maintain itself over billions of years? This would depend on how large a greenhouse effect Mars would have established to counteract the lower temperatures that arise at its distance from the sun, and on how well a hypothetical Martian "Gaia" could maintain conditions suitable to its purposes.

On the real, small Mars, the early, denser atmosphere that seems to have permitted the existence of liquid water might once have been almost as thick as our own atmosphere. Made primarily of carbon dioxide, this early atmosphere began to disappear as the carbon dioxide was converted into carbonate rocks through the action of the running water, while the molecules of nitrogen broke apart into atoms and escaped from the planet. Thus, the activity of water on Mars proved self-limiting, for the atmosphere eventually became too thin for liquid water to exist, precisely because liquid water had existed. Without extensive tectonic activity, there was no means to feed the carbon dioxide back into the Martian atmosphere once it was locked up in limestones. Any organic material contributed to the surface layers by meteoritic and cometary impact or by atmospheric photochemistry has long since been oxidized into carbon dioxide by the action of ultraviolet light, unhindered by atmospheric ozone. And so we find Mars as it is today, with a thin atmosphere, with evidence for subsurface water in the form of permafrost, with water and carbon dioxide freezing at the poles, with nitrogen and hydrogen slowly escaping into space—and with no sign of life.

Summary

The two Viking spacecraft that landed on Mars in 1976 have conducted thorough chemical and biological analyses of the soil and atmosphere at the two widely separated landing sites. The results of these studies include evidence that suggests the presence of life on Mars, but a closer examination of this evidence reveals that nonbiological processes are more likely to be the mechanisms that give the observed responses than true life.

First of all, we should recognize the fact that nothing that resembles life has been *seen* by the Viking cameras, so that our hopes for life on Mars rest with microscopic organisms too small to be visible. Second, the analysis of Martian atmosphere and soil shows nothing we would judge "typical" of life; instead, the soil and atmosphere resemble environments drier and colder than the driest deserts on Earth. In particular, no traces of methane gas have appeared, down to a planetwide limit of 25 parts per billion in the Martian atmosphere.

The Viking landers each performed three experiments to test for living organisms directly. The first of these, the gas exchange (GEX) or "chicken soup" experiment, put a sample of Martian soil in a bath of nutrients thought to be favorable to life. The labeled release (LR) experiment dripped compounds labeled with radioactive carbon atoms onto the soil sample, to see whether the soil would produce any (radioactive) compounds typical of life. The pyrolitic release (PR) experiment also labeled carbon atoms, but this time within the Martian atmosphere; after giving any microbes in the soil a chance to interact with this labeled atmosphere, the soil was roasted to see if it now contained any of the labeled carbon monoxide and carbon dioxide.

Impressively enough, all three of these experiments gave results that might signal the presence of life: The GEX showed the release of a large quantity of oxygen; the LR revealed an increase in radioactive compounds above the Martian soil; and the PR showed a positive reaction, similar to that in relatively sterile Antarctic soil. However, when the analysis of Martian soil showed the total *absence* of any organic material down to a level of a few parts per billion and less, the Viking scientists took a harder look at the three life-detecting experiments. They concluded that inorganic, nonbiological chemical reactions, such as those we would expect if Martian soil contains peroxides, could produce the results found by the GEX, the LR, and the PR experiments. This does not prove that no life exists on Mars—not even that no life exists at the two landing sites—but, taken together with the negative results from the chemical analysis of Mars's soil and atmosphere, the fact that peroxides could mimic the effects of organic processes in the GEX, LR, and PR experiments does suggest that we would have to be immensely optimistic to conclude that the Viking experiments have found life on Mars.

The remaining best hope for finding life, or fossil evidence for former life, on the red planet seems to be the dried-up river beds far from the Viking landing sites. Here we might someday find fossil forms of the life we have been searching for on Mars.

Questions

1. How could an intelligent Martian detect life on Earth? Name at least four different methods for such detection.

2. Why do we think that microbes are more likely to exist on Mars than large plants and animals?

3. The GCMS experiment on the Viking landers failed to find *any* organic compounds in Mars's soil. Does this prove that life on Mars does not exist at the landing sites? Why?

4. What does the absence of methane from the atmosphere of Mars down to a detectable level of 25 parts per billion tell us about the possibility of animal life on Mars?

5. Why was the GEX (gas exchange) experiment on the Viking landers called the "chicken soup" experiment? What did this experiment find out about the Martian soil?

6. Why did the LR (labeled release) use radioactive carbon atoms? Did any of these atoms appear in the atmosphere immediately above the soil sample? What does this prove?

7. The PR (pyrolitic release) showed that some carbon atoms in carbon monoxide and carbon dioxide gas became "fixed" in the Martian soil. Does this suggest that life exists on Mars?

8. How would you estimate the chance that the two Viking landing sites happen to lie on parts of the planet that are particularly hostile to life?

9. Why might the regions at the edge of the Martian polar caps be more favorable to life than regions closer to the Martian equator?

10. Why do we think that Mars would have a thicker atmosphere if it were a larger planet? What effects would a thicker atmosphere have for the chances of life on Mars?

Further Reading

Horowitz, Norman. 1977. The search for life on Mars. *Scientific American* 237:5, 52.

Klein, Harold. 1977. Where are we in the search for life on Mars? *Mercury* 6 (Mar./ April):2.

Sifting for life in the sands of Mars. 1977. *National Geographic* 151:1, 9.

Washburn, Mark. 1977. *Mars at last!* New York: G. P. Putnam's Sons.

CHAPTER 16

The Giant Planets
and Their Satellites

The four giant planets in our solar system, Jupiter, Saturn, Uranus, and Neptune, differ greatly from the rocky inner planets with which we are more familiar. Because of their enormous masses, the giant planets can maintain atmospheres rich in hydrogen and the other light elements, and thus are similar in composition to the primordial solar nebula from which the entire solar system condensed. This seems especially true of Jupiter, the largest and nearest of these bodies. Jupiter thus provides an enormous natural laboratory in which we can test our ideas about the chemical reactions that we think occurred in the early solar system.

The large masses of these bodies also help them maintain extensive satellite systems, rings, and starlike interiors. If we could penetrate the atmospheres of the giant planets with probes to take us to their deep interiors, we would find that these planets do not have solid surfaces. Instead, the enormously deep atmospheres become denser and denser until they gradually merge into solid matter.

At least in the case of Jupiter, this strange interior structure generates a powerful magnetic field. When astronomers first began to observe the planets with radio antennas, they were startled to discover that Jupiter emits intense radio waves, especially at long radio wavelengths (low frequencies). At wavelengths of tens of meters, similar to the shortest-wavelength radio used for terrestrial communication, Jupiter sometimes emits more photons than any object in the solar system, including the sun. The explanation of Jupiter's intense radio emission comes from the planet's strong magnetic field, more than ten times stronger than the Earth's. Jupiter rotates more rapidly than any other planet, in less than 10 hours, and as it rotates, its magnetic field sweeps charged particles along, close to the planet. Some of these particles produce short-wavelength radio emission through the synchrotron process (see page 58), while others produce long-wavelength bursts through interaction with the planet's ionosphere. These bursts seem to be triggered by the motion of Jupiter's inner large satellite, Io, in ways that we do not yet understand. During its passage close to Saturn in 1979, the Pioneer 11 discovered that this planet also has a magnetic field, though not as strong as Jupiter's.

The Composition of the Giant Planets

To show how primitive the giant planets are in their elemental abundances, we can perform a simple computer experiment: Consider a mixture of the chemical elements in the same proportions that exist in the sun and in other stars, and let these elements combine with one another to form molecules in every possible way. If we specify an approximate pressure and temperature, we can model the conditions in the atmospheres of the giant planets as they formed from the original cloud of gas and dust that made the solar system. The chief compounds that we predict from this exercise match the ones that dominate the atmospheres of Jupiter and its neighbors: methane, ammonia, water vapor, and everywhere an excess of hydrogen. (Helium must also be highly abundant, and neon should be present in about the same proportion as ammonia, but because these two gases do not participate in chemical reactions, we do not need to pay them much attention in this discussion.)

These compounds reveal how the outer planets differ from the four inner planets. Venus, Earth, and Mars have *secondary* atmospheres, produced by the outgassing of volatile compounds from the rocky material that formed the planets. Jupiter, Saturn, Uranus, and Neptune appear to have *primordial* atmospheres, made from the primitive matter in the solar system, which they have retained since the time of their formation 4.6 billion years ago. At least for Jupiter, the best-studied of the giant planets, the atmosphere may even be identical in its elemental composition with the sun (Figure 16-1).

Many scientists have considered this hydrogen-rich environment to be representative of conditions on primitive Earth. But we must emphasize that the analogy between Jupiter and early Earth suffers from several important differences between these planets. First, Jupiter's enormous mass, 318 times Earth's mass, keeps hydrogen from escaping, so Jupiter will *always* have a huge number of hydrogen atoms to combine with any elements or molecular fragments. In contrast, Earth rapidly lost its hydrogen, either the original hydrogen from its formation or the hydrogen produced by the dissociation of molecules, a loss which produced an environment more suitable for the evolution of a complex biochemistry. Second, Jupiter does not have a solid surface, and thus has no likely microenvironments,

Fig. 16-1 Jupiter, 11 times the diameter of Earth, shows banded atmospheric patterns parallel to the planet's direction of rotation. The Great Red Spot, 25,000 kilometers long, has persisted for at least the past few centuries.

such as Earth's tidal pools or transient ponds, in which the products from chemical reactions in the atmosphere could become concentrated. Nor does Jupiter offer the opportunity for chemical reactions to be catalyzed by soil surfaces, such as the clay minerals that apparently played this role on Earth.

In addition to these key differences between Jupiter and Earth, another important problem for the origin of life arises from the vertical convection in Jupiter's atmosphere (Figure 16-2). This convection creates a circulation pattern between the relatively cool upper regions and the lower atmospheric levels, where temperatures stay hot enough (above 700° C) to destroy complex molecules. The circulation most likely occurs in a time much shorter

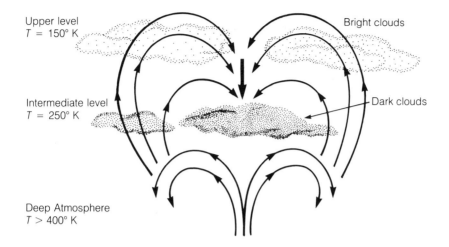

Upper level
$T = 150°$ K

Bright clouds

Intermediate level
$T = 250°$ K

Dark clouds

Deep Atmosphere
$T > 400°$ K

Fig. 16-2 Vertical convection currents carry material upwards from the warmer layers of Jupiter's atmosphere to form the bright clouds. Where the currents descend, we can see darker clouds at lower levels in the atmosphere.

than a year, so any large molecules formed in the upper atmosphere would be broken apart by collisions at relatively high temperatures within this span of time. An intermediate level exists where conditions are better. Scientists estimate that water vapor will condense to form clouds in a region with a temperature of about 27° C and a pressure just a few times greater than the surface pressure on Earth. But the gas in this region will be in constant motion, circulating between the upper and lower regions, so at this level, water vapor continually condenses and evaporates. Higher in the atmosphere, ammonia or ammonia-sulfur compounds take the place of water vapor, forming the clouds we see from Earth. Wherever liquid droplet clouds are present, thunderstorm activity is likely because of the strong vertical convection that can lead to charge separation. Indeed, lightning discharges within the Jovian clouds were detected by the Voyager spacecraft in 1979.

The Great Red Spot offers an interesting possible exception to the rule of rapid destruction of the complex molecules that might form in Jupiter's upper atmosphere. If this enormous feature turns out to be some type of giant vortex—a Jovian analogue of a terrestrial hurricane—as some scientists have suggested, then it may contain a region of large-scale, long-lived updraft, within which particles of the proper size could remain suspended for decades. If so, molecules could avoid destruction by the heat of the lower atmospheric levels. This intriguing possibility adds to the general interest in this enigmatic feature of Jupiter's atmosphere and suggests that the Great Red Spot will repay close-up scrutiny from spacecraft. Suitable images have become available within just the last year (Figure 16-3).

Why does the temperature rise as we descend into Jupiter's ever-thickening atmosphere? The reason, remarkably enough, is that the planet has an internal source of heat, which generates more energy each second than the planet receives from the sun. This heat probably arises from the slow contraction of Jupiter's interior, similar (though on a much smaller scale) to the process that eventually made the sun's interior hot enough for nuclear fusion reactions to begin. But Jupiter, with a thousandth of the sun's mass, will *never* grow hot enough inside to become a star; Saturn, Uranus, and Neptune, though capable of generating some internal heat, have less mass than Jupiter and produce still less heat. Pluto is apparently a small, icy body, more like one of the larger satellites of Saturn than a giant planet.

Chemistry on the Giant Planets

Despite the extreme differences between Jupiter and Earth, we remain interested in this giant of our solar system as we consider the ways in which

Fig. 16-3 The Great Red Spot of Jupiter was photographed by Voyager 1 in March 1979. The smallest details in the picture are about 50 km across.

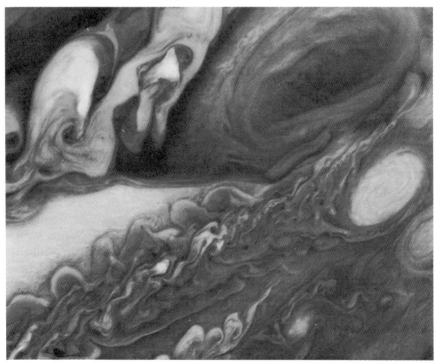

life can begin. Jupiter gives us a chance to study conditions similar to those that existed 4.6 billion years ago as the solar system formed. This is as if we were to encounter a primitive tribe of humans just as they were developing language. What sounds would be identified with what objects? How would such decisions be made? How and when would concepts such as time, space, and love be expressed? If we could observe without being noticed, we could expect to learn about our own history by watching the experiments of these people. In much the same way, we hope to learn something about possible pathways for prebiological chemistry in the primordial solar nebula by studying chemical reactions on Jupiter.

We know that such reactions are presently occurring for several reasons. First, astronomers have discovered that in addition to the gases that we have already listed, Jupiter's atmosphere contains traces of substances that our computer model does not predict, including carbon monoxide (CO), acetylene (C_2H_2), and ethane (C_2H_6). These gases cannot continue to exist in Jupiter's environment unless they are continually produced by reactions among the other atmospheric constituents, since they rapidly interact with hydrogen to form methane (CH_4). Ultraviolet solar photons provide the energy to form C_2H_2 and C_2H_6 from CH_4 in Jupiter's upper atmosphere, while thermal energy from the planet's interior forms CO from CH_4 and H_2O at deeper levels, far below the visible clouds. Are there more complex molecules as well?

Our second clue suggests an affirmative answer: Jupiter's clouds do not have the white or gray colors that we expect to see when sunlight is reflected from frozen water or ammonia. Instead, the clouds show various subtle shades of color, the most famous example being the salmon-colored tint of the Great Red Spot (Figure 16-3). What substances produce these colors? How were they formed? Do these chemical reactions resemble the prebiological reactions that must have occurred on the primitive Earth?

These key questions remain unanswered. At this time, two competing theories strive to provide an explanation of Jupiter's colors. On one side, we find those who believe that all of Jupiter's coloration can be explained by *inorganic* compounds that arise from atmospheric chemistry. The yellows and browns could be caused by sulfur atoms, either in combination with ammonia or combined with each other to form a pure substance. The Great Red Spot might owe its redness to the presence of red phosphorus that could be produced by chemical reactions from the gas phosphine (PH_3), which is known to be present on Jupiter. The opposing side of the color controversy points to laboratory simulations of Jupiter's atmosphere in which compounds with the colors of Jupiter were produced by shining ultraviolet light on an appropriate mixture of methane and ammonia, or by subjecting these gases to some other source of energy such as an electric spark, simulating Jovian lightning. These experiments invariably produce a wide range of organic compounds, some of which have the colors of

Fig. 16-4 This photograph of Saturn, taken by the Pioneer 11 spacecraft in 1979, distorts the planet slightly, owing to the preliminary computer processing of the image. Saturn's magnificent rings appear dark, because the sun and the spacecraft were on opposite sides of the rings, thus giving a view of the rings rarely seen from Earth.

Jupiter's clouds.[1] It seems quite possible that both types of chemistry are actually occurring on Jupiter, and each may contribute compounds that cause the colors we observe. What other compounds are being formed we can only guess at present, but it certainly seems likely at this stage of our knowledge that we shall find at least the preliminary stages of prebiological organic chemistry on Jupiter as we explore this planet more thoroughly.

We have concentrated on Jupiter in this discussion because it is the giant planet about which we know most. But real differences seem to exist among these giants. Beautiful as it is, Saturn does not have colorful clouds like those of Jupiter, and certainly nothing like the Great Red Spot (Figure 16-4). Instead, we seem to confront a thick layer of ammonia cirrus, which completely covers the lower, warmer atmosphere that might in fact have characteristics like Jupiter's. These dense clouds arise from the lower temperature of Saturn's upper atmosphere, the result of the planet's greater distance from the sun. The visible levels of the atmosphere of Uranus and Neptune are so cold that we can't even find evidence of ammonia, which is probably frozen out at low altitudes. What happens at still lower levels we do not know, but radio emission from these regions shows that they are warmer than 0° C.

Could Life Exist on the Giant Planets?

We have seen that the outer planets offer interesting environments for the study of prebiological chemistry. This study, however, seems a long way from the discovery of life itself. Can we be sure that life does not exist on the giant planets? No! So say Carl Sagan and Edwin Salpeter, two well-

[1]The famous Miller-Urey experiments to simulate conditions on the primitive Earth (page 174) resemble what must be taking place on Jupiter now. In the laboratory reactions, various colored compounds emerged from the gases that modeled Earth's primitive atmosphere.

known astronomers from Cornell University. Sagan and Salpeter argue that since we really don't know how life began on Earth, we can't specify the necessary conditions for life to appear on a planet as different from us as Jupiter. Once life begins, living organisms themselves can regulate their environment. Quite possibly, therefore, living organisms on Jupiter or Saturn would be able to overcome some of the obstacles to life that we have described. Giant gas-bag creatures that use hydrogen to maintain their buoyancy *might* be drifting or flying in Jupiter's upper atmosphere as you read this, participants (perhaps) in a Jovian ecology within which some species prey on others. Jupiter furnishes the most likely home for such life forms, because Jupiter appears to have the most chemical activity of the four giant planets. At the simplest level of speculation, however, we can imagine similar creatures floating in the atmospheres of any of the giant planets.

Unfortunately, this speculation does not advance our knowledge. Our increased understanding of life on Earth, and our inability to find life on Mars, should leave us a bit more conservative in speculating on the likelihood of finding life on every celestial object, life that would have adapted uniquely to that particular environment. In close-up studies of the sun's planets, we are so far one for three in finding life. With this record, we cannot argue that *every* planet should have life upon it, or within it. And yet, in the face of our basic ignorance concerning both Jupiter's environment and life's universality, we ought not discard these intriguing possibilities for life on Jupiter too quickly.

Rings and Satellites

While we are still in a speculative mood, let us not restrict ourselves to the planets alone. Each of the giant planets has an extensive satellite system, ranging in number from Neptune's two moons to Jupiter's fourteen (or more!). Saturn, Uranus, and Jupiter have systems of rings; Saturn's are easily visible with a small telescope, while the rings of Uranus remained undiscovered until 1977, when they blocked the light of a star just before and after the occultation by the planet itself. They are much darker than the rings of Saturn, and they share the unusual orientation of Uranus itself, which keeps the planet's poles of rotation nearly in the plane of the orbit (Figure 16-5). The rings of Jupiter were discovered by the Voyager spacecraft in 1979. They are also much harder to see than the rings of Saturn (Figure 16-6).

Each of these ring systems lies close enough to its planet that the difference in the gravitational attraction exerted by the planet on two adjacent particles in the ring is greater than the gravitational attraction of the particles for each other. The boundary within which this condition exists is known as the *Roche limit,* after its discoverer, Edward Roche. The rings

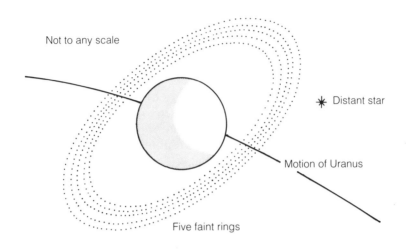

Not to any scale

✳ Distant star

Motion of Uranus

Five faint rings

Fig. 16-5 The rings of Uranus, much thinner than Saturn's, were discovered when Uranus "occulted" a star. As the planet passed in front of the star, astronomers saw the star disappear and reappear five times before the planet's disk hid the star. The same effect was observed when the star reappeared on the other side of Uranus. More than the original five rings have now been discovered.

thus consist of material that could not accrete to form a satellite because it was too close to the planet, whose disruptive force kept any agglomeration of matter from growing. The millions of tiny satellites—rocks and boulders—that constitute the rings we see today are kept in stable orbits by the interaction of the gravitational forces from the planet and its satellite systems. These boulders don't disintegrate into dust for the same reason that our artificial satellites can survive in their orbits within Earth's Roche limit: The forces that hold together a silicate lattice (or stainless steel!) are far stronger than the disruptive effects of gravity.

Our knowledge of the number of objects that circle the outer planets, not to mention their structure and composition, remains incomplete. Neptune may have a faint ring system undetected by human beings. Jupiter and Saturn almost surely have more satellites than we know of now, since one or two have been glimpsed and lost again during recent years of observation. The same may be true of Uranus and Neptune. An asteroidal object called Chiron, with an orbit that lies mostly between that of Saturn and Uranus, was discovered by Charles Kowal in 1977, while James Christy presented evidence in 1978 that Pluto has a satellite whose orbital period equals the planet's period of rotation.

Let us consider the satellites that we do know. The four largest moons in Jupiter's family were first recorded by Galileo, whose contemporary

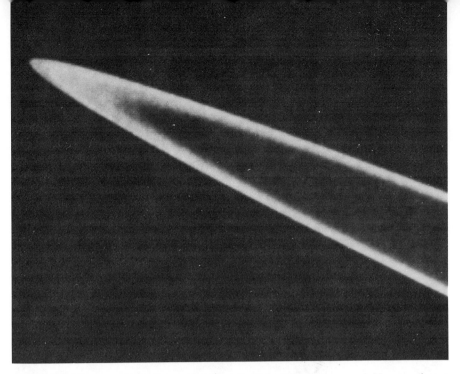

Fig. 16-6 Voyager also discovered that Jupiter, like Saturn and Uranus, has a system of rings, tiny particles in orbit around the planet.

Simon Marius named them after four of Jupiter's illicit loves: Io, Europa, Ganymede, and Callisto. These four satellites are large objects with solid surfaces (three are larger than our moon), and they imitate the gross characteristics of the solar system rather well: The innermost moons, Io and Europa, have densities of matter similar to those of the inner planets, while Ganymede and Callisto represent the outer planets, with low densities implying a high abundance of ice in their interiors (Table 16-1). Amalthea, even closer to Jupiter than Io, remained undiscovered until 1892. It fits this pattern, being evidently a small, rocky object about the size of Long Island (Figure 16-7). A still closer, smaller moon was discovered by Voyager 2.

TABLE 16-1

THE FOUR LARGE ("GALILEAN") SATELLITES OF JUPITER

Satellite	Distance from Jupiter (km)	Diameter (km)	Average Density of Matter	Inclination of Satellite Orbit to Jupiter's Equator
Io	422,000	3640	3.53	0°.0
Europa	671,000	3130	3.03	0°.5
Ganymede	1,070,000	5280	1.93	0°.2
Callisto	1,885,000	4840	1.79	0°.2

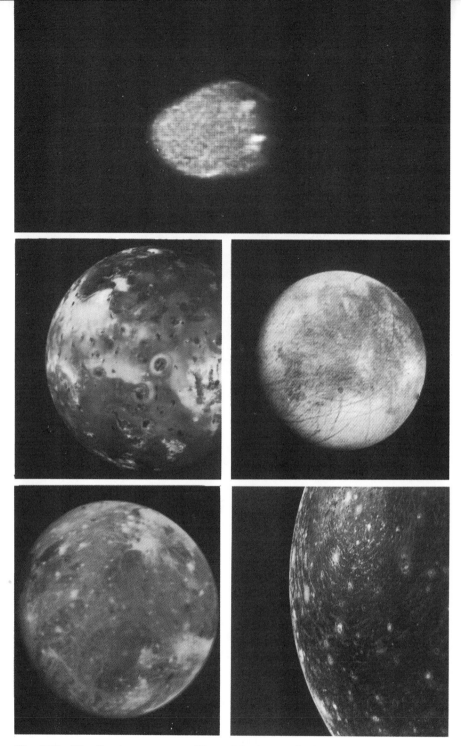

Fig. 16-7 The five innermost satellites of Jupiter each showed a different appearance to the Voyager cameras in 1979. Amalthea, the innermost satellite, is far smaller than the next four moons, Io (left center), Europa (right center), Ganymede (bottom left), and Callisto (bottom right). The moons are not shown to the correct relative scale.

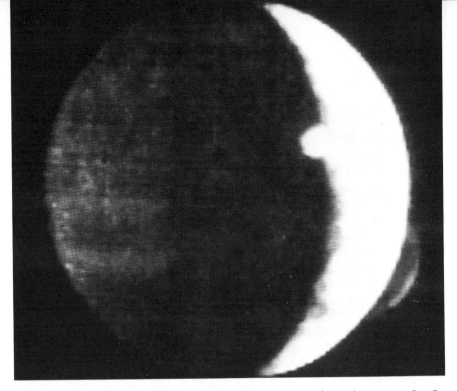

Fig. 16-8 This photograph from Voyager 1 shows two active volcanoes on Io. On the lower right, just at the satellite's limb, ash clouds are rising more than 260 kilometers above Io's surface. The second volcano appears as an irregular extension of the line between light and dark on Io, where a volcanic cloud is catching the rays of the rising sun. We can see the dark side of Io dimly because Jupiter reflects sunlight onto it.

All five of these objects received intensive scrutiny with the Voyager spacecraft in 1979. Surprises abounded, including the lack of large impact craters on Callisto, evidence for plate motions on the crust of Ganymede, a complex network of lines on Europa, and above all, the active volcanoes of Io (Figure 16-8). Callisto and Ganymede have an appearance consistent with a high proportion of ice in the interiors of these objects, while the lines on Europa may be caused by the cracking of the surface in response to internal pressure. This pressure may come from the same cause as the volcanoes of Io—a molten interior produced by the dissipation of tidal forces from giant Jupiter. On Io this tidal force, the result of Io's proximity to Jupiter, is so extreme that the crust of the satellite may be just a few tens of kilometers thick. Dark spots on the surface have been interpreted as lakes of liquid sulfur. This moon is surrounded by a cloud of sulfur, sodium, and potassium atoms that can be seen from Earth.

These satellites are fascinating little worlds which we would be sure to visit on a trip to Jupiter, but we must admit that they do not help us much in our search for life and life's precursors. Despite its volcanic activity, Io is disappointing in its lack of a thick atmosphere; it must have lost all its water and the only volatile now found to emerge from its volcanoes is sulfur dioxide. The other moons of Jupiter are so much smaller than the big four that they arouse little interest for our specific purposes.

Saturn's ten known satellites are a varied group. The six inner moons have low densities and high surface reflectivities that once again suggest the presence of large amounts of water ice. But then we encounter Titan, largest of all satellites, which exceeds the planet Mercury in size (though not in mass). Titan is the only moon in our solar system with a substantial atmosphere, and it also has an unusually red color. Here we have an object with an atmosphere *and* a solid surface, just what we have been looking for in our search for life in the outer solar system. Unfortunately, despite intensive efforts, we still know precious little about this satellite.

We do know that Titan's atmosphere contains methane, that it is unusually warm in its upper reaches (but still a chilly $-113°$ C), and that some clouds or haze obscure the view of Titan's surface. Measurements made with the Very Large Array of radio telescopes (see cover photograph) suggest that Titan's surface temperature is about $-190°$ C. We don't know how much methane exists, why the atmosphere is warm, or what other gases (if any), besides traces of acetylene and ethane, make up Titan's atmosphere. Some other compounds must be present because Titan's reddish color suggests the occurrence, as in Jupiter's case, of interesting chemical reactions. This color may arise from a photochemical smog whose composition, if we only knew it, could tell us more about the kinds of chemical reactions that occur in primitive environments. Titan's special appeal from this viewpoint comes from the fact that products of the atmospheric chemistry will accumulate on the satellite's surface, whereas in the giant planets such compounds are destroyed by the high temperatures of the lower atmosphere. We expect to obtain much more information about this intriguing object from Earth-based studies during the next few years. But our best hope for advancing our knowledge of Titan will come in 1980 and 1981 with the visits from the Voyager spacecraft.

Saturn has at least nine other satellites besides Titan, but only one of these draws our special attention: Iapetus. This unusual moon is the only object in the solar system which we might seriously regard as an alien signpost—a natural object deliberately modified by an advanced civilization to attract our attention, and to leave us no rest until we have deciphered its meaning.[2] What is this apparent modification of Iapetus? Take a look at Figure 16-9. Iapetus changes its brightness by a factor of *seven* as it

[2]This idea was incorporated in Arthur C. Clarke's novel and screenplay *2001: A Space Odyssey.*

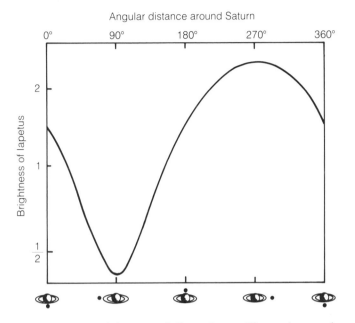

Fig. 16-9 Iapetus, second largest of Saturn's satellites, changes its apparent brightness by a factor of 7 during the course of each orbit around the planet.

moves around its orbit. In other words, one side of this satellite appears to be seven times darker than the other. No other natural satellite comes close to this variation in light, not our moon with its familiar mottled surface, nor Io with its volcanoes, nor Titan with its methane atmosphere. One explanation for this brightness variation suggests that frosts are more common on one side of Iapetus than on the other, because encounters of the satellite's leading hemisphere with debris, or with charged particles, have worn the frosts away, exposing dark rock. This would certainly explain the observed variation in brightness but we can't help wondering whether the signpost hypothesis deserves checking, so that a trip to Iapetus would reveal an obelisk with . . . ? Speculation is fun, but getting answers can be thrilling, and we may be able to solve this enigma when spacecraft fly past Iapetus in 1980 and 1981, sending us pictures of both the bright and dark hemispheres of Saturn's ninth moon.

Spacecraft to the Outer Solar System

During the early 1970s, the United States sent two spacecraft, Pioneer 10 and Pioneer 11, on journeys of several years to Jupiter and (in the case of

Pioneer 11) on to Saturn. The two Pioneer spacecraft are continuing to travel away from Earth on trajectories that will make them the first objects made by humans to leave the solar system. In 1979, the Voyager spacecraft, launched from Earth in 1977, passed through Jupiter's satellite system, measuring the planet's magnetic field and mapping the concentration of charged particles around it, all the while taking pictures and making spectroscopic observations of the planet and its satellites. The best pictures show details less than 1 kilometer across on the satellites, and about 10 kilometers across on the planet. Project Voyager has thus opened new worlds to us, for we have never before seen such detail on Jupiter or its moons. We now have our first opportunity to study small regions of Jupiter carefully, regions where the colors are most intense, and we may at last begin to resolve the riddle of the Great Red Spot's origin, structure, and composition. In 1987, the Galileo spacecraft will reach Jupiter. This advanced instrument package will consist of an orbiter to study the satellites and atmosphere of the largest planet, and a probe to descend deep into the atmosphere, measuring composition, pressures, temperatures, and lightning discharges.

Two years after the Voyagers sail past Jupiter, the spacecraft will arrive in the vicinity of Saturn (Figure 16-10). Far out from our home, almost 10 times Earth's distance from the sun, the Voyagers will survey the planet's satellites and its rings, as well as the great ball of Saturn itself. If all goes well, one of the two spacecraft will then head still farther outward to Uranus, and by 1986, eight full years after leaving the Earth, this spacecraft will study the seventh planet, 20 times farther from the sun than our blue-green haven of life.

Past Uranus, there may be an opportunity to visit Neptune in 1989. Then both Voyagers will continue to coast outward from the sun, eventually leaving the solar system completely. Like their predecessors, Pioneer 10 and Pioneer 11, these emissaries from Earth carry messages from the creatures that made them. Instead of the Pioneer plaques, each Voyager spacecraft carries a long-playing phonograph record, anodized in gold for protection against erosion by interstellar dust particles. If another civilization should someday find the spacecraft and realize that the records carry not only sounds but pictures as well, coded into the equivalent of audio signals, they could listen to sounds from Earth and view pictures of our planet and its inhabitants (see page 322).

The two identical records each present about a hundred scenes from Earth, showing landscapes, human activities and development, the oceans, various animals and plants, and a few astronomical objects for easy recognition. The sounds on the record span an array of ethnic music, as well as greetings in eighty different languages, and selections from Bach, Beethoven, Stravinsky, and Chuck Berry. With the Pioneer and Voyager messages into space, we seem to be repeating ancient cultural patterns by creating artifacts that will preserve our accomplishments beyond the time span of a human life. Despite the tiny probability that such objects will

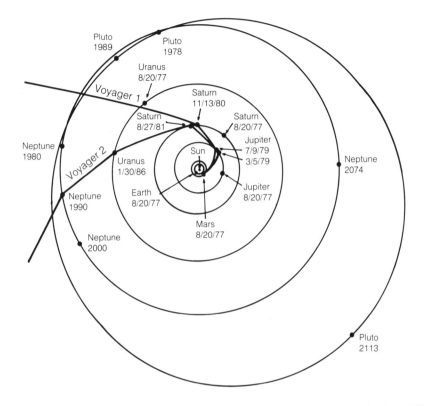

Fig. 16-10 Because of a favorable line-up of the giant planets during the late 1970s and early 1980s, the Voyager spacecraft can pass close to Jupiter, Saturn, Uranus, and Neptune with only minor course corrections.

ever be recovered, we send our monuments into space like bottles cast adrift in a celestial ocean. Do we hope that some beachcomber will find our spacecraft and read their messages? And are we ready for a reply?

Summary

The four giant planets, Jupiter, Saturn, Uranus, and Neptune, each far exceed the mass and size of Earth, the largest of the four inner planets. Unlike the rocky composition of Mercury, Venus, Earth, and Mars, the four giants are spheres of gas, mostly hydrogen and helium, which resemble the stars and the rest of the universe in their composition far more than the inner planets do.

Jupiter, Saturn, and Neptune each have so much mass that their slow contraction, the result of their self-gravitational forces, releases significant amounts of heat. Uranus may also exhibit this effect, but to a much smaller degree. The giant planets owe their gaseous, hydrogen-helium composition to the fact that they have retained much of their original matter. In contrast, the smaller inner planets have relatively thin secondary atmospheres, released by volcanic activity after the hydrogen and helium had escaped. The giant planets' initially great masses, together with their greater distances from the sun, have allowed this fundamental difference to persist.

Jupiter, being both the largest and the nearest of these objects, has received the closest scrutiny. In addition to hydrogen and helium, its atmosphere contains methane, ammonia, and water vapor, just the gases we would predict for an object that had a composition determined by the ingredients in the primordial solar nebula. But we also find colored material in the cloud layers, indicating that more complex substances are being formed by ultraviolet light, lightning discharges, and thermal energy. Some of these compounds could be similar to those formed on the early Earth or delivered to our planet's surface by comets and meteorites before the origin of life. The most famous example of a colored region is the Great Red Spot, whose true nature remains unknown. The lower atmosphere of Saturn may exhibit similar phenomena, but they are hidden by a thick ammonia cloud layer.

Each of the giant planets has an extensive system of satellites, ranging in numbers from Jupiter's fourteen moons, three of which exceed our own moon in size, through Saturn's ten, Uranus's five, and down to the mere two moons of Neptune—one of which, however, approaches the size of Mercury, larger than any other satellite save Titan, the largest moon of Saturn. These satellites formed along with the planets, except for the small, outer moons of Jupiter and Saturn, which may be captured asteroids. The rings of Saturn, Uranus, and Jupiter represent swarms of minimoons, millions of tiny particles in satellite orbits. These rock- or pebble-sized fragments represent satellites that never managed to form, because the material is too close to the planet.

Titan is especially interesting to us because its unusual red color probably comes from a smog or haze of particles produced by chemical reactions in its atmosphere. Here, too, we look for examples of the kind of spontaneous chemistry that might have taken place in the early solar system before the origin of life. Unlike the giant planets, Titan has a surface on which these products might accumulate, and it is small enough so that any hydrogen released in the process would escape.

Questions

1. Compare the amount of sunlight that falls on each square meter of Earth's surface with that which falls on each square meter of Jupiter (five times Earth's distance from the sun) and of Saturn (9.5 times Earth's distance).

2. Why are Jupiter and Saturn warmer than we would expect from the small amount of sunlight that reaches these planets?

3. What does it mean to say that the giant planets have primordial atmospheres, while the inner planets have secondary atmospheres? What has happened to the primordial atmospheres of the inner planets?

4. What would happen to a spacecraft that tried to land on Jupiter?

5. What produces the various colors of Jupiter's clouds? What is the Great Red Spot?

6. Could life exist on the giant planets? Where would be the most likely spots for such life to appear?

7. What is unusual about Saturn's largest satellite, Titan? What do we hope to learn by studying this object?

8. What kinds of markers might be left in the solar system by an intelligent civilization to attract the attention of emerging intelligent species? Why is Iapetus a possible example of such a marker?

Further Reading

Cruikshank, Dale P., and Morrison, David. 1976. The Galilean satellites of Jupiter. *Scientific American* 235:108.

Fimmel, Robert, Swindell, William, and Burgess, Eric. 1974. *Pioneer odyssey: Encounter with a giant.* Washington, D.C.: NASA, U.S. Government Printing Office.

Hunten, Donald. 1976. The outer planets. In *The solar system.* San Francisco: W. H. Freeman and Company.

Ingersoll, Andrew P. 1976. The meteorology of Jupiter. *Scientific American* 235:46.

Sagan, Carl, Drake, Frank, Druyan, Ann, Ferris, Timothy, Lomberg, Jon, and Sagan, Linda. 1978. *Murmurs of Earth: The Voyager interstellar record.* New York: Random House.

Soderblom, Laurence. 1980. The Galilean Moons of Jupiter. *Scientific American* 242:1, 88.

Science Fiction

Clarke, Arthur. 1968. *2001: A space odyssey.* New York: New American Library.

KONZERT F-dur

Brandenburgisches Konzert Nr. 2

Joh. Seb. Bach (BWV 1047)

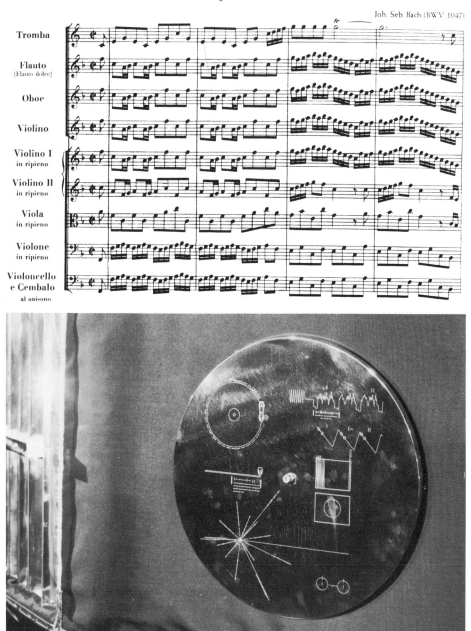

PART FIVE

The Search for Extraterrestrial Intelligence

And for the soul
If it is to know itself
It is into a soul
That it must look.
The stranger and the enemy, we have seen
  him in the mirror.

—GEORGE SEFERIS

We stand now at the brink of the age when human beings for the first time possess the capacity to talk with other beings at interstellar distances. But to turn this potential into reality, we must do more than wait for those others to call us; rather, like the multitudes of civilizations that may well exist in our galaxy, we must analyze the means by which civilizations could contact one another to determine the best way to proceed. If we judge this analysis to be the fanciful pastime of astronomers alone, we are unlikely to establish interstellar contact; if we instead undertake a carefully paced plan to look for our neighbors, we seem likely to meet with success, perhaps not in the first year of our effort, but within a span of time that human civilizations would judge a reasonable one.

J. S. Bach's Brandenburg Concerto No. 2 (top) begins the music section of the information carried by each of the gold-anodized records aboard the two Voyager spacecraft (bottom). These records are protected by a cover that explains how to play them.

Fig. 17-1 The two stars Alpha Centauri A and B appear as a single, overexposed source of light in the top photograph. The double set of diffraction spikes reveals the presence of the two stars close together. The lower photograph shows an enlarged, short-exposure image of A and B only. Each of these two stars could have one or more planets similar to the Earth.

CHAPTER 17

Is Earth Unique?

Our age will appear in terrestrial history as the first great era of solar system exploration, for at last we are making the acquaintance of the cosmic community in which we live. As yet, however, we have found life nowhere but on Earth, despite our best efforts to study Mercury, Venus, our moon, Mars, Jupiter, and Saturn with instrumented spacecraft and with human landings. On our own planet, we have made detailed chemical investigations of meteorites that have arrived from interplanetary space.

Our failure to find life in any of these environments may seem regrettable, but this very failure signals that the search is well underway. Furthermore, our knowledge about how life may have originated on Earth and on other planets has advanced considerably from the great efforts that went into the Viking landings and the analysis of the data returned from Mars. In the outer reaches of the solar system, we have found environments that resemble those some scientists have postulated for the primitive Earth: hydrogen-rich atmospheres, in which thermal energy and ultraviolet light convert simple molecules into more complex compounds. In some meteorites, we have found evidence for the products of such reactions, since we have discovered organic material that includes several types of polymers and even some of the amino acids and nucleotide bases that are used in our kind of life. These discoveries encourage us because they support our ideas about how life began on Earth, and also because they suggest that the fundamental processes that can lead to the origin of life occur throughout the cosmos.

But closer to home, as we have learned more about Mars and Venus, we have come to appreciate the mysterious combination of conditions that a planet must achieve if it is to change from an object with interesting chemical reactions into a world that can develop and sustain *life*. Why, in our solar system, has this balance occurred only on the Earth? What are the special properties of our planet that have singled it out among the other planets and their satellites as the home of abundant life? Is chance alone responsible for the fact that this book was printed on Earth rather than on Venus? Is it even possible that the Earth is unique in the galaxy as the *only* abode of intelligent life?

Distinguishing Characteristics of the Earth

If we confine ourselves to the most basic characteristics of planets, those that are not the result of the others, we can easily identify three facts that make the Earth unique in the solar system: its *distance from the sun,* its *size,* and, perhaps less important, the relatively *large mass of its natural satellite.* These three characteristics of Earth each have an important impact on the possibilities for the origin and evolution of life.

Suppose that the Earth's orbit changed, so that we circled the sun where Venus does. We can calculate what would happen: The increased intensity of solar radiation would raise the average temperature of our planet, thus producing greater evaporation of water and hence more water vapor in our atmosphere. The water vapor would make it harder for the surface of the Earth to radiate its heat into space, and this would raise the temperature still further, which would produce more evaporation of water, and so on and on. The Earth would therefore undergo the same runaway greenhouse effect that has led to the drying out of Venus. The high temperature would also lead ultimately to the production of a thick carbon dioxide atmosphere on Earth similar to that on Venus.

Suppose instead that the Earth exchanged its position with Mars. Then the *decrease* in incoming solar energy, arising from our greater distance from the sun, would cool the oceans and increase the size of the polar caps of ice; this would leave *less* water vapor in our atmosphere and would also increase the reflectivity of our planet. Hence the temperature would decline still further, and so on and on until we had a runaway refrigerator. In the extreme case, the Earth would acquire a complete covering of water ice. We can imagine intermediate situations, however, in which a thick carbon dioxide atmosphere maintained a warmer climate, or in which a temperate environment might exist only near the equator. But in an orbit still farther out, the Earth would lose even this possibility. We are clearly better off at our present distance from the sun.

We have already discussed the importance of a planet's having the right *size.* Large planets can't lose their hydrogen, while the smallest planets cannot retain an atmosphere. This constraint actually goes beyond a simple ability to hold an atmosphere once it exists. As we have seen in Chapter 12, a planet the size of the Earth has a better chance to develop a temperate environment than a planet as small as Mars or Mercury. The larger object has a greater probability of producing a dense atmosphere, because it had a greater accumulation and outgassing of volatile elements during its formation and later history.

Thus, *size* and *location* form two key aspects of a planet's suitability for life. What about the moon? The existence of a large natural satellite seems rather incidental to the problem of life's development and continued existence, despite the familiar evidence that the moon influences such widely varied activities as human romance and fish spawning. But the moon apparently does provide two benefits of great importance to us. First, the moon produces large *tides,* which may have been a key element in making microenvironments within which life could begin. The moon brings a second benefit: It *stabilizes the orientation of the Earth's rotation axis.* We have seen that the variations in the inclination of the Martian poles to the plane of the planet's orbit can produce large changes in the climate of Mars (see page 273). These variations in the inclination of Mars arise from a combination of the gravitational pulls of the sun and Jupiter. Recent calculations have shown that the Earth's ice ages, some of which have made many species of life extinct, probably arise from much smaller changes in the inclination of the Earth's rotation axis and in the eccentricity of the Earth's orbit around the sun. These changes could be much *more* dramatic if we did not have the nearby gravitational pull of a large satellite to suppress them. Greater variations than ice ages in our planet's climate would probably be fatal to developing life.

On the other hand, it is not *essential* to have a large satellite to bring about the desired stability. If the rotational period of the Earth were much faster or much slower, the same effect would occur. The 243-day period of Venus, for example, produces a stable inclination of that planet's rotation axis. So what may be special about the Earth-moon system is that it allows a 24-hour day with stability. How would life develop on planets with shorter or longer periods of rotation? We can only guess, but we can suggest that with the wide variety of accommodations to different daylight periods exhibited by the various forms of life on Earth (e.g., penguins and polar bears which live through long days and long nights), there seems to be no intrinsic drawback to the development of life on a planet with a dramatically different day-night cycle than ours. The absence of the large tides caused by the moon would also not be fatal. Solar tides and the fluctuations of water levels caused by weather might well be sufficient to produce suitable microenvironments.

The Planetary Systems of Other Stars

We expect to find systems of planets around stars of spectral type F5 to K8, stars that resemble our sun in their lifetimes and absolute brightnesses, since there is nothing in our theories for the origin and evolution of our sun and its planets that would not apply to planetary systems around similar stars. Thus, we would expect to find a set of rocky inner planets with atmospheres produced by degassing, weathering and escape, for the same reasons that our own rocky inner planets have atmospheres. Judging from our own example, the chances seem good that one of these inner planets will orbit its star at the ''right'' distance. In our own solar system, the Earth is ''right'' and Mars and Venus are not far away. Furthermore, we have two planets (Earth and Venus) that are the right size, and one of them is in the right place. This suggests that something like one in every two planetary systems will be similarly favored with a planet of the right size in the right place. We say one in every two to be conservative; the optimistic guessers would estimate that since we exist, almost *every* planetary system should have a planet of the right size in the right place! Such optimism requires either that Earth-sized planets will always be present, or that a planet the size of Mars could support life if it were closer to the sun. Either or both of these hypotheses *may* be true, but, lacking evidence, we do best to be cautious.

The remaining worry, oddly enough, is the moon! We don't understand why our satellite is so large (relative to ourselves), or even how it formed, so we cannot make accurate predictions for other planetary systems. But as we have seen, the need for a massive satellite is probably not as great as the other two requirements. We can therefore estimate that perhaps one in every four planetary systems should have an Earthlike planet in the right position, with sufficient stability of climate for life to develop. This estimate rests on the idea that half of all planetary systems will have a planet of the right size in the right place, and half of *those* will either not need a massive moon to damp the inclination changes, or will have such a moon, as the Earth does.

Do not think that this deduction must be rigorous simply because numbers appear! Other features, as yet unrecognized by us, may make the Earth unique, or nearly so. For example, we have not dealt with the fact that only the Earth among the sun's planets has oceans of water. A large abundance of liquid water clearly helps to nurture life as we know it, and we have assumed that water will follow naturally once a planet has the right size in the right place. But other factors, such as the distribution and abundance of volatile-rich meteorites and comets in the primitive solar system, could be significant in determining whether or not liquid water appears. Finally, we must emphasize that the existence of a planet identical to the Earth does not guarantee that life will develop on that planet, nor are we in a position to state that only an Earthlike planet can harbor life. As we have emphasized in earlier chapters, we may yet find evidence of life on Mars, and some scientists have suggested that the warm regions of Jupiter's atmosphere may also be populated by living organisms. But our experience thus far tells us that Earthlike planets are the most probable habitats, and probabilities are what we must work with when we search for life outside the solar system.

To carry this argument further, we would clearly like to be able to investigate *other* solar systems to see whether our predictions are valid, or whether our own system of planets has some unusual property that is not shared by the others. Unfortunately, we can't carry out such a comparative study now, since we know of only one solar system, just as we know only one example of life. But our intuition suggests that other such systems exist. The growth of human knowledge about the universe has been accompanied by a steady erosion of our sense of being "special." Early astronomers thought that the Earth was the center of the universe. The discovery that the Earth orbited the sun still left an impression that the sun must be the center of the stellar system. Once the sun turned out to be located in a spiral arm far from the center of the galaxy, our galaxy appeared to be one of the largest in the universe. We now know that even this is not true. It is not simply that *we* are not special, but that the astronomical universe contains very few, if any, truly unique objects. There are many quasars and pulsars, although at first only one or two were known; many supernovae, many europium-rich stars, and many dense interstellar clouds. This perspective suggests to us that there are probably many other solar systems, despite the fact that at the present time we have no direct observations of them.

Support for this assumption comes from studies of multiple-star systems. Most stars in our own galaxy, and presumably in other galaxies as well, come in doubles, triplets, quadruplets, and even higher-number combinations. Most double-star systems show the two stars quite close to each other, with a typical separation approximately equal to the distance from the sun to Neptune. This relatively small distance, thousands of times less than the average separation of neighboring star systems, is a clue that the formation of stars from interstellar clouds is unlikely to result in a single object. The abundance of *stellar* systems with dimensions on the same order as our *planetary* system suggests that in those cases where we find a star that does not have a close, visible companion, we might expect a small, dark star or a series of planets to exist instead, as a result of the fragmentation of the original cloud that produced the star itself.

We see evidence for this tendency for fragmentation during the formation of massive objects in our own solar system. The giant planets Jupiter and Saturn are each accompanied by an extensive retinue of satellites, and the innermost five satellites (of Jupiter) or six (of Saturn) formed with the planets themselves. Indeed, with the recent discovery of a satellite for Pluto, only two of the nine known planets in our solar system have no satellites at all. Mercury and Venus may simply be too close to the sun.

How Can We Detect Other Solar Systems?

These indirect arguments are encouraging, but we would obviously prefer to have some observational evidence for the existence of other planetary systems. Unfortunately, this evidence has proved very difficult to obtain. Planets shine by reflected light, far more weakly than the stars that illuminate them. Furthermore, the relatively close nestling of planets around stars means that we must look for a weak point of light at almost the same place on the sky as a bright star (Figure 17-1). A planet of Jupiter's size, and at the same distance from Alpha Centauri A (a member of the nearest star system) as Jupiter is from our sun, would have an apparent brightness *1 billion times* less than the apparent brightness of the star! With this tremendous difference in brightness, even the best telescopes we have on Earth could not reveal a planet as large as Jupiter in orbit around Alpha Centauri A or Alpha Centauri B, assuming that the planet-star distance is the same as the distance from Jupiter to the sun.

But can we detect planets around other stars if we can't *see* the planets? Four good possibilities exist: First, build a better telescope, and place it outside the Earth's atmosphere to avoid the blurring effects caused by the air. Second, try to find planets not by their weak reflected light, but by the perturbations that their gravitational force produces on the motion of their parent stars. Third, look for Doppler-effect changes in the spectrum of the parent star as the planet orbits around it, pulling it first one way and then the other. Fourth, try to detect the radio waves emitted near a Jupiter-like planet, since in our own solar system Jupiter often outperforms the sun as an emitter of radio power.

Each of these possibilities might yield positive results within the next few years. Astronomers hope to have the Space Telescope in operation by the early 1980s. This 2.4-meter reflecting telescope will orbit the Earth, high above the disturbing effects of our planet's atmosphere. By use of an optical technique called apodization, the telescope may be able to obtain enough resolving power to find dim points of light, Jupiter-sized planets, close to the glare of nearby stars. Another possibility would be to use our moon as a screen as it passes in front of a star, momentarily reducing the star's light and permitting the light from accompanying planets to be detected (Figure 17-2).

The second technique for finding planets relies on the fact that planets exert a gravitational force on stars, exactly equal and opposite to the gravitational force that stars exert on planets. The fact that the stars have far more *mass* than the planets makes the planets move in big orbits and the stars in little ones, in both cases around the center of mass of the entire system. The center of mass follows a smooth orbit around the center of the galaxy, while the gravitational pull of the planets causes the star to appear to "wiggle" as it tries to follow this orbit (Figure 17-3).

Scale greatly exaggerated

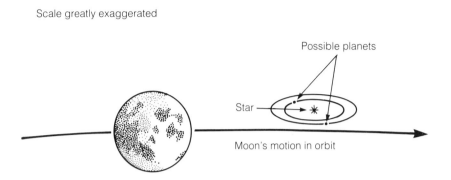

Fig. 17-2 As the moon passes in front of a distant solar system, there will be a tiny instant of time when most of the star's light is obscured and the light from any planets orbiting the star at sufficient distances can be observed.

About a dozen cases of wiggles in the motions of nearby stars have been carefully studied. Some of these stars turn out to have companions with masses not much less than that of the sun; these objects are thus clearly not "planets" but are more likely to be faint white dwarf stars. However, for at least one star—called "Barnard's star" after its discoverer, E. E. Barnard—such studies have suggested the presence of a companion with a mass of one thousandth the sun's mass (about the same mass as Jupiter) and an orbital period of 11.5 years. In addition, there may be a second companion with a mass half as great as the primary companion and an orbital period of about 20 years. If these results can be verified beyond doubt (they are currently extremely controversial), they would provide the first proof of the existence of planets in orbit around other stars, and the similarity of the masses and orbital periods of the apparent planets around Barnard's star with the masses and orbital periods of Jupiter and Saturn would be striking.

The third technique we have mentioned relies on the Doppler effect, the change in photon energy that occurs when a source of photons is in relative motion towards us or away from us (see page 27). If a planet orbits around a star, then, as we have just discussed, the star in fact makes a small orbit around the center of mass of the star-planet system. Instead of trying to observe the star's orbit directly, we can attempt to detect the changes in the star's *velocity* relative to ourselves. These changes arise because in part of its orbit the star is moving towards us, while at the other side of the orbit it is moving away from us, relative to its average velocity in space (Figure 17-3). Orbiting around a star like our sun, a planet with Jupiter's

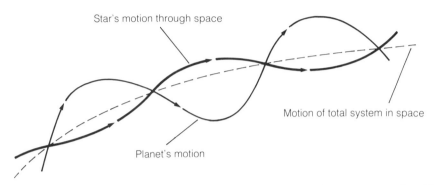

Fig. 17-3 If a star has planets around it, then the star will make a small orbit around the center of mass of the system of star plus planets. This orbit will add a wiggle to the basic "proper motion" of the star, which would otherwise be a straight line. The star's motion in orbit around the center of mass of the star-planet system will change the star's *velocity* relative to ourselves in a systematic way. The velocity will show a periodic increase and decrease as the star moves first towards us and then away from us in its small orbit.

mass would produce a change of only about a hundredth of a kilometer per second in the star's velocity. Velocity changes of this amount cannot be detected at present, but instruments are being developed that may permit such precise measurement within the near future from the Space Telescope in orbit above the atmosphere, or even from telescopes on Earth.

The fourth possibility we mentioned for detecting planets involves the use of radio techniques. Much of the speculation concerning *civilizations* that may exist on other planets centers around the use of radio waves to exchange messages (see Chapter 20). A much more direct possibility of the radio detection of other planets deserves immediate discussion, since it has nothing to do with the possible existence or nonexistence of technologically advanced creatures. The planet Jupiter emits more radio power than the sun does, at certain times and at certain radio frequencies. In other words, astronomers from another civilization with a sensitive radio antenna who happened to observe our own solar system under the right conditions would detect mainly Jupiter, and only secondarily the sun. If we turn our radio antennas toward another star and find bursts of radio photons similar to those that we receive from Jupiter, we could conclude with some assurance that the star does have a planet something like Jupiter in orbit around it. Our antennas and receivers are already almost sensitive enough to perform this experiment for the closest stars; a modest additional investment (say, 100 million dollars) would yield a system quite capable of finding a Jupiter anywhere within a radius of 30 light years from the sun.

The Likeliest Stars

While we are waiting for these techniques (and others that will undoubtedly be developed) to find other solar systems, we can speculate about which stars among our neighbors would be good candidates for such searches. Many scientists are sufficiently confident of the abundance of solar systems to suppose that any apparently single star with a spectral type similar to the sun's *must* be accompanied by planets. We are most interested in the possible planets of stars near the sun, since they are the ones with which communication—by either rockets or radio—would be easiest. We therefore begin our search with a consideration of the 23 star systems that lie within 4 parsecs of the sun (Table 17-1).

Note that half of these systems are multiples. Only three of the stars (Alpha Centauri, Sirius, and Procyon) have a greater true brightness than the sun does; that is, only these three stars lie above the sun on the main sequence. The other stars either lie far down the main sequence, typically in spectral classes K and M, or (for the companions of Sirius and Procyon) they have evolved to the white dwarf stage (see Chapter 6). It seems that although the sun does provide us with a generally average star, we can count ourselves lucky in having a source of energy that exceeds 90 percent of all stars in mass and thus in true brightness. Furthermore, a much brighter star such as Sirius (which has 23 times the sun's luminosity) cannot last as long as the sun; in fact, Sirius must be less than a billion years old, or it would have already evolved from the main sequence, as its white dwarf companion clearly has done.

We have already stated our reasons for wanting stars to be on the main sequence for at least 5 billion years to improve our chances of finding a planet inhabited by an intelligent civilization. This requirement eliminates Sirius and makes Procyon A a bad risk. If a star *has* planets, we also want at least one of the planets to be warm enough for life to exist on it. Because of the expectation that either water or ammonia will provide the necessary solvent for life (see page 212), we seem to require a temperature that stays at least within the range of 0 to 100° C (if water is the solvent), or of −108 to −33° C (if ammonia does the job). So we are looking for temperatures in the range of −108 to +100° C, with a preference for the higher values.

In our own solar system, we know that only the planets Earth and Mars have surfaces that include temperatures in this range. Venus, which would have a temperature of about 45° C if it had no atmosphere, has instead a carbon dioxide atmosphere that keeps its surface temperature at 475° C. If we include Venus in the proper temperature zone, considering its atmosphere to be an exceptional case (perhaps if it had been a smaller planet, its development would have differed), we find that the sun, or another star like it, can produce the proper temperatures on planets that orbit between 0.7 and 2.0 times the Earth's distance from it. This relatively

TABLE 17-1

Boldface type indicates those stars most like the sun in this list.

Star Name	Distance (parsecs)	Spectral Type	Luminosity (sun = 1)	Mass where known (sun = 1)
Alpha Centauri A	1.3	G2	1.53	1.1
B		K4	.44	.88
Proxima Centauri	1.3	M5	.00006	.1
Barnard's star[a]	1.8	M5	.00044	
Wolf 359	2.3	M8	.00002	
Lalande 211385[a]	2.5	M2	.0052	.35
Sirius A	2.6	A1	23.0	2.31
B		w.d.	.0020	.98
Luyten 726-8 A	2.7	M5	.00006	.044
B		M6	.00004	.035
Ross 154	2.9	M4	.0004	
Ross 248	3.2	M6	.0001	
Epsilon Eridani	3.3	K2	.30	.8
Luyten 789-6	3.3	M6	.00012	
Ross 128	3.3	M5	.00033	
61 Cygni A[a]	3.4	K5	.082	.63
B		K7	.038	.6
Epsilon Indi	3.4	K8	.14	
Procyon A	3.5	F5	7.6	1.77
B		w.d.	.0005	.63
Sigma 2398 A	3.5	M4	.003	.4
B		M5	.002	.4
Groombridge 34 A	3.6	M1	.006	
B		M6	.0004	
Lacaille 9352	3.6	M2	.012	
Tau Ceti	3.7	G8	.47	
BD +5°1668[a]	3.8	M5	.0015	
Luyten 725-32	3.8	M5	.0003	
Lacaille 8760	3.8	M0	.027	
Kapteyn's star	3.9	M0	.004	
Krüger 60 A	3.9	M3	.0015	.27
B		M4	.0004	.16

[a]These stars are thought to have unseen companions.

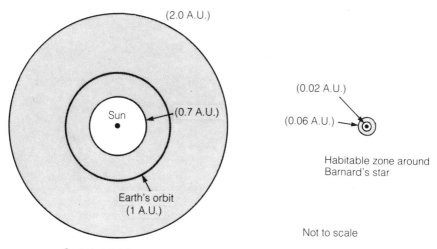

(2.0 A.U.)

Sun

(0.7 A.U.)

Earth's orbit
(1 A.U.)

Sun's habitable zone

(0.02 A.U.)

(0.06 A.U.)

Habitable zone around
Barnard's star

Not to scale

Fig. 17-4 We define the habitable zone around the sun as the spherical shell of space within which a planet will have the "proper" temperature for life to exist. If we define this temperature range as −108 to +100° C, then a star like the sun has a habitable zone that extends from 0.7 to 2.0 times the Earth's distance from the sun. The habitable zone around the much fainter Barnard's star extends only from 0.02 to 0.06 times the Earth's distance from the sun.

narrow zone (Figure 17-4) brackets the habitable regions: Farther away from the sun, temperatures tend to fall too low for life, while closer to the sun, excessive heat ruins our hopes for life's existence. But these limits will vary somewhat with the composition of the atmosphere.

If we turn to the list of stars in Table 17-1, we find that our desire to duplicate the proper *temperature* conditions means that to be habitable, planets must orbit far closer to the faint K and M stars that comprise the majority of the list than to stars like the sun. These low-luminosity stars do possess tremendously long lifetimes, so *if* habitable planets orbit around them, they can expect many tens of billions of years of steady starshine rather than the mere 10 or 11 billion our sun will provide the Earth.

But how close must a planet be to, say, an M5 star such as Barnard's star? This second-closest star system consists of a single dull red star that emits less than a two thousandth of the sun's luminosity each second! For a planet to receive as much starlight energy from Barnard's star as we receive from the sun per square meter per second, the planet must be 45 times closer to Barnard's star than we are to the sun. In other words, even Mercury's distance from our sun far exceeds the maximum distance a planet could have from Barnard's star and yet stay warm enough for life to exist (Figure 17-4).

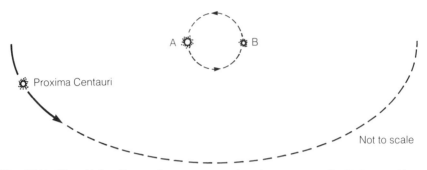

Not to scale

Fig. 17-5 The Alpha Centauri system contains three stars, of which two (A and B) closely resemble the sun. All three stars orbit around the center of mass of the system, but A and B are separated by about 25 times the Earth's distance from the sun, while Proxima Centauri's distance from the center of mass is 2000 times the separation of the A and B stars. Figure 17-1 is a photograph of Alpha Centauri A and B.

At such close distances, another worry comes into play: the possibility of a gravitational lock between the orbital and rotational motions of the planet that keeps the same hemisphere always facing the star. This has happened for our own moon and the five inner satellites of Jupiter; the eccentricity of Mercury's orbit has led to a 3:2 resonance instead (see page 243). How hostile to life would such a rotation lock be? We would like some alternation of conditions—wetting and drying, freezing and thawing—to help the initial chemistry that starts life along its way. But perhaps there are other pathways that lead to the same results. The presence of a sufficiently thick atmosphere on one of these rotationally locked planets might provide adequate modulation of the perpetual day or perpetual night conditions. We simply can't rule these M dwarfs out *a priori* as possible centers for inhabited planets; but they do seem less likely objects than their less numerous but more luminous brethren.

The *habitable zone* around stars, the place where planets would have the right temperatures for life to exist, often receives the title of *ecosphere*. Because most ecospheres are so small—since most stars shine less brightly than the sun—we can reduce the length of the list in Table 17-1 considerably when we look for the stars most likely to have habitable planets. In fact, since Sirius and Procyon can be eliminated because of their short lifetimes, the best candidates within four parsecs of the sun with reasonably large ecospheres are Alpha Centauri, Tau Ceti, and Epsilon Eridani. Let us take a brief look at each of these possibilities.

The Alpha Centauri system (Figure 17-5) consists of two fairly bright stars—A with 1.5 times and B with 0.44 times the luminosity of the sun—and a third, much fainter star, Proxima Centauri, with less than a ten thousandth of our sun's true brightness. Astronomers once thought that multiple star systems were very unlikely places to look for planets because the gravitational attractions of the various stars would prevent the existence of stable planetary orbits. Recent calculations have shown, however, that this viewpoint is too pessimistic. Instead, planetary orbits appear to be stable in a double-star system if one of two possibilities occurs. Either the planets must be in orbits close to one of the two binary stars (the stars themselves orbit around their common center of mass), or the planets must orbit at a large distance from the double star system.

Both Alpha Centauri A and Alpha Centauri B, with spectral types G2 and K5, are quite capable of supporting life on planets in orbits around them, provided that the planets' orbits are sufficiently close to one or the other of these two stars. Since the separation of A and B is about 25 times the distance from the Earth to the sun, we could imagine planets around both stars with orbits similar to those of the inner planets in our system. In this case, inhabitants of a planet moving around one star would see the other star as a sort of a giant moon 1000 times brighter than our moon though still 1000 times fainter than the star which kept the planet in orbit. This "supermoon" would be out in the daytime for half the year, and then out at night for the other half of the year. The two stars have slightly different colors, yellow and orange, which would add to the beauty of the effect (see Figure 17-6).

The other case, in which a planet orbits at great distances from both of the double stars, makes the planet follow an orbit similar to that of Proxima Centauri (Figure 17-5). Although such orbits are stable, they do not help

Fig. 17-6 A planet in orbit around either Alpha Centauri A or B would see two bright suns, one yellow and one orange. Proxima Centauri would appear as a dim red object, barely visible without a telescope.

us in our search for life, because the great distance of Proxima Centauri from either Alpha Centauri A or Alpha Centauri B at any time means that both of these stars appear no brighter than our moon. Hence, to hope for life on a planet that occupied such an orbit would be similar to hoping that life would develop by moonlight, which appears unreasonable. We could, however, imagine that a planet orbits extremely close to Proxima Centauri itself. But this M5 star is so faint and so cool that it presents us with the same conditions we faced in considering Barnard's star: Habitable planets must have very small orbits. If the Earth orbited around Proxima Centauri at our present distance from the sun, we would see our new "sun" as a small red disk 60 times brighter than the full moon (as illuminated by our present sun) though only a tenth as large on the sky. With this as our parent star, our hopes for survival would be dim indeed, although we would certainly enjoy the celestial spectacle, with nearby A and B shining brightly in our heavens.

Farther out in space than the Alpha Centauri system, the next good candidates we encounter are Epsilon Eridani and Tau Ceti. Epsilon Eridani, a K2 star, has only 30 percent of the sun's luminosity, about the same as Alpha Centauri B, but a planet in orbit around this star at the distance of Mercury or Venus from the sun would still fall within the habitable zone. Tau Ceti, a G8 star, resembles the sun a bit more than Epsilon Eridani does, for its true brightness equals 47 percent of the sun's. Again, we can expect a reasonably large ecosphere around Tau Ceti, larger than Epsilon Eridani's but smaller than the sun's. Epsilon Indi and 61 Cygni A are also likely candidates, since they have spectral types not far below Epsilon Eridani's. But with intrinsic luminosities that are respectively 53 and 74 percent lower, they do not rank as high on our list.

Both Tau Ceti and Epsilon Eridani drew early attention from scientists who speculated about life around other stars, and from the first radio search to the present day, astronomers have pointed their antennas in the direction of these two stars with the hope, so far unfulfilled, of detecting civilizations there. The Alpha Centauri system has not received comparable attention, primarily because the radio telescopes located on the northern hemisphere of our planet cannot observe it, and these are the principal instruments that have been used to search for indications of artificial signals up to the present time.

As our discussion of the likelihood of finding another civilization on another planet will show (see Chapter 19), we should not expect life to be so easy that the first two or three stars to be studied would reveal another civilization, especially after only a short period of observation. But we should not abandon the idea that even one of these nearby stars may be circled by a planet carrying a civilization far more advanced than ours, to be discovered when we learn the proper techniques for making contact. What a splendid way to prove the existence of other solar systems!

TABLE 17-2

APPROXIMATE ABUNDANCES OF THE CLASSES OF MAIN-SEQUENCE STARS

Spectral Type	Approximate Mass (sun = 1)	True Brightness (sun = 1)	Percent of Total[a]	Number of Stars[a]
O	32	50,000	0.00002	55,000
B	6	300	0.09	360,000,000
A	2	10	0.6	2,400,000,000
F	1¼	2.0	2.9	12,000,000,000
G	0.9	0.9	7.3	28,000,000,000
K	0.6	0.2	15.1	60,000,000,000
M	0.2	0.005	73.2	293,000,000,000

[a]We assume the total number of stars in the Milky Way to be 400 billion. Stars not on the main sequence (and thus not included in this list) form 0.8 percent of the total.

Meanwhile, our consideration of the 23 star systems within 4 parsecs of the sun has indicated what an average region of our galaxy is like. We can see that most stars have too low a luminosity to be promising candidates for planet-based life, if we confine ourselves to the kinds of environments we are familiar with in our own solar system. But with so many stars in a galaxy such as ours, we need not be discouraged even if we take the conservative view of eliminating the small, cool stars from consideration. As Table 17-2 shows, there are billions of stars with characteristics comfortably close to those of our sun. We may simply need to extend the boundaries of our search to greater volumes of space before we encounter the tendrils of an alien intelligence unfolding in our direction at the speed of light.

Summary

We began this section with a worry that some extremely rare characteristic of our planet, our solar system, or our star might lead us to the conclusion that we are the only intelligent civilization in the galaxy. This worry was triggered by the realization that at this stage in our exploration of our solar system, our planet is the only environment where we have been able to find evidence for the existence of life. But there seem to be good reasons for this anomaly, the Earth's *size* and *distance from the sun*. If Venus and the Earth changed places, it would be Venus that was inhabited, provided that Venus had formed at the Earth's present position so that it would have retained the same mixture of volatile elements. Our moon has played an important role in the origin of life by stabilizing the Earth's inclination and increasing the magnitude of oceanic tides, but faster or slower rotations might provide similar stability and more gentle tides could provide adequate environments for the early stages of life.

Other solar systems should have rocky inner planets like ours, and the chances of finding a planet of the right size in the right place with the right rotational period seem to be about one in four, if we can assume that our system is typical. Unfortunately, for the time being we have no direct observational evidence of any other solar systems. Indirect evidence abounds, however, that points to the frequent formation of planets as a natural accompaniment to star formation. Several techniques for detecting solar systems around nearby stars are becoming available and may provide the evidence we seek within the next decade.

As we look about us in the galaxy, we find that the volume of space within just 4 parsecs of the sun contains three stellar systems that are good candidates to have planetary systems with ecospheres suitable for life. Two other stars in our list are marginally acceptable, so even this tiny sample of the stars in our galaxy may contain another planet like the one on which we live. This does not mean that such a planet must harbor an intelligent civilization, or even that life must have originated on its surface. But it does suggest that the Earth is almost certainly not unique. To find other life, we must begin a more sophisticated search that may extend many parsecs away from our solar system.

Questions

1. What are the basic characteristics that distinguish the Earth from the other planets in our solar system? What is their relative importance?
2. What would happen if the Earth were suddenly moved to the position of Venus?
3. Why do astronomers think that solar systems are relatively common in our galaxy?
4. What methods have been proposed to search for solar systems associated with nearby stars? Can you think of some other ones?
5. What is the most common type of star in the galaxy? Why do these stars seem less likely than others to be circled by Earthlike planets?
6. What is an ecosphere? How do its dimensions vary with the spectral type of the star it surrounds?
7. Among the nearest stars, which ones are most likely to have Earth-like planets? Why?
8. Describe the Alpha Centauri system. What would it be like to live on a planet orbiting one of the brighter components of this system?

9. Suppose that in our Milky Way galaxy, most stars have planets in orbit around them, but only one star in a thousand has planets within a habitable zone around the star. If each such star, on the average, has one planet in the habitable zone, how many planets within habitable zones are to be found around the 400 billion stars of our galaxy?

10. Consider the nearby star Lacaille 9352, which has a true brightness only a hundredth of the sun's. How many times closer to *this* star would a planet have to be for each square centimeter of the planet's surface to receive the same amount of starlight energy each second as occurs in *our* solar system?

Further Reading

Cameron, Alastair. 1976. The origin and evolution of the solar system. In *The solar system*. San Francisco: W. H. Freeman and Company.

Dole, Stephen. 1970. *Habitable planets for man*. 2nd ed. New York: Elsevier Scientific Publishing Co.

Sagan, Carl. 1973. *The cosmic connection*. New York: Dell Books.

Sagan, Carl. 1976. The Solar System. In *The solar system*. San Francisco: W. H. Freeman and Company.

van de Kamp, Peter. 1971. The nearby stars. In *Annual Reviews of Astronomy and Astrophysics* 9:103.

van de Kamp, Peter. 1975. Unseen astrometric companions of stars. In *Annual Reviews of Astronomy and Astrophysics* 13:295.

CHAPTER 18

The Development
of Extraterrestrial Civilizations

Still more exciting than the prospect of discovering extraterrestrial life is the possibility of finding extraterrestrial civilizations with whom we can communicate ideas. For the purposes of considering this tremendous opportunity, we can define "civilizations" as groups of beings with self-conscious, complex interactions who share information among themselves and could share their music, poetry, and learning.

Although civilizations, like the creatures that form them, could have many different structures, we are most likely to communicate with societies that are relatively like our own. In particular, we can expect to encounter civilizations with *curiosity* about the universe far sooner than those which, for one reason or another, have no such curiosity.

In our search for other civilizations, four questions seem paramount:

1. How many civilizations exist?
2. How long do they persist?
3. How eager are they to talk to us?
4. How does communication proceed?

We must attempt to answer these four questions as best we can to estimate the difficulty of making contact with other civilizations. Then we can decide whether or not we want to invest this amount of effort for the chance of a fascinating, though unknown, reward.

This process must occur in any civilization that confronts the issue of how much time and energy should be spent in looking for its neighbors. As a civilization, however, we ourselves must rank among the youngest in the universe, for human beings have existed for only a few million years and have had the ability to communicate by radio for only the past 70 years. In contrast, we may guess that other civilizations developed not thousands or millions but billions of years ago. If these civilizations have not yet disappeared, their development must surely have so far surpassed our own present status that we can hardly imagine their abilities and ideas.

How Many Civilizations Exist?

In all our speculation about other planets with life and other civilizations, we have been forced to rely on our single example of a living planet, our parent Earth. If we could locate even one other civilization, we would take a giant step toward proving that Earth's conditions represent a fair sample of those which occur in planetary systems. Two examples combined have far more impact than a single, possibly unique, case. Thus, before men sailed from Europe to America, they imagined strange and varied races of man-apes in different regions of the Earth. But once they found human beings in America, too, the conclusion that our planet has a single race of hominids grew immensely stronger.

For the time being, we on Earth remain in pre-Columbian ignorance of our neighbors on other planetary systems, and we have no convincing evidence that our civilization does *not* represent an extremely rare event. When we examine the different factors that appear to enter the development of a civilization, however, we soon conclude that within our own galaxy, many civilizations should exist. The number of these civilizations and the distance to the nearest civilization depend on the actual values of the factors that determine whether or not a civilization will develop around a particular star. Since this number governs the ease with which we can discover another civilization, we should take the time to look carefully at how we estimate the number of civilizations and at how reliable this estimate may be.

In compiling our estimate of the number of civilizations in our galaxy (and, later, in the entire universe), we must consider all the factors that influence the result. To reach an intuitive understanding of the basic problem of estimating this number let us first consider a problem closer to our experience—the search for the perfect restaurant.

The Search for the Perfect Restaurant

Suppose we ask ourselves a question: How many restaurants exist that are just like our favorite restaurant? We all know that our favorite owes its perfection to a combination of many factors. First of all, its location must be reasonably close to where we live or to where we spend our day. Second, it must have the right sort of decor: If we like old-time mahogany furnishings with inlaid mirrors, our restaurant must have them to please us; if we prefer art nouveau, a different interior design is required. Third, the restaurant personnel must be our kind of people. Fourth, the food must be to our taste, and cooked the way we like it. Fifth, the price of meals must not exceed what we enjoy paying. Sixth, the restaurant must be open when we want to eat there.

We can ask ourselves what number of restaurants similar to our own perfect restaurant now exists in the United States. To estimate this number, we first consider the total number of cities. We think that cities are the place to find perfect restaurants, because experience has shown that small towns do not have a great variety of restaurants of the sort that might be able to produce one as perfect as our favorite. The number of cities in the United States equals about 5000, if we count cities with more than 5000 inhabitants. Next, we must multiply this number of cities by the fraction of cities that have restaurants. Some cities have an interesting set of laws that prohibits true restaurants and substitutes "clubs" instead. Since most cities do have restaurants, we shall set the fraction of cities with restaurants equal to 0.99. Multiplying this fraction by the 5000 cities gives us a total of 4950 cities with restaurants.

Next we must consider the number of restaurants per city. Careful sampling by the authors has shown that this number equals approximately 200. Naturally, the larger cities have far more restaurants and the smaller cities have far fewer. If we multiply the number of cities with restaurants by the average number of restaurants per city, we find the total number of restaurants in the United States in cities. This number, though interesting, does not provide us with the number of perfect restaurants. We must first multipy the total number of restaurants by the fraction that have the location which we consider right. This fraction might be one in twenty, or 0.05. Next, we must multiply our result by the fraction of those restaurants that have the right decor, perhaps one in ten, or 0.1. And the fraction of restaurants that meet these tests and also have the right personnel might be one in two, or 0.5.

We must still pass through three more steps in our search. We first estimate the fraction of restaurants that meet our qualifications so far that have the right sort of food—say, one in five, or 0.2. Then we must consider the fraction of these restaurants with the right prices, which could be one in ten, or 0.1. Finally, we must guess the fraction that have the right hours of opening, probably as high as 0.9.

To find the number of restaurants in the entire United States that match our idea of the perfect restaurant, we must multiply together the factors we have estimated. We have seen that the number of restaurants in the cities of the United States equals the number of cities (5000) times the fraction of cities with restaurants (.99) times the average number of restaurants per city (200). We must multiply this number by the estimated fraction with the right location (0.05), then by the fraction of these with the right decor (0.1), and then by the fraction of these restaurants with the right personnel (0.5). This number in turn must be multiplied by the fraction with the right food (0.2), then by the fraction with the right prices (0.1), and finally by the fraction with the right hours of opening (0.9).

If we write an equation that expresses these multiplications, thus estimating the number of perfect restaurants in the United States, we have

Number of perfect restaurants	=	number of cities	×	fraction of cities with restaurants	×	number of restaurants per city
	×	fraction with right location	×	fraction with right decor	×	fraction with right personnel
	×	fraction with right food	×	fraction with right prices	×	fraction with right hours

Upon performing this multiplication, we find:

Number of perfect restaurants
$$= (5000) \times (0.99) \times (200) \times (0.05) \times (0.1) \times (0.5) \times (0.2) \times (0.1) \times (0.9)$$
$$= 44.55$$

Our result that there are 44.55 perfect restaurants serves to remind us of a few key facts about estimates such as the one we have just performed: First, the final result inevitably contains some error; and second, the result depends on the product of various factors, some of which are just educated guesses. If any one of the factors that enters our multiplication turns out to be far wrong, the final product will be wrong by the same proportion.

We can summarize our estimate by saying that the number of perfect restaurants in the United States seems to be about 45, but we would not be surprised if the number turned out to be 50, or even 20 or 100. For example, if the fraction of restaurants with the right food were 0.02 instead of 0.2, the number of perfect restaurants would be 4.5 instead of 45. This sort of analysis is very similar to our attempts to estimate the number of civilizations in the Milky Way galaxy.

The Number of Civilizations

To estimate the number of civilizations, we need only perform a series of multiplications analogous to those that enter our search for the perfect restaurant. In the search for civilizations, we begin with the number of stars in our galaxy rather than with the number of cities in the United

States. We must then multiply this number by the fraction of stars that last long enough for life to develop around them. Next, we must multiply by the average number of planets per star, and these three terms together will give us the total number of planets in the galaxy in orbit around stars that last long enough for life to develop. We must multiply this number by the fraction of planets with conditions favorable to life, and then by the fraction of *those* planets upon which life actually develops. This product gives us the number of planets in the galaxy with life on them at some time, and we must multiply this result by the fraction of planets with life upon which intelligent civilizations arise. The final factor is the length of time during which a civilization has both the ability and the desire to communicate with other civilizations in the galaxy: We must multiply our product by the ratio of the lifetime of a civilization with both the ability and the desire to communicate to the total lifetime of the galaxy.

Our basic equation that estimates the number of civilizations in the galaxy *now* with which we can communicate thus becomes:

$$
\begin{aligned}
\text{Number of communicating civilizations in our galaxy now} = {} & \text{number of stars in the galaxy} \times \text{fraction of stars that last long enough for life} \\[2ex]
\times {} & \text{average number of planets per star} \times \text{fraction of planets suitable to life} \\[2ex]
\times {} & \text{fraction of those planets where life actually develops} \times \text{fraction of planets with life where intelligent civilizations arise} \\[2ex]
\times {} & \frac{\text{lifetime of civilization with ability and desire to communicate}}{\text{lifetime of the galaxy}}
\end{aligned}
$$

This equation is called the *Drake equation* by astronomers, because Frank Drake first devised it. Previous chapters in this book have helped us in producing estimates of the factors that enter the Drake equation. As we read from left to right in the equation, we go from the relatively sure numbers to the less sure, finally arriving at the totally unknown. For example, the number of stars in our galaxy, about 400 billion, is known to within a factor of 2, so we think. This is not perfect accuracy, but it is far better than our estimate of, say, the fraction of planets with life where intelligent civilizations develop. Here we have only ourselves to guide us, together with whatever conclusions we can draw from the way that we believe we evolved. Finally, when we come to the last term in the equation, we shall encounter great difficulties in estimating the length of time that a civilization survives with communications ability and desire. Nonetheless, this equation has such great importance that we must proceed to fill in the numbers as best we can and thus estimate the number of civilizations now in existence with whom we could hope to exchange messages.

First of all, for the number of stars in our galaxy we have relatively little doubt of the 400 billion that we discussed in Chapter 3. Chapters 5 and 6, which examine the ways stars evolve, have shown us that the fraction of stars that live for at least 5 billion years, the minimum time that we think is needed for intelligent life to evolve, equals at least half of all stars, and probably a greater number. Let us, however, use 0.5 as the fraction that fills the second place of our equation.

The average number of planets per star remains a mystery to us, but if we use our solar system as an example, which we think reasonable because the sun seems a perfectly average star, we come out with a number of about 10 (in our case 9).

The next factor, the fraction of planets with conditions suitable to life, can be partly answered by examining our solar system, to the extent that we believe our solar system to be a representative planetary system. We find one planet, Earth, that appears eminently suitable to life, and one other, Mars, that appears very close to being suitable, if not exactly so. Our discussion in Chapter 17 suggests that a conservative estimate gives to one planetary system in every four a planet with suitable conditions for life. If planetary systems average 10 planets per system, this figure implies that one planet in 40 has conditions that favor the origin of life.

We also estimated in Chapters 10 and 17 that the fraction of planets suitable to life on which life actually develops rises close to unity, but in line with our conservative approach, we set this fraction at 0.5, with the risk of underestimating the number of planets with life by a factor of 2.

How Long Do Civilizations Last?

The next fraction we encounter in our equation is the fraction of planets with life where intelligent civilizations develop. Again, if we were to extrapolate from the sole example we know of a planet with life, our own, we would set this fraction at unity; again, to be conservative, we shall take this fraction to be 0.5. Finally, we must consider the *lifetime* of a civilization once it has achieved the ability to communicate and the desire to do so.

With this last term, we face an almost insurmountable difficulty in estimation. Our own civilization has achieved the ability to send messages across interstellar space within the last few generations, and we shall discuss the ways in which we might do so in Chapter 20. We still have not shown much desire to do so, however, for only a few such messages have been sent (see page 402). The most important question of all—how long we shall last as a civilization with the ability and the desire to communicate with other civilizations in our galaxy—continues to be shrouded in mystery, for we cannot look into the future. Since this lifetime plays a crucial role in determining the number of civilizations we can expect to find in our galaxy, always provided that our lifetime as a civilization reflects the average, we shall leave this quantity as an unknown in our equation, and simply designate it L.

If we do this, and if we combine the numbers in our equation together with the fact that the lifetime of our galaxy is approximately 10 billion years, we get the following result:

$$\text{Number of civilizations in our galaxy} = (400 \text{ billion stars}) \times (0.5) \times (10) \times (0.025) \times (0.5) \times (0.5)$$

$$\times \frac{L}{10 \text{ billion years}}$$

We must bear in mind that we are measuring the lifetime L of a civilization with communications ability and desire in *years*, just as we are measuring the lifetime of our galaxy in years.

We can see that upon inserting these numerical values, the big numbers in the equation tend to cancel each other. The 400 billion stars in our galaxy overwhelm the 10 billion years in the denominator of the last fraction, leaving only a factor of 40 after canceling the first number with the last denominator. Most of the other numbers in the equation are rather close to 1. This fact has great significance, for most of the numbers represent a fraction of planets or stars with some quality that leads toward the establishment of communications possibility. If any of these fractions were tremendously less than 1, for example, 0.0001 instead of 0.5, the total product would fall far below the number we shall reach if we assume that the fractions do *not* fall far below one. In a similar fashion, our search for the perfect restaurant would be greatly handicapped if, for example, the fraction of restaurants with the right kind of personnel were not 0.5 but 0.0001 instead.

When we multiply all the factors in our equation together, we find that the result for the number of civilizations in our galaxy *now* with communications ability and desire equals 1.25 times L, where L is the lifetime, measured in years, of a civilization once it has developed the ability and desire to communicate with others. Given the uncertainty in the factors that went into this equation, we can make an accurate statement by saying that our *estimate* puts the number of civilizations in our galaxy now with the ability and desire to communicate equal to the lifetime of such civilizations, *L,* measured in years.

Table 18-1 recapitulates the numbers that have entered our equation, together with other values for the appropriate fractions suggested by Carl Sagan in 1974. We can see that although Sagan's estimate of the different fractions differs from our own, a certain balance has kept the best-guessed result approximately the same: The number of civilizations in our galaxy now approximately equals the lifetime of an average civilization with communications ability and desire, measured in years.

In short, a civilization's lifetime seems to determine just about everything. If *we* represent an average civilization, and if we destroy ourselves in, say, 100 years after we developed radio techniques capable of interstellar communication, then L should equal 100 years, and the number of civilizations in our galaxy now, or at any other time, should be about 100, with our chances of finding one another very slight. If, on the other hand, civilizations discover ways to maintain themselves indefinitely after they have reached our level of technological advance, then L could equal approximately 2 or 3 or 5 billion years, the average lifetime we can expect for the suitable stars in our galaxy. In this case, the number of civilizations now existing in our galaxy that can communicate with one another would reach into the billions. Most likely, an intermediate value is closer to the truth. If L equals, say, 1 million years, then about a million civilizations now exist in our galaxy with whom we might communicate.

Let us never lose sight of the fact that this number refers only to our Milky Way galaxy, and that billions and billions of galaxies, if not an infinite number, exist in the universe. We restricted ourselves to our own galaxy because local communication remains a far easier matter than extragalactic communication, though the latter cannot be judged impossible. We must merely be prepared to wait a longer time between messages. The nearest stars in our galaxy would require several years for a round-trip message between ourselves and them, while the farthest stars require many thousands of years for a round-trip radio message. In contrast, to exchange a message with a civilization in the Andromeda galaxy demands almost 4 million years, while to exchange a message with one of the galaxies in, for example, the Coma cluster would require 250 million years (Figure 18-1). But we would not have to have a dialogue to be excited by contact with extraterrestrial intelligence. Receiving messages from some civilization that might have vanished long before its wisdom, poetry, or music reached us would still be thrilling, as we know when we read the works of Plato or listen to Bach. Whatever we may think about the likelihood of receiving such messages, we can agree that our chances for real conversations are much more likely if we limit ourselves to our fellow civilizations in the Milky Way galaxy, who, as we have just estimated, number anywhere from a hundred to several billion.

TABLE 18-1

ESTIMATES OF THE PROBABILITY FACTORS THAT DETERMINE THE NUMBER OF CIVILIZATIONS IN OUR GALAXY AT ANY GIVEN TIME

	Sagan (1974)	Our Best Estimate	Most Favorable Case	Least Favorable Case
Number of stars in our galaxy	100 billion	400 billion	600 billion	100 billion
Fraction of stars with sufficiently long lifetimes	1	$\frac{1}{2}$	1	1/40
Average number of planets per star	10	10	20	4
Fraction of planets suitable for life	1/10	1/40	$\frac{1}{4}$	1/100
Fraction of suitable planets where life does occur	1	$\frac{1}{2}$	1	1/1000
Fraction of planets with life that develop civilization	1/100	$\frac{1}{2}$	1	1/1,000,000
Ratio of civilization's lifetime to galaxy's life	$\dfrac{L}{10 \text{ billion years}}$	$\dfrac{L}{10 \text{ billion years}}$	$\dfrac{L}{10 \text{ billion years}}$	$\dfrac{L}{10 \text{ billion years}}$
Product of previous factors: the number of civilizations in the galaxy now or at any average time	L/10	1.25 L	300 L	$\dfrac{L}{100,000,000,000}$

Fig. 18-1 The Coma cluster of galaxies lies about 110 million parsecs (360 million light years) away from us. The thousand or so galaxies in this cluster contain several hundred *trillion* stars, so at any given moment, many billion civilizations may exist in the Coma cluster.

How Eager Are Civilizations for Contact?

We have acquired some feeling for the number of civilizations that may exist in our galaxy, and for the importance of estimating their lifetimes if we hope to gain even an approximate idea of this number. What about the *desire* of civilizations for communication? Do human beings, for example, want to contact other civilizations? Yes and no; some do and some don't.

If the average lifetime of a civilization capable of interstellar communication far exceeds our 70 years or so, then our own civilization cannot now serve as a good representative. Most of a civilization's lifetime would then be spent not in our first flush of technological advance but in the enjoyment of a long period of time during which the question of whether or not to search for others could be debated.

The chief forces that favor interstellar communication are *curiosity* (the desire to find out what's there), *gregariousness* (the urge to talk and to listen), and what might be called *social avarice* (the hope of obtaining "valuable information"), to which we might add the more nebulous urge to try simply because we *can*. In opposition to attempts at communication are *fear* (the worry that hostile aliens will enslave, devour, or destroy us, either in person or through the shock of discovering who they are), *inertia* (the judgment that we just don't feel like exerting ourselves on such a project), and the press of *other priorities* for our time and resources, if these commodities begin to be in very short supply.

Curiosity and gregariousness seem to be human characteristics that have allowed our society to exist and to dominate our world, while fear and inertia, though always with us, seem to have failed consistently in their confrontations with what we may call "positive" characteristics for social success. With a small leap of faith, we may conclude that "successful" civilizations on other planets should have gained their dominant positions by possessing much the same characteristics. If our curiosity had not tended to overcome our fear, we would not now be able to stare into the night skies and wonder about the wisdom of contact with other civilizations, who may have trod the same basic road of development.

Some civilizations will attempt to contact others, and some will not. Of course, the average civilization's lifetime as a seeker after contact may fall far below the average total lifetime with communications *ability*. We may find that most civilizations capable of long-term existence will pull in their antennas and concentrate on themselves. Indeed, it may be true that *only* those civilizations that do *not* seek contact can last for long; the urge to search could be closely related to various self-destructive impulses. In either case, the key lifetime in the master equation would then turn out to be far less than we might hope.

On the other hand, even a civilization not interested in contact may betray its presence through the "leakage" of radio signals into space just as we do now (see page 395). Its communicative lifetime would then be correspondingly lengthened. Finally, there is the interesting possibility suggested by Sebastian von Hoerner that once a civilization has made contact with another more advanced than itself, its chances for long-term survival will increase, because it can learn from the experience of the society it has discovered. This "positive feedback" will therefore increase L and lead to longer average communicative lifetimes.

We have now provided, as best we can, the answers to the first three questions that we posed on page 342: How many civilizations exist? How long do they persist? And how eager are they for communication? We have seen that the answer to the first question depends directly on the answer to the second, and have agreed simply to forget about civilizations not eager for communication as the best answer to the third question. If a significant fraction of all civilizations that exist do possess the desire to communicate, then our best estimate sets the number of such civilizations in the galaxy *now* equal to the lifetime of such a civilization, measured in years.

The "least favorable case," which we have also included in Table 18-1, is totally arbitrary. (The reader is invited to experiment with even less favorable cases.) Here we have assumed that only G-type stars very similar to the sun are suitable, and we have reduced the figures for the average number of planets and the fraction suitable for life. But the large changes in our numbers arise from the assumption that the origin of life might be an exceedingly improbable event and that the evolution of intelligent life with a capacity for interstellar communication is even less likely. If our numbers are correct (and we have no way of proving that they are not), then we could expect only four civilizations to exist in the Milky Way at any given time, and then only if the mean lifetime of civilizations were equal to the lifetime of the galaxy itself! For the reasons we have given in the previous parts of this book, we do not regard this case as likely. But we can use these different estimates of the number of advanced civilizations in the galaxy to confront the last of our four questions.

How Does Communication Proceed?

In order to consider the ways in which interstellar communication might proceed, we must use our estimate of the number of civilizations in the Milky Way to determine how widely separated these civilizations ought to be. To estimate the average distance among civilizations, we must calculate the fraction of all the stars in our galaxy that now have communication-oriented civilizations nearby. We do know fairly well the average separation between stars, so the fraction of stars with likely conversationalists will provide us with the key distance we are seeking.

How far away, then, is the closest civilization with which we might communicate? Since stars in the sun's region of the Milky Way are spaced about 2 parsecs apart, a sphere 20 parsecs in radius, centered at the sun, will contain 33 thousand cubic parsecs of volume and about 4000 stars (Figure 18-2). A sphere around the sun 200 parsecs in radius will have 1000 times this volume and will contain some 4 million stars, or about 1/100,000 of all the stars in the galaxy. For greater distances from the sun, we would have to consider the fact that stars in our galaxy are not distributed uniformly in all directions (Figure 18-3).

Now suppose that the average lifetime of a civilization equals 20,000 years, and that there are 20,000 civilizations in our galaxy. The galaxy's number of stars would then be 20 million times the number of civilizations, so that only one star in 20 million should now have a civilization capable of communication on one of its planets. If these civilizations are located at random in the Milky Way, we should be able to find the nearest such civilization by examining the 20 million or so nearest stars. This would require looking at all the stars within 350 parsecs from the sun—about a hundredth of the distance across the galaxy.

If, on the other hand, most civilizations last for 20 *million* years, then 20 million civilizations ought to exist now in the galaxy. Then one star in every 20,000 should have a civilization on an orbiting planet, so to find the

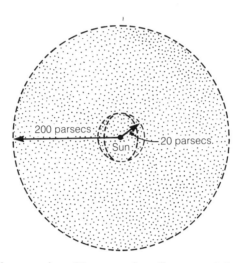

Fig. 18-2 If we draw a sphere 20 parsecs in radius around the sun, it will contain about 4000 stars. A sphere with a 10 times larger radius (200 parsecs) would contain about 1000 times more stars.

nearest civilization, we would have to search only among the 20,000 nearest stars, or out to a distance of 35 parsecs. (Notice, by the way, that even a lifetime of 20 million years for a civilization barely exceeds a thousandth of the lifetime of most main-sequence stars.) When we consider that the strength of any radio or television signal decreases in proportion to the *square* of the distance, we can see that to find a civilization 10 times closer to us means a saving of 100 times in the amount of power needed to produce a detectable signal in either direction. Unfortunately, we simply cannot make an accurate estimate of the number of civilizations—and thus of the distance to the nearest civilization—until we determine the average lifetime of other civilizations, by finding them or their echoes.

Fig. 18-3 The Milky Way galaxy is about 30,000 parsecs across its long dimension but only a thousand parsecs thick. About 1 percent (4 billion) of all the stars in the galaxy lie within 5000 parsecs of the sun; most of the galaxy's stars appear closer to the galaxy's center, especially within the 5000 parsecs closest to the center.

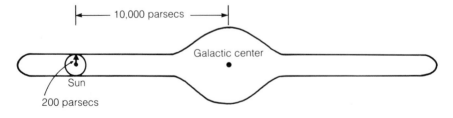

One more fact ought to draw our attention when we estimate the distance to the nearest civilization. If the average lifetime of a civilization turns out to be relatively short—say, only 1000 years—then the number of civilizations in the galaxy becomes quite small (about a thousand) and the average distance between civilizations increases. If only a thousand stars—one star per 400 *million*—has a civilization near it, then we must search through 400 million stars to find the nearest such civilization, which would take us out to distances of 1500 parsecs (4900 light years) from the sun. But now the time for any messages to travel between neighboring civilizations must, on the average, exceed the lifetime of civilizations, so that two-way contact would become unlikely. As Frank Drake first pointed out, in this case most civilizations would have *disappeared* by the time that a return signal arrived, even from their nearest neighbor. The cutoff point for this to occur corresponds to an average lifetime of about 3500 years per civilization. This would put neighboring civilizations about 630 parsecs (2000 light years) apart. If the average lifetime falls below 3500 years, then no messages can be exchanged with even the nearest civilization before it (and our civilization, too) will vanish. Conversely, if the average lifetime of civilizations exceeds 3500 years, then the nearest civilization should be less than 630 parsecs away, and several or, if we are lucky, many messages could pass back and forth before one civilization or the other passed into history.

We have based our calculations on our best estimate of the number of civilizations in the galaxy, which we have taken to be equal to a civilization's lifetime (in years). In less favorable cases, civilizations will be more widely spaced at any time, while in more favorable cases, the average distance between neighboring civilizations will be less. Figure 18-4 shows a graph of the number of civilizations, N, estimated in our galaxy, plotted against the average lifetime of a civilization. The heavy lines show the relationship between the number and lifetime that we expect if $N = L$; that is, if the number of civilizations in the galaxy equals an average lifetime, and also for the two extreme cases of $N = 300 L$, the most favorable case we care to imagine, and for $N = L/100$ million, an unfavorable case.

The dashed lines in Figure 18-4 which cut across the heavy lines show the number of two-way message exchanges that can occur for a given combination of separation and lifetime, with a given choice of how N depends on L. Let us suppose, for example, that we believe that N equals L. Then if L equals 100,000 years, the average distance between neighboring civilizations will be about 200 parsecs (650 light years), and two neighboring civilizations could exchange about a hundred messages back and forth before one of them vanished.

The graph shown in Figure 18-4 sums up our best answers to the four questions we posed: How many civilizations exist? How long do they last? How eager are they for contact? And how can we communicate with them?

Let us pause to notice that we have considered these questions as if the present moment in time were an entirely representative one. In fact, our galaxy has a finite age (just 10 billion years or so), and the ever-increasing age of the galaxy should see an ever-increasing number of planets where civilizations appear for the first time. The effect of this increase would be to make our use of the Drake equation (page 346) a more complex

Fig. 18-4   This graph shows the relationship between the average lifetime of a civilization with communication ability and desire (plotted on the horizontal axis) and the number of civilizations in the Milky Way galaxy (on the vertical axis). The graph has been drawn for three cases: $N = L$ (that is, the number of civilizations in our galaxy equals the lifetime of an average civilization, measured in years); $N = 300$ L; and $N = L/100$ million. The dashed lines show the number of two-way messages that could be exchanged in a given situation.

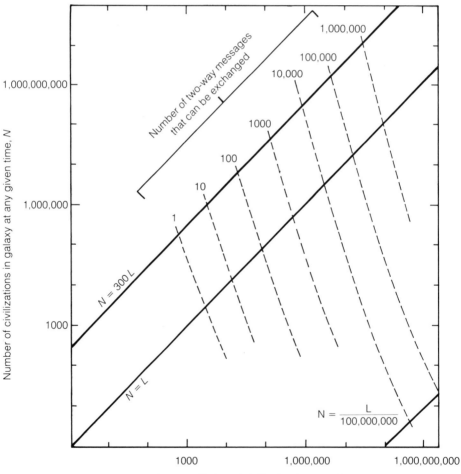

procedure than we have outlined, as we would have to take into account the fact that many stars have not yet existed for enough time to produce intelligent civilizations.

And how long is the lifetime L? How long does it take for an average civilization to vanish from the face of our galaxy? Although no one on Earth knows the answer to this question, we can imagine some possibilities for the development of a civilization once it has passed the stage that human society now represents.

Further Advances of Earthlike Civilizations

Our speculations about civilizations far more advanced than our own cannot have much accuracy, since we must extrapolate from our own experiences into an unknown future. One possible outcome for human society, its thorough destruction through war, can hardly lead to what we call a more advanced civilization, unless many cycles of destruction and rebuilding will, after millions of years, finally produce a stable society. In speculating about our future advances, we can guess what the immediate future may bring, and we can also see certain unavoidable limitations that physical laws impose on any society. Thus, for instance, we can speculate that the first contact with another civilization tends to increase a civilization's lifetime, as its members gain a new respect for their abilities. We can also state that no civilization, no matter how advanced, can build a spacecraft that will travel faster than the speed of light. But it is always possible that realms of physical experience as yet unknown to us may reveal extensions of our physical laws that will show us entirely new concepts, still beyond our imagination.

The prediction most often made concerning human interaction with the rest of the solar system states that our society will attempt to use the spaces that surround us in much the same way that we have used our Earth. That is, we shall try to extract materials of use to us; to build homes, offices, and sports arenas; to manufacture products for other humans to enjoy; and to make pleasure excursions from place to place. In short, we shall colonize space.

The chief proponent of space colonization now, Gerard O'Neill, of Princeton University, has led study groups that predict a population in space colonies of several billion people within 40 years after the time of initial construction. O'Neill says that the time to begin is now, and that within a decade matter mined from the moon, launched into space by superconducting slingshots and gathered into construction areas at the moon's distance from the Earth, could yield the first few space habitats (Figure 18-5). The preference for mining the moon instead of the Earth rests on the fact that to eject a given mass from the moon into space requires

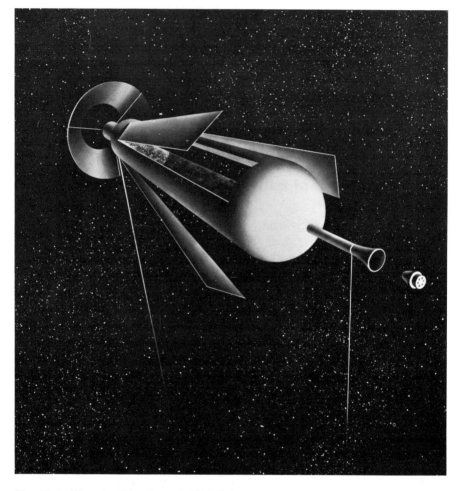

Fig. 18-5   The physicist Gerard O'Neill has suggested that we could mine matter from the moon's surface, fling it into space, and use this material to build cylindrical space habitats, each of which would have mirrors to reflect sunlight into the cylinder. Each habitat would be a few kilometers across and a few tens of kilometers long.

only 5 percent as much energy as launching the same mass from the Earth, because the force of gravity on the moon falls far below that on Earth.[1]

Since the moon's crust contains great amounts of oxygen, silicon, magnesium, aluminum, iron, and copper, O'Neill sees no great difficulty in

---

[1]The force of gravity at the moon's surface is one sixth as great as the gravitational force at the Earth's surface. The ratio of energies needed to escape from the moon and from the Earth shows a still greater effect, since we need only $1/22$ as much energy to escape from the moon's surface.

refining out the useful elements (while retaining the residue for shielding against ultraviolet light and cosmic rays), and then in building the cylindrical habitats, each a few kilometers long and half a kilometer across. The space cylinders would rotate to simulate the effect of gravity (Figure 18-6), and each colony would hold a few thousand people. Later versions, 10 times longer and wider, might each support a million inhabitants, using solar power and the material from the moon's surface to supply their needs.

Even the immediate neighborhood of the Earth-moon system contains plenty of room for thousands of space colonies, filled with billions of humans who might pass their entire lives in space. O'Neill suggests that an

Fig. 18-6   Inside the space habitats, each of the "land" strips would see the other two strips high in the "sky." In fact, each of the strips of land would be equivalent to the other two, and all would have the same sort of view. The rotation of the cylindrical habitats around the long axis of the cylinder would simulate the effects of gravity by what is familiarly called "centrifugal force."

initial goal for these colonies will be the manufacture, again from lunar materials, of immense power-generating stations that would convert sunlight energy into microwaves, which would be beamed to Earth to supply electrical energy wherever needed. Once made, the solar power stations would be placed in *synchronous* orbits around the Earth, at a distance of 36,000 kilometers above the Earth's surface (Figure 18-7). Objects orbiting at this distance from the Earth take the same time to complete one orbit as the Earth takes to rotate once, so if they move around the Earth in the same direction as the Earth rotates, they will appear stationary to an observer on the Earth's surface.

For our purposes, the key question raised by O'Neill's proposals refers, as always, to the average civilization, which ours may or may not be. Do most civilizations start to colonize the space around their home planets? And how far does such colonization proceed?

Confronted once again with questions that we really cannot answer, scientists have managed to produce some definitions and some reflections based on our knowledge of physics. Suppose that some civilization *does* begin to build colony after colony, until most, if not all, of their population lives in space habitats rather than on the starting planet. Why would they do this?

First of all, because life in space colonies can be more easily manipulated than life on a planet. A space habitat can, at least in theory, produce the climate, the atmosphere, and the political situation most desired by its inhabitants. (Whether some colonies would try to conquer others remains open to speculation.) Second, because space provides far more living room

Fig. 18-7 At a distance close to 42,000 kilometers from the Earth's center, an object will take just about 24 hours to orbit once around the Earth. If we place such an object in an orbit around the Earth's equator, in the same direction as the Earth rotates, the object will appear to remain stationary as seen by an observer on the rotating Earth. These synchronous orbits, used by weather satellites that constantly photograph the Earth, could also become useful "parking" orbits for satellites that generate electrical power from solar energy and then beam the power to receiving stations for use on Earth.

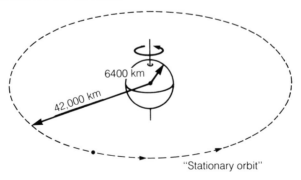

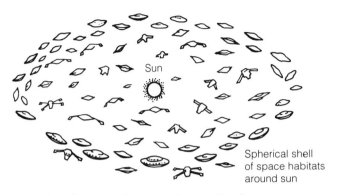

Fig. 18-8   We can imagine some future society on Earth, or some more advanced society elsewhere, that set out to use *all* of the energy that the sun liberates. To do this, we would have to cover an entire sphere around the sun, perhaps at the Earth's distance, with space habitats or other collecting areas. (The habitats could all be in orbit around the sun, but since the orbits would be unstable in the long term, small corrections would have to be made with special rockets from time to time.) Since the Earth intercepts about one billionth of the sun's energy output, we can see that we would need a billion habitats with the same cross-section as Earth to capture all of the sun's energy.

than a planet's surface. Most of the Earth lies far *inside*, while we live on its outermost shell and can mine only the first few kilometers of its 6400-kilometer radius. If we could spread out a planet like the Earth at will, we could obtain the material not for just a few thousand or a few million space habitats, but for trillions upon trillions. These colonies would spread all the way around the sun, thus making the maximum possible use of the photon energy that the sun emits (Figure 18-8). The Earth in its orbit now intercepts only about one billionth of the sun's energy.[2] That is to say, we would need a billion Earths around the sun to catch all of the sun's energy. Though we do not have a billion Earths, we could build space colonies that *could* capture all the light and heat that the sun produces. These space habitats could, at least in theory, support about a billion times the Earth's present population—4 billion billion (4×10^{18}) people, all using solar power to grow crops, to run their machines, and to enjoy life!

This speculation about the eventual growth of space colonies from Earth parallels an earlier idea of Freeman Dyson, another physicist from Princeton. Dyson suggested that an advanced civilization might wish to use the matter from one of the planets in its system to surround its star and "harvest" its energy. In this case, the civilization might remain small

---

[2]Even this one billionth of the sun's total energy output exceeds our present energy requirements on Earth by a factor of about 10,000. Therefore, if only we could convert solar energy into more "useful" forms at a reasonable cost, we could resolve the energy crisis at once.

in population and confined to the planet of its origin. But now it would have an enormous amount of energy at its disposal. If advanced civilizations have developed as either O'Neill or Dyson has speculated they might, then if we tried to find their central stars, we would not be able to see them. Instead, we would detect a glow of infrared light from the surrounding shells of matter. If space colonies or some type of radiation collectors catch a star's energy output and thus maintain themselves at temperatures above absolute zero—say, at the 300° Kelvin that characterizes the Earth's surface—they must radiate infrared light into space. We might hope to detect these civilizations by their infrared radiation, but we might, ironically, confuse this photon emission with that from a star in its initial stages of formation, when the protostar slowly heats itself by contraction and emits infrared light (see page 102).

Thus, we cannot count on being able to detect a civilization that has wrapped its parent star in a cocoon of colonies, unless we can share in their radio communications (see Chapter 20). For purposes of easy reference, astronomers call civilizations that use all the energy radiated by their parent star Type II civilizations. This terminology, invented by the Soviet astronomer N. S. Kardashev, designates as Type I those civilizations which, like ourselves, understand the basic laws of physics, can attempt interstellar communication, and have not yet enveloped their star in a nest of colonies designed to use all of the star's photons. Such a civilization—our own, for example—may use one trillionth as much energy per second as the parent star liberates. Therefore, if we go by energy usage, Type I civilizations fall about a trillion times below Type II civilizations.

A Type III civilization uses about a trillion times *more* energy per second than a Type II civilization. Fantastic? Only from our point of view. Kardashev uses the term "Type III" to describe civilizations that have not simply enveloped a star but are using an entire *galaxy* of stars as a source of energy. Since a giant galaxy may contain almost a trillion stars, Type III civilizations go as far beyond Type II as the Type II civilizations surpass the Type I. As we shall see in Chapter 19, travel between stars presents great difficulties but not impossibilities, provided that we set aside thousands or millions of years for the journeys. Our own Type I civilization has no urge to invest in these great voyages, but a Type II civilization could well take a longer view and decide to take a billion years or so to colonize an entire galaxy.

Our galaxy does not contain a Type III civilization, unless we occupy one of the last corners to be filled in,[3] and we cannot well estimate the number of Type II civilizations in the Milky Way. In terms of energy use,

---

[3]Of course, we can see hundreds of billions of stars in our galaxy, which tells us right away that a "pure" Type III civilization does not exist here. But we cannot exclude the possibility that for every star we see shining in our galaxy, another star has been wrapped in an energy-catching cocoon by an expanding civilization, well on its way to Type III stardom!

a Type II civilization would stand as far beyond a Type I as we (a Type I civilization) outdo a colony of bees. Since Type II civilizations would need a fairly long time—at least several thousand years—to wrap up their parent star, we can conclude with some confidence that only those civilizations with a large degree of stability will evolve from Type I to Type II.

Once again we face a crucial lack of knowledge—in this case, of the average lifetime of a civilization once it has achieved Type I status. If this average lifetime falls below, say, 500 years, then we would expect few Type II civilizations to exist in our galaxy. If, on the other hand, the average lifetime far exceeds a few thousand years, then conditions appear ripe for many Type I civilizations to evolve toward Type II. But what sorts of plans or devices does a Type II civilization have for interstellar communication?

We won't know until we talk to one. It does appear certain that any Type II civilization will take a longer view of life than we do, and it is possible that all the Type II civilizations in our galaxy are now in communication with one another. Perhaps they have adopted a common policy toward would-be members, and, much as the state of Oregon seeks to discourage overdevelopment of the natural beauty of its countryside, so too the galactic club of Type II civilizations may have decided that further members cannot be accommodated.

But we must not allow speculation about modes of development for advanced civilizations to restrict our attempts to answer our fourth question: How does interstellar communication proceed? The tentative answers are large enough to fill a chapter, the following one in this book, despite the fact that no interstellar communication has definitely been detected.

Summary

When we attempt to estimate the number of civilizations with which we might communicate, we face a formidable problem of making estimates that are not fanciful guesswork. If we limit ourselves to estimating how many civilizations capable of communication exist *in our own galaxy at the present time,* we can construct a formula that presents this number as the product of seven key terms.

The first number in our formula is the number of stars in the Milky Way galaxy, which we must multiply by the fraction of stars that last long enough for life to develop nearby. If we multiply this result by the average number of planets per star, and then by the fraction of these planets with conditions suitable for the origin of life, we obtain the number of planets in our galaxy with conditions suitable to life, in orbit around stars that last long enough for life to develop. Next, we must multiply by the fraction of *these* planets upon which life actually *does* develop at some point in the planet's history, and must then multiply by the fraction of those planets with life which see the emergence of a civilization with the ability to communicate. Our multiplication of the first six terms has then provided us with the number of planets in the Milky Way upon which intelligent civilizations have appeared at some time during the history of our galaxy. To find the number of civilizations in our galaxy *now,* we must multiply by one last factor, the ratio of a civilization's lifetime, once it acquires the ability and the desire to communicate, to the total lifetime of the galaxy.

The seven-term multiplication gives us an estimate of the number of civilizations in our galaxy now with whom we might exchange messages. When we insert our best guesses for the relevant numbers, we find that this key number approximately equals L, where L is the lifetime (in years) of a civilization that has achieved the ability and the desire to communicate. If L equals 1 million years, then we expect that about a million civilizations capable of interstellar communication now exist in the Milky Way. If, in contrast, L equals just 100 years, then only about a hundred such civilizations exist at any particular time.

Larger values of L imply more civilizations, which in turn imply a *lesser* value for the average distance between neighboring civilizations. If L equals, for example, 3500 years, then neighboring civilizations should be about 630 parsecs apart. Smaller values of L than this imply that most civilizations disappear before they can exchange a single round-trip radio message with their closest neighbor! On the other hand, if L equals 20 million years, then the average distance between neighboring civilizations should be a mere 35 parsecs, and civilizations could exchange thousands of messages back and forth during their millions of years of coherent existence. Which of these two possibilities more closely mirrors reality in our galaxy remains a problem that we can answer only by finding other civilizations.

Questions

1. Why does the lifetime of a civilization enter directly into our estimate of how many civilizations now exist in our galaxy?
2. Why do scientists feel more confident about estimating the fraction of planets on which life develops than the average lifetime of an extraterrestrial civilization?
3. Why are we interested in finding the average distance between civilizations in the Milky Way galaxy?
4. If the average lifetime of civilizations turns out to be less than 5000 years, we may be unable to establish two-way communication with other civilizations. Why?
5. Using the approximation $N = L$, where N is the number of civilizations in our galaxy at any time and L is the average lifetime of a civilization, how many round-trip radio messages could neighboring civilizations exchange if $L = 1,000,000$ years (see Figure 18-4)?

6. Why might an advanced civilization's parent star become totally invisible? How could we detect such a civilization?

7. What is a Type III civilization? What would we expect to see when we look for the galaxy in which such a civilization exists?

8. If we use the approximation $N = L$, how long would an average civilization's lifetime have to be for the civilization to exchange at least 10 round-trip messages with its closest neighbor?

Further Reading

Asimov, Isaac. 1979. *Extraterrestrial civilizations*. New York: Crown.

Drake, Frank. 1962. *Intelligent life in space*. New York: Macmillan.

Ponnamperuma, Cyril, and Cameron, A. G. W., eds. 1974. *Interstellar communication: Scientific perspectives*. Boston: Houghton Mifflin.

Sagan, Carl, ed. 1973. *Communication with extraterrestrial intelligence*. Cambridge, Mass.: The M.I.T. Press.

Sagan, Carl, and Shklovskii, Josef. 1966. *Intelligent life in the universe*. San Francisco: Holden-Day.

Sullivan, Walter. 1966. *We are not alone*. Rev. ed. New York: McGraw-Hill Book Company.

CHAPTER 19

How Can We Communicate?

When human beings have thought about meeting living creatures from other parts of the universe, they have used their intuition and experience to picture a physical interchange, a face-to-face confrontation that may end in friendship or in violence. Scientific analysis, however, suggests that the best means of interstellar communication consists of radio messages, not spaceflight. Thus, we should anticipate an exchange of television programs, not a showdown in space, as the way in which we shall first encounter other civilizations.

The Superiority of Radio and Television

What makes space travel so difficult, and radio messages so easy? The answer lies in the problem of accelerating particles with mass, such as protons, people, or spaceships, to large velocities. Particles with *no* mass, such as the photons that form light waves and radio waves, always travel at the speed of light, 299,793 kilometers per second. Since the speed of light appears to be absolutely the greatest velocity that any particle can reach, light waves and radio waves always travel through space as fast as is universally possible.

In addition, photons can be produced in large numbers cheaply and easily. To send a radio message of, say, 5 minutes' duration from the Earth to the moon, or from the moon to the Earth, requires about 1 kilowatt-hour of electrical energy, which costs about 10 cents. The transmitters, antennas, and receivers used in sending and receiving such messages cost thousands of dollars but still many times less than the cost of space travel. For comparison, to build a Saturn rocket capable of sending men to the

Fig. 19-1 The Apollo 11 spacecraft was sent to the moon by a Saturn V rocket, the most powerful ever built in the United States.

moon (Figure 19-1) cost many hundreds of millions of dollars, plus billions more for developing the systems that supported this effort.

When we consider larger distances than that from the Earth to the moon, we find that the ratio of costs between a manned spacecraft and a simple radio message grows steadily larger as we think about greater distances and greater spacecraft speeds. Radio waves keep on traveling at the speed of light, whatever the distance they must travel. Spacecraft, too, will coast at a constant velocity, but only after we accelerate them to that velocity, which may require immense amounts of energy. To travel to other star systems at the speed of the Apollo journey to the moon would take tens of thousands of years, but radio waves can make the trip in only a few years. If we try to decrease the spacecraft's travel time by increasing its speed, we must expend far more energy, and build far more sophisticated spacecraft, than we have done so far.

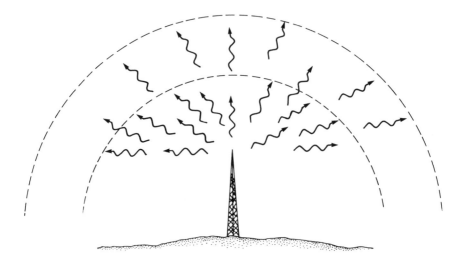

Fig. 19-2 Any signal carried by photons, such as those in radio waves, will diminish in intensity by a factor of 4 each time the distance from the source doubles. At twice the distance, the radio signal must pass through an area four times greater than before.

Let us therefore consider the exchange of messages by radio waves. Although radio photons move through space at the speed of light, the *intensity* of a beam of radio waves (the number of photons that cross one square centimeter of area each second) decreases with distance. In fact, for simple geometric reasons, the intensity of a beam of photons varies as 1 over the *square* of the distance from the photon source (Figure 19-2). Thus, to produce a signal of the same intensity at a thousand times the original distance, we must provide *1 million times* more photons each second from the source, since 1000 squared equals 1 million.

To produce more photons requires more energy, so that communication over ever-increasing distances does require ever-increasing amounts of energy; in fact, the energy cost rises as the square of the distance to be covered. Hence, communication over interstellar distances, millions of times greater than the Earth-moon distance, would require *trillions* of times more photon energy, *if* we used the same equipment and the same sorts of messages as were employed by the Apollo astronauts. Since these costs would rise into the hundreds of billions of dollars just to generate the photons that carry the messages, it is clear that we must instead use shorter messages and much, much more sensitive antennas if we seek to communicate with other star systems. In the next chapter we shall discuss the kind of antenna arrays that are envisioned for interstellar communication. But we can be sure that if ordinary *photon* communication were to cost billions of dollars, then the expense of human space travel would be so far out of sight as to be unattainable, at least in the foreseeable future.

Interstellar Spaceships

We have seen that photons can carry information far more efficiently than can any "physical" object, namely objects with mass. We recognize this fact early in our lives when we listen to our radios and televisions rather than waiting for the governmental courier or the town crier to come and give us the news. Nonetheless, our biological heritage and much of our social training tell us that personal visits have a special importance. Thus, the Soviet Premier and the U.S. President like to meet each other, even though they have special telephone circuits for their individual communication by radio.[1] The very effort of making such personal visits seems to underscore their significance. We find, therefore, for every report of a strange, perhaps an extraterrestrial, signal on radio or television broadcasts many thousands of reports of extraterrestrial visitors here on Earth in person.

Because of this urge to travel, which arises from the curiosity and gregariousness shared by humans everywhere, we shall pause to examine how interstellar spacecraft might work. This investigation has a value beyond that of simply convincing *ourselves* that photons will better serve our intentions, for the consideration of such spaceships shows the physical principles that we think must apply throughout the universe, and these principles should lead other civilizations to similar conclusions.

Let us imagine, then, what a reasonable interstellar spaceship might be like. First of all, we must deal with the overwhelming fact of life in space: *Distances are enormous.* A simple ratio to remember is that the distance from the sun to the next closest star is *100 million times* the distance from the Earth to the moon. If we imagine the journey from the Earth to the moon (400,000 kilometers) to be equivalent to a trip to the refrigerator, then the distance to Alpha Centauri would itself be like the voyage to the moon!

We often hear people say that because no one a hundred years ago dreamed that we could build flying machines, let alone rockets to the moon, so too in another hundred years we shall travel to the stars by means yet undreamt of. Although pessimistic thoughts make few friends, this analysis founders on the rock of reality. The fact is that the mechanics of gliding and of rocket propulsion had been studied and fairly well understood for many centuries. What was lacking was the power plant or engine capable of making into a useful vehicle what had only been a toy. In an analogous manner, the mechanics of space flight have been understood in outline for three hundred years and in detail for sixty, but *even if some*

[1]Notice that no one bothers to say how often the two leaders use these circuits: The excitement lies in personal visits.

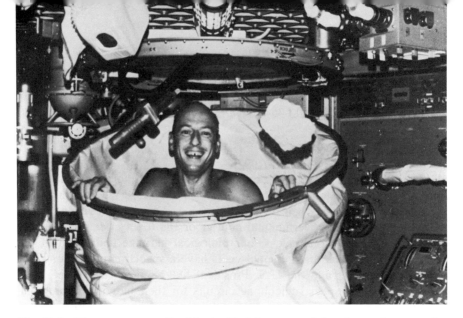

Fig. 19-3 Three astronauts lived in the Skylab spacecraft for almost three months, confined to living quarters that measured about 20 meters by 8 meters by 8 meters.

marvelous new source of energy should be discovered, the constraints imposed by the human requirements and by the time of flight would remain. Let us examine these limiting effects in greater detail.

Think about the space ships that you see in science-fiction movies such as *Star Wars* and on television in programs such as "Star Trek," and you may be struck by two key facts, each worthy of some thought. First of all, the spaceships tend to be enormous. Why? Because people need room in which to function, and the ship must carry all the fuel, food, and other items they require. The Skylab vehicles (Figure 19-3), in which astronauts lived with moderate discomfort for several months, measured some 60 meters in length and had a mass of about 10^{10} grams (10,000 tons). The *Star Trek* ship *Enterprise* must be at least 10 times longer, and about 1000 times more massive, than Skylab. Amazingly large as the *Enterprise* may be in terms of present-day technology, we can easily show from an example developed by Edward Purcell that this ship is actually too small for interstellar travel!

Suppose that a spacecraft liberates energy of motion by bringing together matter and antimatter to convert all of the energy of mass into kinetic energy, and suppose that this process occurs with 100 percent efficiency. How much matter and antimatter do we need for a journey to another star? We can calculate the answer rather easily, if we assume that the spaceship aims to achieve a velocity of 99 percent the speed of light. At this velocity, interstellar journeys between neighboring stars will require several years; at a lower velocity, the travel time would of course be longer.[2]

[2]To travel at speeds greater than 99 percent of the speed of light would not significantly lessen the travel time as seen from outside, though the space crew would seem to have spent less time on the journey (see page 373).

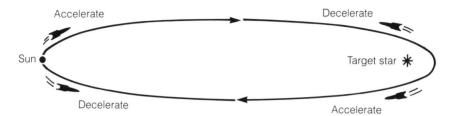

Fig. 19-4 To travel to a nearby star and back at nearly the speed of light, we must make four separate accelerations. First, we must reach nearly the speed of light; then, we must slow down at our destination. Third, we must accelerate to nearly the speed of light toward the Earth, and finally we must again decelerate to land on Earth.

To reach a velocity of 99 percent the speed of light by letting matter and antimatter annihilate each other, we must start with *14 times more mass* than we shall have when we reach our cruising velocity. If we aim to slow down at the far end of the journey, we need another factor of 14 to provide the deceleration. Thus travel from here to another star, at 99 percent the speed of light, requires *196 times* as much matter-antimatter fuel as the final payload! Finally, of course, we may wish to return home, in a reverse analogue of the journey out; this will require yet another 196 times as much fuel, starting out there, as the payload's mass upon our return (Figure 19-4). Hence, if we hope to make a journey in a self-contained spaceship out to another star, decelerate there, start back home, and stop when we return, we need 196 × 196 or almost 40,000 times as much mass in the form of matter-antimatter fuel as we use for the payload—crew's quarters, bridge, and so on. To carry a total payload of, say, a modest 10,000 tons, like the Skylab, we must begin with 400 million tons of fuel, half of which is anti-matter! This much fuel would fill a volume of a cubic kilometer, much larger than the entire volume of the starship *Enterprise*.

Thus, the matter-anitmatter annihilation that the *Enterprise* uses has significant drawbacks, since we need a fantastic amount of fuel to make things work. Yet this is *the best source of energy we can imagine,* though at present we have no way to use it. It is time, however, to turn our attention to the second difficulty that science-fiction plots must ignore: the long, not to say tedious, travel times before we arrive anywhere interesting.

The second striking illusion about space flight as seen on television and in movies is that travel is made to occur almost instantaneously. In real life, voyages between stars would require many years, if not many centuries. We must distinguish at least two sorts of imaginary spacecraft: those that could travel to the nearest stars at about the same speed as our best rockets can now, and those that could travel tens of thousands of times faster, at speeds close to the velocity of light. Table 19-1 shows the distances that various speeds will take us in one year, and we can see that

TABLE 19-1

DISTANCES TRAVELED IN 1 YEAR BY PHOTONS AND BY VARIOUS PARTICLES WITH MASS

Traveler	Distance Traveled in 1 Year (km)	Number of Years Needed to Cover the Distance that Photons Travel in a Year
Human being (0.0003 km/sec)	10,000	1,000,000,000
Automobile 0.03 km/sec)	1,000,000	10,000,000
Jet aircraft (1.2 km/sec)	38,000,000	263,000
Pioneer 10 (10 km/sec)	333,333,333	30,000
Fusion spacecraft (3000 km/sec)	100,000,000,000	100
Photons (300,000 km/sec)	10,000,000,000,000	1

the fastest-moving object that we have made so far, Pioneer 10, still travels outward from the sun at one twenty-thousandth of the speed of light.

With our minds made up, we could probably build the first sort of interstellar spacecraft, capable of travel at Pioneerlike speeds, within the next century or so. Such a spaceship would take tens of thousands of years to reach even the nearest star to the sun. Hence, those who support these missions may face serious problems in convincing others to support them. For who wants to receive news of another civilization a hundred generations from now? Yet this kind of mission remains the only technologically feasible one within the foreseeable future. If a choice had to be made between spending money to search for extraterrestrial signals *now* and spending money to send out a space crew to return after, say, 200,000 years, then an immense prejudice in favor of human journeys would be needed to choose the second alternative.[3]

As we mentioned earlier, we cannot plan on building a spacecraft that could travel anywhere close to the speed of light with our present technology. Yet *only* by traveling at nearly the speed of light can we hope to cover interstellar distances in tens of years. The need to imagine such a spacecraft is in fact so great—at least as far as the consideration of human travel to the stars is concerned—that we shall for the moment imagine that we *could* build a ship capable of traveling almost as fast as photons. Then the crew aboard the spacecraft could profit greatly from an additional, fascinating aspect of travel at this enormous velocity: *For the crew, time would slow down as the ship moved through space with nearly the velocity of light.*

[3]To grasp the significance of such an interval of time, try to imagine what your ancestors were interested in 200,000 years ago!

Fig. 19-5 The Stanford Linear Accelerator passes beneath Highway 280 in Menlo Park, California. Every working day, physicists routinely use this accelerator to accelerate charged particles, mostly electrons and protons, to speeds in excess of 99 percent of the speed of light.

Time Dilation

The theory of relativity predicts that time will slow down or dilate, and this effect has been repeatedly measured in particle accelerators, where physicists accelerate elementary particles to almost the speed of light every day (Figure 19-5). But how can time slow down? And how do we know *which* object has a speed close to the speed of light: Aren't all velocities relative? These questions have answers which are directly relevant to the problem of high-velocity space travel.

First of all, how can time slow down? We measure the flow of time in various ways; for example, by the length of time it takes for a clock to tick, for the Earth to rotate or to orbit around the sun, or for a human being to grow old. All such seconds, days, years, or lifetimes have a standard relationship to one another; thus, for example, each non-leap year contains 31,536,000 seconds. Whatever *units* of time we employ on Earth, we still think we are measuring the same flow of time, and that two identical clocks will tick at precisely the same rate, even if one clock is in San Francisco and the other in Rome.

What Einstein predicted, however, and what experiments have verified, is that the clocks will tick at the same rate only if neither clock moves with respect to the other. If in fact we put the clock in Rome aboard a jet airplane and fly with it at a speed of 1000 kilometers per hour, then an observer in San Francisco who measures the clock's rate of ticking will find that *the clock in motion ticks slightly more slowly than the clock at rest.*

For speeds much less than the speed of light, such as 1000 kilometers per hour, the effect can barely be measured, as it amounts to only a tiny fraction of a percent. But for speeds close to the speed of light, 300,000 kilometers per *second,* the slowing down of time can assume great importance. A clock that passes by an observer at 99 percent the speed of light will tick 7 *times more slowly* than a clock at rest with respect to that observer.[4] This time dilation will occur whether the clock in question is mechanical, atomic, or biological, so we must conclude that the flow of time within a moving system does indeed slow down, in comparison with the flow of time in a system at rest with respect to the observer.

Now comes the truly amazing part: How can we tell which clock is "at rest" and which is "in motion"? Isn't all motion relative? The answer to this is *yes,* all motion is relative, provided we are talking about *unaccelerated* motion (motion in a straight line at a constant speed). But accelerations (changes in speed, in direction, or in both speed and direction) are *not* simply relative and can be recognized as accelerations by any observer. A detailed analysis of the time dilation effect, as well as experimental results, shows that particles that are accelerated to almost the speed of light will experience time passing more slowly than they would if they remained at rest. We may conclude that a person who travels through space at nearly the speed of light and then returns to Earth will age less during the journey than a person who stays at home. This slowing down of time has a progressively greater effect for speeds closer and closer to the speed of light, 300,000 kilometers per second. Thus, if an astronaut travels to Alpha Centauri (4.3 light years away) at 95 percent of the speed of light and returns at the same speed, the astronaut will age by only 3 years, while people on Earth age by 9 years during the astronaut's voyage. If the astronaut travels at 99 percent the speed of light, people on Earth will now calculate that the journey takes 8.7 years, but the astronaut will age by a year and three months! If we imagine speeds still closer to the speed of light—say 99.999999 percent of the speed of light—we find that a journey that covers 10,000 light years (3000 parsecs) would seem to take 10,000 years to the people left on Earth but would take just over a year from the life of those making the trip. Table 19-2 shows the distances a spacecraft could cover in various intervals of time, *if* it could accelerate to speeds closer and closer to the speed of light.

---

[4]The algebraic expression for the time dilation, if v is the velocity of the moving clock and c is the velocity of light, comes out as:

$$\text{Time elapsed in system at rest} = \frac{\text{time elapsed in moving system}}{\sqrt{1 - (v/c)^2}}$$

TABLE 19-2

Round-Trip Times for Journeys at an Acceleration of 1 g[a]

Time as Measured by Spacecraft Crew (years)	Time as Measured on Earth (years)	Greatest Distance Reached (parsecs)	Farthest Object Reached
1	1	0.018	Comets
10	24	3	Sirius
20	270	42	Hyades
30	3100	480	Orion Nebula
40	36,000	5400	Globular star cluster
50	420,000	64,000	Large Magellanic Cloud
60	5,000,000	760,000	Andromeda galaxy

[a]Following an example by Sebastian von Hoerner, we imagine a spacecraft that accelerates and decelerates at 1 g; that is, the force of acceleration or deceleration is equal to the force of gravity at the Earth's surface.

The Difficulties of High-Velocity Spaceflight

It seems evident that spaceflight at speeds near the speed of light is highly advantageous, if we could just devise a way to do it. Suppose we tried another approach. Part of the difficulty we face in interstellar rocket travel is the need to propel the fuel as well as the payload. Robert Bussard has suggested instead that we should consider an interstellar "ramjet" that could scoop up the interstellar gas along its path with some type of magnetic funnel that would require an entrance area hundreds of square kilometers in area. This gas would then serve as the fuel for a fusion reaction engine that would use the same nuclear reactions that make the sun and the stars shine to drive the spaceship. As we discussed in Chapter 4, interstellar gas consists mainly of hydrogen atoms and hydrogen molecules, with some helium (about 10 percent of the hydrogen, by number) and a trace of all the other elements (about 1 percent of the total for all of them). The average density of matter in interstellar space, an incredibly tiny 10^{-24} gram per cubic centimeter, provides just about 10 atoms in every cubic inch of the interstellar medium. Such a low density of matter can be modeled by placing one tennis ball inside the state of Missouri. Nonetheless, these incredibly rarefied interstellar particles could provide both the potential fuel for an interstellar spaceship and also an almost certain *obstacle* to travel at speeds close to the speed of light.

Following Bussard, we can easily imagine, if not construct, a rocket ship designed to scoop up hydrogen atoms and molecules over an enormous area as the ship plunges through space (Figure 19-6). We must worry about the energy required to maintain the magnetic field that collects the particles. This energy would have to be added to the energy requirements for moving the spaceship itself. At present, we must admit that we lack the technology to accomplish either objective, but even if we could, our problems would not be over. As the spaceship's speed increased, the incoming particles of interstellar matter would appear more and more formidable until they became so dangerous as to prohibit further increases in the ship's speed.

Why? The same theory of relativity that predicts the slowing down of time at high speeds also predicts that *a particle's kinetic energy will increase dramatically as the particle's speed increases.* We all know that in familiar situations, faster-moving particles have more kinetic energy than slower-moving particles do. At velocities far less than the speed of light, a particle's kinetic energy varies in proportion to its mass times the *square* of its velocity. Thus, for instance, an automobile moving at a speed of 100 kilometers per hour has *16* times its kinetic energy at a speed of 25 kilometers per hour, which makes collisions at higher speeds far more destructive. Einstein showed that as particles near the speed of light, the simple rule of proportionality stated above does not hold, and that particles' kinetic energies increase *much more rapidly* than they do at lower

Fig. 19-6   We can imagine a rocket ship designed to scoop up interstellar atoms, molecules, and dust grains for fuel, thus using a "Bussard drive," named in honor of the man who first suggested it. The ship, however, would see oncoming particles as bullets of ever-increasing energy as it moved closer to the speed of light.

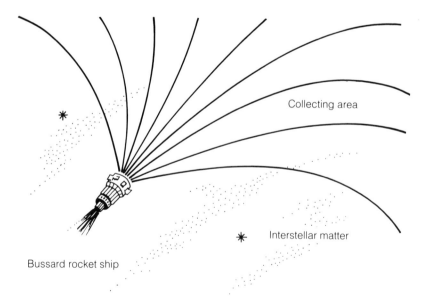

speeds. A particle moving at 99.9 percent of the speed of light has 3.2 times the kinetic energy it has when it moves at 99 percent of the speed of light. At 99.99 percent of the light velocity, the particle gains another factor of 3.2 in its kinetic energy. In order to move at exactly the speed of light, any particle with some mass must have an *infinitely large* kinetic energy. This is one reason why physicists do not believe that any particle with mass can move as fast as the speed of light.

Consider the effects of this enormous increase in kinetic energy. First of all, we can see that in order to accelerate a spaceship to almost the speed of light, we must supply the ship with an enormous amount of energy as the ship speeds closer and closer to the velocity of light. This creates a problem of supplying such huge energies, which prohibits the use of ordinary fuel carried along with the spaceship. Hence our reliance on interstellar hydrogen for rocket fuel seems ever more necessary.

No matter what propulsion system we use, however, we encounter still another serious problem as we try to travel at almost the speed of light. As the ship moves faster and faster, all of the particles in the interstellar medium become progressively more dangerous projectiles. From the spaceship's point of view, these particles are now approaching at speeds close to the speed of light. Thus, measured in the frame of reference of those on board the spaceship, each atom, molecule, or tiny dust particle will possess a kinetic energy that grows enormously as the particles approach the ship at, say, 99.99 percent of the speed of light. These tiny particles through which the ship must pass, and which it is trying to collect for fuel, take on the energies of bullets and bombshells. If the ship must expend a considerable fraction of its available energy to deal with the onslaught of these projectiles (so they appear from the ship), that much less energy will be left for propulsion. As the spacecraft's speed rises closer and closer to the speed of light, there may be no effective way to be sure that none of the oncoming particles can penetrate the spacecraft with devastating effect.

What this means is that the highest velocities suggested for future spacecraft may prove to be unattainable in practice, either because of the lack of propulsion systems or because of the danger of traveling through the interstellar medium at speeds within a fraction of a percent of the speed of light. We may be able, someday, to travel through space at perhaps 99 percent of the speed of light, so that a journey of 100 light years would age us only 14 years and thus would return us to the world of our great-grandchildren, but we shall probably never be able to reach the speeds of, say, 99.999999 percent of the speed of light that would enable us to travel through most of our galaxy within a human lifetime.

To keep all these plans in perspective, we would do well to recall that the greatest speed yet attained by a humanmade spacecraft equals not 99 percent nor 0.9 percent of the speed of light, but a mere 0.005 percent of the light velocity. Great as our achievements may have been during the last few centuries, we cannot place much confidence in a straightforward extrapolation that suggests to us that within another few centuries we shall be able to travel 20,000 times faster. For the time being, we have no way to do so. Furthermore, it is an ironic fact that the same relativity theory that makes travel at nearly the speed of light important—because of the time dilation effect—also has the effect of making the interstellar medium, through which any spacecraft must travel, an ever more dangerous hail of bullets as the spacecraft's velocity nears the universal speed limit.

Automated Message Probes

In the year 1972, the spacecraft Pioneer 10 set out from Cape Kennedy, accelerated away from the Earth, flashed past the planet Jupiter, and by 1979, coasting at about 10 kilometers per second (36,000 kilometers per hour), was about 2.3 billion kilometers away from us, almost as far from the sun as the planet Uranus (Figure 19-7). Pioneer 10 will continue to sail through space at a fairly constant velocity (relative to the sun) until the spacecraft happens to come close to a source of gravitational force, for example, a star. If the star in question were Alpha Centauri, the closest star to the sun, then this encounter could occur in about 100,000 years.

Pioneer 10's trajectory, however, points nowhere near the direction of Alpha Centauri; instead, it is headed for a point in space close to the boundary of Taurus and Orion. Hence, any encounter with another star will take longer than 100,000 years to occur—far longer, in fact, when we consider how empty space is of stars. Because the Pioneer spacecraft has no guidance system, it can encounter other stellar systems only by chance, and the chance of its coming close to another star within, say, 1 billion years remains infinitesimal. We can calculate that this spacecraft, which takes 100,000 years to cover the average distance between stars, must traverse 100 trillion times this distance if it is to intersect a planetary system like our own. But such travel will require 100 trillion times 100,000 years, or 10 billion billion years, almost a billion times the age of the universe! This enormous waiting period, or, more honestly, the vanishingly small chance that Pioneer 10 will ever encounter another planetary system (let alone one that supports intelligent life) reminds us, first, that space is mostly empty, and, second, that the message plaque on Pioneer 10 seems destined to be worn away by interstellar dust before any other civilization sees it. Likewise, the gold-coated copper records on the Voyager spacecraft (see page 318) will sail onwards through our galaxy, carrying messages and pictures from Earth, but without much of a chance that any other intelligent species will ever admire them.

Still, we have begun our attempts at communcation. The Pioneer 10 plaque represents humanity's first effort to send interstellar messages by rocket, for possible recognition by other civilizations. Once the idea of

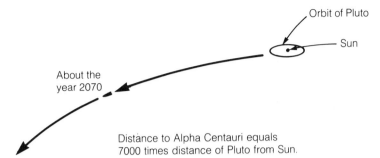

Orbit of Pluto

Sun

About the
year 2070

Distance to Alpha Centauri equals
7000 times distance of Pluto from Sun.

Fig. 19-7 The Pioneer-10 spacecraft, which left Earth in 1972, has passed the orbit of Uranus, 19 times farther from the sun than our Earth. At its speed of 10 kilometers per second, however, the spacecraft will require 100,000 years to travel a distance equal to Alpha Centauri's distance from us.

communication has arisen, better means can be devised. We might conclude from the vanishingly small chances of the Pioneer and Voyager messages that we would be foolish to send spacecraft without human control into space, hoping that such messengers could eventually encounter another civilization. A closer look, however, shows that the right kind of spacecraft could perform far better than those we have sent so far. What we need are spacecraft that send radio messages as they pass close to stars. Ronald Bracewell has suggested that sophisticated robot scouts of this type could be in common use by advanced civilizations for routine exploration of our galaxy. There is even a possibility that one or more of these interstellar robots could be in our solar system now, lost among the asteroids, or slowly circling the sun beyond the orbit of Jupiter.

An automated message probe, launched from Earth now, could reach the vicinity of Alpha Centauri after about 100,000 years. Similar probes could be directed, for relatively little cost, at other nearby stars similar to our sun.[5] These spacecraft, in their most sophisticated versions, would each contain a radio receiver and transmitter. The receiver could detect any radio or television messages emitted by a civilization near another star; that is, a small computer aboard the spacecraft could analyze any radio output to see whether the photons arrive in random or in coded patterns. If the spacecraft did detect civilization-made radio and television, it would then deliver a prerecorded message at exactly, or nearly, the same frequency as the local transmission.[6] This matching of frequency would maximize the chance that the spacecraft's signal would be detected and analyzed.

[5]If we mass-produced such probes and the rockets to send them into space, a reasonable estimate might be $100 million per probe, so that for the cost of putting a man on the moon, we could send off several hundred automated messengers to other stars.

[6]It has been suggested that by selling advertising time in this prerecorded message, we could finance part of the costs of sending the automated spacecraft!

Although the basic purpose of automated probes would be simply to announce that we exist, even a small spacecraft could broadcast a message of several billion "bits" of information, enough to transmit all the information in the *Encyclopedia Britannica*. With allowances for the difficulty of teaching another civilization how humans communicate their thoughts, our current technological capability assures that we could transmit just about everything we could think up about the Earth with a rather small automated spacecraft in the near future.

Included in the message aboard such a spacecraft would be, of course, the *origin* of the message, if we seek two-way communication. Other automated spacecraft, perhaps deliberately modeled on our own, could be sent back to us by any civilization as advanced as ours, so we might be in danger of exhausting all that we have to say with a single exchange of messages! The chief drawback in setting this scheme in operation remains the long travel times—hundreds of thousands of years to the closest stars, millions or tens of millions of years to those somewhat farther away. Hence, the transfer of information would occur, at least to begin with, at an extremely slow rate. In contrast, messages sent directly by radio can travel more than a thousand times faster, and would take proportionately less time for the journey.

In actuality, only the *first* message would have to travel so slowly. Afterwards, with exact knowledge of where to look, photons could carry the information at the speed of light, and could thus reduce the round-trip travel time to decades or centuries. Automated message probes make sense as a one-time affair (per star to be investigated). In comparison with beams of radio waves, message probes compensate for their greater construction cost by traveling in one piece to the star in question. Radio beams spread out as they travel, so that we must invest more and more power to reach farther and farther stars and planets. On the other hand, once we have launched a spacecraft past the solar system, it can coast through space essentially forever without diminishing its ability to function. Once the probe nears another star, solar (that is, starlight) energy cells can provide power to the receiver, transmitter, and computer aboard the spacecraft.

If we chose to, we could launch one such spacecraft each week with no great strain on our budget. The reason that we don't do so, assuming that we do have the urge to contact other civilizations, comes mostly from the long travel times. Few of us support wholeheartedly the sending of messengers to tell their story a million years in the future. Interstellar communication makes a splash in our minds when we think of communication *now;* communication efforts that may pay off for the ten-thousandth or fifty-thousandth generation can safely be left to our descendants.

But who, then, will begin the process? The answer to this conundrum may come from the possibility of "eavesdropping" on radio waves not produced with interstellar communications as their goal. Or we may find that other civilizations less timid, better established, and more secure than ourselves have performed the same analysis but have reached a positive conclusion. Thus, any night now our televisions could light up with an unfamiliar signal—not WABC, KTVU, or KNXT, but EXO, OBOY, or YUHU, signals of far-out programming that will treat us to a real 90-minute spectacular. Then we would know that automated message probes do indeed make sense.

Summary

Although many people assume that interstellar travel must soon become a human reality, analysis using the laws of physics shows that interstellar spaceflight must remain incredibly difficult, if not downright impossible, for the foreseeable future. The best spacecraft humans have built until now travel at one twenty thousandth of the speed of light, and would take 100,000 years to reach even the closest star to the sun. If we seek to build much faster spacecraft, we must find new ways to accelerate the craft to greater velocities; such acceleration requires tremendous amounts of fuel, even if we assume that we devise a way to use a 100 percent-efficient annihilation of matter and antimatter.

If we could somehow build a spaceship that traveled at nearly the speed of light, then the time dilation effect, first discussed by Albert Einstein, would allow space travelers to age more slowly than those who stay behind, since time appears to slow down for those who travel close to the velocity of light, in comparison with those who stay at home to await their return. The same theory that predicts the slowing down of time, however, also states that at speeds close to the speed of light, every tiny particle of interstellar gas or dust would effectively become a bullet of enormous energy, as perceived by the spaceship or by those inside it. Hence, we would also have to devise a way to deflect these bullets, and such deflection adds to the energy problem we encounter in accelerating the spacecraft to near-light velocities.

In contrast to the manifold difficulties we meet in trying to move particles with mass at greater and greater speeds, *photons,* because they have no mass, always move at the speed of light, greater (at least by a tiny bit) than the speed *any* particle with mass can *ever* achieve. Photons do not cost much to produce, and even for the relatively short distances we meet on Earth, photons have come to predominate over particles with mass for sending messages. When we consider the enormously greater distances that exist between neighboring civilizations, we find that photons have a tremendous advantage over "nuts and bolts" spacecraft. Even though we may take pleasure in thinking about interstellar space travel in our own flesh, the physics of the situation—the fact that civilizations are spaced many parsecs apart—argues in favor of photons as the best way to communicate whatever has to be said. If our conclusion has merit, then we ought to try to figure out what sort of interstellar messages are passing by us right now, and how we can join the network of intercommunicating civilizations, should we choose to do so.

Questions

1. Why do radio waves seem far superior to spacecraft for interstellar communication?

2. Why are radio waves more difficult to detect at greater distances from the radio transmitter?

3. Why must any interstellar spaceship carry an enormous amount of fuel, if it is to reach speeds that are a significant fraction of the speed of light?

4. What difficulties stand in the way of a spacecraft that scoops up fuel from interstellar gas and dust as it travels through space at 99 percent of the speed of light?

5. What is the time dilation effect? How could this help with the problem of the immense amounts of time that interstellar journeys require?

6. For velocities, v, that are nearly equal to the speed of light, c, the energy needed to reach a given velocity varies in proportion to the ratio $c/(c-v)$. How much more energy is needed to accelerate to a velocity of 99.9 percent of the speed of light than is needed to reach a velocity $v = .99\,c$?

7. About how long should it take for the Pioneer plaque to be carried into another planetary system? What does this imply about the usefulness of sending such messages into space?

8. What advantages could we design into an automated message probe that would render it far superior to the Pioneer plaque and the Voyager record?

Further Reading

Bracewell, Ronald. 1975. *The galactic club*. San Francisco: W. H. Freeman and Company.

Mermin, David. 1972. *Space and time in special relativity*. New York: McGraw-Hill Book Company.

Purcell, Edward. 1963. Radio astronomy and communication through space. In *Interstellar Communication,* ed. by A. G. W. Cameron. New York: W. A. Benjamin, Inc.

Science Fiction

Anderson, Poul. 1970. *Tau zero*. New York: Lancer Books.

Heinlein, Robert. 1957. *The door into summer*. New York: Doubleday, Inc.

CHAPTER 20

Interstellar Radio and Television Messages

We have seen that interstellar travel requires large amounts of energy, and the more rapidly we wish to travel, the more energy we must supply to our spacecraft. We do not now have the means to build an effective interstellar voyager, and even as our technology improves, we may remain unwilling to spend so much energy if less expensive means of communication are available. If we can agree that informational contact rather than face-to-face dialogue represents our primary goal, and if other civilizations have a similar attitude, then we can get down to business and start communicating.

We should note that the question of how interstellar communication begins has almost certainly been resolved, at least in its general outlines, by civilizations in our galaxy more advanced than ourselves. These (supposed) advanced civilizations acquired interstellar communication abilities not within the last few years, but thousands, hundreds of thousands, or perhaps many millions and even billions of years ago. Unless such civilizations are extremely rare—either because they are unlikely to develop, or because they destroy themselves rather quickly—most of these civilizations that care to contact others should have done so by now. If this is true, then our focus becomes not how developing civilizations can contact one another, but how a new civilization can join the network. Is there a test? Or do we simply call "information"?

These questions are not simply fanciful if we think that our development as a civilization has followed broadly universal lines. If we are average, then we can hardly be the first, or anywhere near it, in the Milky Way, just as we can't be among the very last civilizations that will appear in galactic history. Thus, we must try to figure out how other radio and television messages can be detected. If we can, courtesy and caution suggest that first we listen, and then we send our own messages.

We might think that if everyone listens and no one transmits, no one will hear anything. We are, however, considering this problem, or so we believe, long after many other would-be communicators have already dealt with it, perhaps by establishing a galactic network. In addition, we know that our own terrestrial communications leak radio, radar, and television waves into space. These stray photons could be detected at interstellar distances by a sensitive array of antennas and receivers. If the same sort of leakage occurs in other civilizations, we might be able to eavesdrop on their radio communications. Before we can consider these suggestions in detail, however, we must consider four questions: (1) Where should we look? (2) Which frequency channels are most likely to be used for interstellar communications? (3) What total range of frequencies might be involved? (4) What sort of messages might be considered "standard" for opening conversations with new members of the galactic network?

Where Should We Look?

On page 334, we listed the nearest stars, some of which seem likely to have planets. Nearby stars basically similar to the sun provide the best chance for finding other civilizations, because the intensity of radio photons emitted by any civilization decreases in proportion to 1 over the square of the distance from it. However, radio telescopes directed toward the apparently "best" of these high-priority stars, Tau Ceti and Epsilon Eridani, have failed to reveal any signals that might provide evidence of another civilization.[1] We must, therefore, settle down for a long effort, prepared to search star after star, before we have a good chance of finding the nearest civilizations (see page 399). Which stars should we examine first?

All else being equal, the radio waves from nearby civilizations will outshine those from faraway civilizations. Each time we double our distance from a given source, we reduce the intensity of the radio signal—the number of photons that reach any particular antenna each second—by a factor of 4. This dominant fact causes us to look first at the nearest stars, later at the more distant ones. As human technology continues to advance, however, this limitation no longer restricts us to the short list on page 334: Our ability to detect a civilization whose television, radio, and radar transmissions roughly equal our own, *if* we choose to invest in the equipment to do so, embraces a sizable fraction of our own galaxy (Figure 20-1), one that includes many million stars. Thus, even though closer civilizations require less effort for detection, we do not need to consider only the very

---

[1]These searches for signals, however, tested only a few of the millions of possible frequencies and bandwidths (see page 386) for very short periods of time.

Fig. 20-1 The 7 million nearest stars to the sun lie within about 250 parsecs (800 light years). This volume stretches across about 1 percent of the diameter of the Milky Way galaxy, which contains about 400 billion stars.

nearest stars, but rather may include the nearest *millions* of stars in the Milky Way as candidates for radio signal detection. The question then springs back to *which* stars among these millions merit special attention.

In plain fact, the answer comes out: none in particular. Single stars seem better candidates than double- and multiple-star systems, but not by much (see page 337). Stars whose luminosity resembles the sun's seem superior to low-luminosity M stars, such as Barnard's star, because of the tiny ecospheres that surround faint stars (see page 335). If we restrict ourselves to stars whose true brightness equals at least 1 percent of the sun's, we must discard 80 percent of our galaxy's stars, but the remaining 20 percent of all stars will still provide 80 billion stars in the Milky Way, and many hundred thousand within range of the receivers we could build. We certainly must reject the brightest stars, those whose lifetimes fall short of the billion years that we think represents the minimum time in which a civilization can develop. These most luminous stars number only about 1 percent of the stars in our galaxy, so we do not lose much in numbers when we choose to ignore them in our search.

We arrive at the conclusion that the best search for extraterrestrial signals should proceed star by star, among single and multiple systems, so long as the star has a spectral type from F5 through K8; that is, so long as the star lasts for at least 5 billion years and has a large enough true brightness to maintain a reasonably large ecosphere. Our discussion in Chapter 12 has suggested to us that these stars have a high probability of possessing planets, and we can see no fundamental objection to the thought that, on the average, at least one planet in every four planetary systems should have conditions favorable to the development of life, and thus of intelligent civilizations. When we examine the numbers, as we have in Chapter 18, we see that we should prepare to search through at least several thousand stars to have a reasonable chance of finding another civilization that exists now (see page 355). This prospect, although difficult, should cause no despair, for if we examine one star per hour, we could work through a thousand stars every six weeks, and would soon reach the level of many thousands that could produce the reward of interstellar contact. The problem thus reduces to one of deciding to commit the resources to construct the computers and antenna systems required to conduct a sufficiently sensitive search.

What Frequencies Should We Search?

How can we hope to determine the frequencies at which we are likely to have the best chance for discovering another civilization? Aren't we facing an insoluble problem in trying to figure out how another civilization would send *its* messages? The answer, we think, is that other civilizations would send messages much as *we* would, so we can in fact expect to determine the best frequencies to search.

We must distinguish between messages that a civilization uses for its own purposes, and which we might overhear or "eavesdrop" upon, and messages sent deliberately to other civilizations, or at least into interstellar space with the hope of detection. Though these two purposes overlap somewhat, we have only to think of our own civilization to see that most radio and television broadcasts have been planned with solely a human audience in mind (unless the broadcasters know more than they let on). Nonetheless, we shall see that radio-television frequencies have such key advantages over other frequencies that we can recommend them for any sort of communication purpose.

Consider the electromagnetic spectrum (Figure 20-2). The tremendous spread of photon energies, more than a billion billion (10^{18}) times in energy from the highest-energy gamma-ray photons to the lowest-energy radio-wave photons, presents a tremendous range of possibility in choosing a photon frequency. What are the most striking "natural," that is, galactic, frequencies of reference? Are there any photon frequencies that would come to mind in any other civilization, whenever they think about how to locate other civilizations for communication by photons? The answer seems to be that *we can find a most likely range* of photon frequencies for interstellar communication, even though we cannot find one single frequency to point to and say: "*That* is the galactic communication channel." In narrowing down the enormous range of possible frequencies, we employ three principles that may be universal: *economy, freedom from interference,* and *cosmic frequency guides.* Let us examine each of these in turn.

The most basic reason for using *radio* photons instead of any other for interstellar communication refers to the economics of message exchanges, the *costs* of communication. If we think of a message as made of individual units, then the smallest unit of information, called a *bit* by scientists who work with the theory of messages, requires at least one photon for transmission. To avoid errors and to ensure detectability, more than one photon per bit may be required, but the number of photons per bit will be about the same for radio messages as for visible-light messages.[2] If we now

---

[2]The efficiency of the detecting system (a radio antenna and receiver, a telescope and a photographic plate, and so forth) will determine how many photons must be detected to obtain one bit of information.

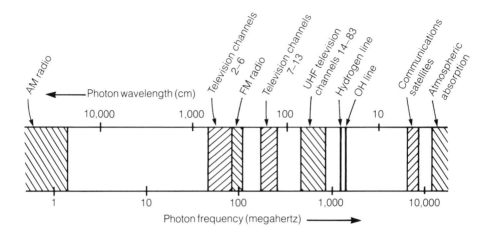

Fig. 20-2 The spectrum of photon energies, frequencies, and wavelengths spreads over an enormous range that is technically an infinite one. The radio region of the spectrum contains the photon energies and frequencies used for radar, television, CB communication, and AM and FM radio broadcasting.

compare radio waves with visible light, an important truth dawns immediately: Each visible-light photon carries about a million times more energy than a radio photon and thus, in energy terms, costs a million times more to send. In summary, radio waves are cheap, visible-light waves expensive.

The second factor in choosing a frequency at which to send or receive messages deals with *freedom from interference*. Some photons cannot travel as freely through space as others. Here again, *radio* waves of all but the lowest frequencies have special advantages. To see why this is so, we must consider the *absorption* of photons as they travel, and the *interference* of other emission processes with our attempt to detect and to locate a given source of information-carrying radiation.

In the astronomical exploration of our galaxy, the advantage of radio waves over visible light for long-distance communication quickly becomes apparent. We cannot photograph the galactic nucleus, or the spiral arms that lie beyond it, because the intervening gas and dust absorbs the light emitted by stars at these distances in the plane of the galaxy. Yet we can map these features with comparative ease with radio waves, except for a few particular frequencies at which absorption by atoms and molecules occurs. In our own solar system, we have found that radar waves penetrate the clouds of Venus easily, while visible light cannot.

The same advantage for radio holds true when we consider interference from other emissions. Visible-light emission from a planet circling a star (perhaps from a powerful laser paired with a giant telescope) must contend with the immense amount of energy radiated at visible wavelengths by the star itself. It is far easier to send messages at radio frequencies, where the energy level of stellar radiation is much lower. Indeed, as we shall see, even our "primitive" civilization already has the capability to send a message at a particular radio frequency with more energy than the sun emits. However, a cosmic "background" of radio emission exists from two separate sources, the radiation left over from the big bang (see Chapter 2) and the myriads of sources of synchrotron radiation caused by electrons spiraling in magnetic fields (see page 58). These two sources define a radio "window" in the electromagnetic spectrum that is relatively free from competing emission, and it is within this window that we expect interstellar communications among our hypothetical advanced civilizations to be taking place (Figure 20-3).

Fig. 20-3   The "waterhole" of radio frequencies happens to fall in the frequency band of least cosmic noise. Fast-moving particles in our galaxy emit radio waves at the lowest frequencies, while the highest radio frequencies are blocked by the absorption of radio waves by molecules in the Earth's atmosphere.

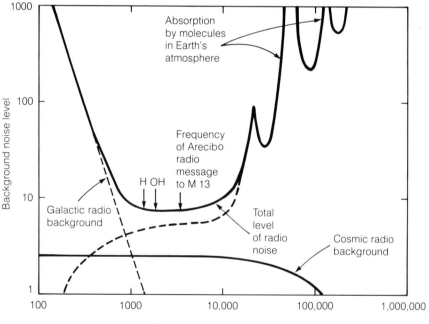

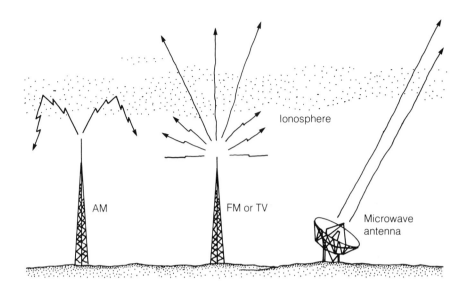

Fig. 20-4 The Earth's ionosphere reflects most of the AM (long wavelength) radio photons that reach it, but lets FM, television, and microwave (e.g., radar) photons pass through it unimpeded.

At still longer radio wavelengths, we encounter the barrier of our planet's ionosphere, which reflects most of the radio emission incident upon it (Figure 20-4). This frequency domain includes the region of commercial AM broadcasts. While the reflecting ionosphere helps to spread these signals around our planet, it prevents signals from penetrating our atmosphere to reach our receivers. FM radio and television, especially the UHF channels, do not suffer from this interference.[3] At the short-wavelength end of the radio window, we encounter absorption by Earth's atmosphere. We could avoid both these impediments by broadcasting and receiving messages from a station in orbit above our atmosphere, but as we have seen, radio absorption in the galaxy itself would keep this effort from yielding much improvement over ground-based facilities.

We have been guided to the radio band of frequencies by considerations of economics and freedom from absorption and interference. Can we look for cosmic frequency guides within the radio band? Can we tell what particular frequencies are likely to attract *any* civilization to a universal station on the radio dial?

[3]Note that the CB band lies just at the edge of the ionospheric cutoff, and thus CB conversations frequently pass unimpeded into the depths of space, together with commercial television and the powerful beams of our defense radars. This information about our current level of intelligence may be the reason why we have not received any signals from advanced civilizations!

In the first discussion of the problem in 1959, Philip Morrison and Giuseppe Cocconi pointed out that the single most important frequency in the universe appears to be 1420 megahertz, the frequency of the radio waves emitted by the spin-flip process in hydrogen atoms (see page 71). This emission arises from the small random velocities of hydrogen atoms in interstellar space. Some of the atoms are continually bumped into the spin-flip state from which they can emit their characteristic radio signals. Hence, throughout the galaxy in which we live and in other spiral galaxies as well, interstellar gas, 90 percent of which consists of hydrogen, constantly emits radio waves with a frequency of 1420 MHz and a wavelength of 21.1 centimeters. Every intelligent being who studies our galaxy knows about these radio waves, which form the most abundant and most widespread emission of radiation at a definite frequency. Furthermore, the 1420-MHz photons have the ability to travel, and to be detected, over great distances. Finally, the band of frequencies around 1420 MHz remains especially clear from interference (Figure 20-3).

Because of the Doppler effect, and the fact that atoms in our galaxy tend to have some motion towards us or away from us, the entire band of frequencies between about 1419 and 1421 MHz contains a great roar of photon emission from the natural processes of hydrogen spin flips. This 1420-MHz emission, affected by subsequent absorption and still later reemission processes, allows us to map the distribution of interstellar hydrogen when we study the frequency range between 1419 and 1421 MHz. But outside this rather narrow band of radio frequencies, much better conditions prevail. In the radio domain somewhat above or somewhat below the 1420-MHz range, relatively few atoms and molecules, as we have seen, provide *natural* interference to radio signals. Humanmade interference is something else again: Only by arranging to protect the few megahertz around 1420 MHz have radio astronomers managed to continue their work on Earth! Suppose that we, or any other civilization, would "naturally" be led to choose a frequency near 1420 MHz for interstellar messages, and suppose that this sort of frequency is now in use for local message traffic, so that we can hope to "eavesdrop" on civilizations at the same frequencies. We still must face a crucial problem of detail: *Which* frequency close to 1420 MHz has the message? How far from 1420 MHz should we tune our dials? And should we look to higher or to lower frequencies?

If we believe that *water* will be essential for most other forms of life as well as for our own (see Chapter 11), then we may find merit in the suggestion made by the physicist Bernard Oliver. Since each molecule of water (H_2O) consists of a hydrogen atom (H) plus a molecule of OH, Oliver pointed to the frequency band between 1420 MHz and 1612 MHz as the most likely channel for interstellar communication. OH molecules produce photons at a series of frequencies—1612, 1665, 1667, and 1721 MHz—by a hyperfine emission process similar to that for hydrogen atoms. The band of frequencies characteristic of photon emission from OH molecules forms a guidepost in the realm of photon frequencies, just as the 1420-MHz frequency of hydrogen-atom emission forms another still more striking marker.[4] If the importance of water occupies a large place in the consciousness of all life forms, then from the fact that H + OH = water, we might indeed find that the gap between 1420 MHz and 1612 MHz gives the frequency domain where interstellar communication occurs. Bernard Oliver calls this gap the *waterhole*—the place (in photon frequency) where galactic civilizations meet (Figure 20-3).

Even if the waterhole idea should prove too much a feature of our terrestrial life, the 1420-MHz emission from hydrogen does provide a galactic reference frequency. We have seen that because of the motions of hydrogen atoms in our galaxy relative to ourselves, we observe great numbers of photons at frequencies from (roughly speaking) 1419 to 1421 MHz. Therefore, we can expect any communication channels to avoid this frequency range, which remains choked with the naturally emitted photons that define this cosmic guidepost in frequency. Let us consider, then, how we could search in nearby frequencies with the hope of discovering radio signals produced by another civilization.

Frequency Bandpass and Total Frequency Range

If we think that we know the most likely frequency at which interstellar communication should occur, we still must determine the size of the *bandpass* of frequencies—similar to the spread in frequency between one radio station and the next on the dial—that the signal covers (Figure 20-5). If we don't know the frequency—and we don't!—we must also decide on the total frequency *range* that we plan to examine in our search. The bandpass is the spread of frequencies that we examine in any individual part of the search, while the range gives the entire spread of frequencies to be covered. Should we divide the range into tiny intervals, of, say, 1 hertz each? This would give us a better chance to detect a signal that had been deliberately held within an extremely narrow frequency channel (Figure 20-5). But to search through a range of, perhaps, 1000 megahertz *1 hertz at a time* would require billions of minisearches in each direction that we choose to look. (Luckily, this problem has a solution, as we describe on page 395.)

We face a difficult problem in guessing how large a frequency bandpass another civilization might use for its messages—both local and interstellar—and in estimating the total frequency range over which we should search for interstellar signals. Several factors enter the selection that *any* civilization will make for the frequency bandpass of its broadcasts, and we

[4]The radio spectrum of water molecules themselves is extremely complicated and presents no single, predominating frequency band.

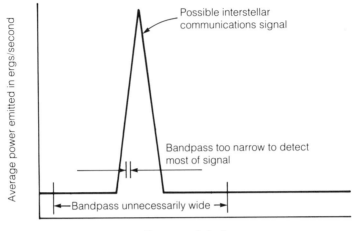

Frequency in hertz

Fig. 20-5 If we attempt to detect a radio signal with equipment that cannot receive radio waves over as wide a frequency bandwidth as the signal covers, then we lose part of the signal and have more trouble detecting it. On the other hand, if we use a much wider bandwidth than the signal covers, we waste much of our effort and again miss the best chance for detection, which occurs when the signal and our equipment cover the same bandwidth.

must think about them in turn. Two key impulses appear to be the desire to send information rapidly, which suggests a relatively *large* bandpass, and the desire to produce a signal that is recognizable against the background of radio noise, which requires a relatively *small* bandpass.

When a message arrives at its (unknown) destination, whoever searches for it must be able to separate the signal from the background of "noise"; that is, from the random emission of photons at nearly the same frequency by various cosmic processes (Figure 20-6). Narrower signals (that is, signals confined to a smaller total frequency range) can be more easily separated from the noise, if we consider signals of the same total radiated power.

Thus, we might think that a civilization with limited resources would say, to minimize cost, let's put the signal in as narrow a frequency band as is possible. In this case, an important minimum size for the frequency bandwidth arises from *interstellar dispersion*. As the radio waves pass through interstellar space, charged particles there (mostly electrons) actually *change* the frequency of the radio waves as they pass by. Furthermore, the *amount* of this change varies in a random way as the electrons oscillate. Therefore, even if we knew the exact frequency at which a radio message had been sent, and even if we knew the precise relative velocity between the source and ourselves so that we could allow for the Doppler shift, we still would find unpredictable changes in the frequency of the arriving radio waves. These changes establish a lower bound on the frequency spread of a small fraction of 1 hertz. Any civilization would in fact be wasting its energy if it tried to confine interstellar messages to a narrower frequency band than this, because the variations in the amount of interstellar dispersion along the path of the message would spread the radio waves in frequency by at least a fraction of 1 hertz. Of course, a civilization with tremendous resources might decide to send as much information as possible, and to use a broad bandwidth. We would need much larger telescopes to detect messages from such a civilization than to find a narrow-bandwidth message.

Let us, therefore, assume that 0.1 hertz represents about the smallest frequency band that any civilization would use for interstellar communication. For comparison, 0.1 hertz is about the frequency bandwidth used by powerful military radars (but not, we may assume, because they are used for interstellar communication). Radio broadcasting stations use bandwidths of 10 kilohertz for AM and 200 kilohertz for FM. Television requires much larger bandwidths, about 6 megahertz for each channel. However, about *half* of the total power radiated by a television station resides in the "video carrier signal," which covers less than one hertz of bandwidth. This spike carries no information about the picture, but it would be far easier to detect (because of its narrow bandwidth) than the 6-MHz-wide signal that fills in the picture. A similar sort of spike, also carrying half the total power, appears in FM radio broadcasting. The trick then becomes how to recognize *which* frequency bandpass, perhaps just 0.1 hertz wide, has the message. Must we search through 2 *billion* frequency channels, each 0.1 hertz wide, to cover the 200 MHz of frequency between 1421 and 1621 MHz? Apparently we must; yet this task may not be as hard as we think at first.

Fig. 20-6 To find a signal against a background of randomly emitted noise will be easier if the signal covers only a narrow range of frequencies.

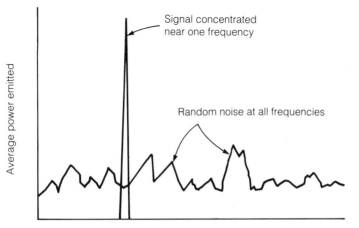

Frequency in hertz

TABLE 20-1

Estimated Power Output of Various Radio-Photon Sources that Operate at Frequencies Greater than 20 MHz

Source	Frequency Range (MHz)	Number of Transmitters	Fraction of Time that Transmitters Emit	Per Individual Transmitter		Total Average Power Radiated (watts per hertz of bandwidth)[a]
				Maximum Power Radiated (watts)	Effective Frequency Bandwidth (hertz)	
Citizen-band radios	27	10,000,000	1/100	5	2	200,000
Professional landmobile radios	20–500	100,000	1/10	20	1	200,000
Weather, marine, and air radars	1000–10,000	100,000	1/100	10,000 to 1,000,000	1,000,000	10 to 1000
Defense radars[b]	400	2	1/10	10,000,000,000	0.1	20,000,000,000
FM radio stations	88–108	10,000	1	4000	0.1	400,000,000
TV stations (for photons that carry picture, not sound)	40–850	2000	1	500,000	0.1	10,000,000,000

[a]The last column shows the power radiated *per hertz of bandwidth*. Systems that cover a wider bandwidth (most noticeably, weather, marine, and air radars) will radiate a greater total power over *all* frequencies than this column would suggest. This table, as well as Figures 20-7, 20-8, and 20-9 follow the results of a study made by W. Sullivan III, S. Brown, and C. Wetherill in *Science*, vol. 199, p. 377, 1978.

[b]We have considered only the most powerful defense radars; these dominate the total power output from all such radar systems.

Radio engineers who are interested in the possibility of interstellar communication, within the constraints we have described, are now starting to design receiver systems that can analyze as many as 1 million frequency channels simultaneously. In other words, instead of having to tune the radio dial to one station at a time to see whether radio messages are arriving from a given direction, we should soon be able to check on a million frequencies at once. This still leaves us short of the 2-billion-frequency channels we just mentioned, but the gain of 1 million times in our efficiency is not to be sneezed at. If we can indeed analyze 1-million-frequency bands, each 0.1 hertz wide, all at the same time, we can acquire a reasonable hope of finding a signal of this width fairly quickly—provided we are looking in the right direction, and that our *total* frequency bandpass does contain the signal frequency. Indeed even one *billion* channel receivers seem feasible at reasonable cost. The problem then will become: How can we be sure we have found another civilization's message and not a source of random cosmic noise?

How Can We Recognize Another Civilization?

Our plans to search for other civilizations by radio rests on the assumption that radio photons have such universal usefulness that any other civilization would be likely to use them, as we do, for sending messages. Table 20-1, which shows the chief sources of radio-wave photons that now exist on Earth, reveals that most of the radio power arises from television broadcasting, at frequencies between 40 MHz and 850 MHz, and from high-power defense radar systems that sweep the skies with intense radio pulses, at frequencies that are constantly changed for security purposes. If we managed to eavesdrop on another civilization that used photons of similar frequencies for their own communications, what could we hope to detect?

With sufficiently sensitive antennas, we could distinguish one television program from another, and could eventually even determine the content of these programs. This analysis would allow us to decide just how eager we would be to establish two-way contact. At greater distances and with less receiver sensitivity, we could tell that radio photons in great numbers arise from a certain location on the sky, but we could not determine what messages these photons carry. Then we might not feel sure that we had another civilization in our beam, rather than a natural source of radio noise. We could, however, go far toward resolving this issue in the following way: We might notice that the strength of some of the signals varies in time in a regular and complex manner.

Planets tend to rotate, as our own does, and this rotation implies that *if* the sources of radio broadcasting are unevenly distributed around the planet, any observer not directly above the planet's north or south poles will detect a cyclical variation in the arriving photon intensity. Figure 20-7

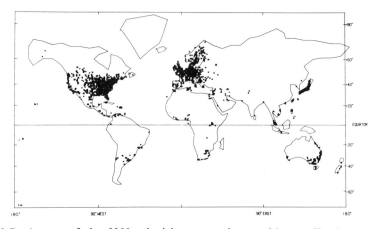

Fig. 20-7   A map of the 2200 television transmitters with an effective radiated power of 50 kilowatts or more shows the great concentration toward the centers of population and wealth on Earth.

shows the 2200 strongest television transmitters on Earth, which cluster heavily in the United States, Europe, and Japan. These television stations send far more radio photons roughly *along* the Earth's surface than straight outward so as the Earth rotates, an observer will see the peak emission from each station as it rises over the curve of the horizon or sets below it (Figure 20-8). The pattern of change will repeat every day, thus suggesting to the observer the existence of some superslow puslar (see page 129) or of a *non*natural source of radio signals.

How could an observer distinguish the Earth, which rotates once each day, from a pulsar, which rotates about once each second, with firm conviction? Any observer would obtain a tip-off that more than a slow pulsar must be causing the radio signals from the fact that the signals show a Doppler shift, caused by the Earth's motion around the sun, that repeats periodically once each year. An observer located in another planetary system would see the radio waves from each television channel—each frequency assigned to television stations of a certain number—shift back and forth by tens of hertz from their average frequency. An extraterrestrial student of the solar system could thus discover not only that radio waves emerge from the sun's vicinity, but also that these radio waves vary in *intensity* on a *daily* cycle, and in *frequency* on a *yearly* cycle. This observer could conclude that the sun has an object in orbit around it that circles the sun once each year, rotating once each day as it does so, broadcasting radio waves that allow these cycles to be determined.

A source of radio photons that showed cyclical variation on anything like the time scale of one day, together with Doppler-shift changes on a time scale of months or years, would be far more likely to be a planet with a civilization upon it than a pulsar of unknown character. But without an accurate analysis of the photons to see if they contained a real *message*, we could never be entirely sure.

Of course, for us to detect radio and television stations on another planet, they must already exist there. On Earth, the rise of radio has been sudden and recent. Figure 20-9 shows the change in the power of Earth's radio and television broadcasts since the year 1940: A thousandfold increase has occurred during the past 30 years, and this increase continues, though at a lower rate than during the 1950s.

The region within 70 light years (21 parsecs) of the sun contains about four thousand stars. Only these stars, and the planets that may exist around them, have had the chance to detect the radio and television photons emitted from the Earth since 1910. These stars lie within the radio

Fig. 20-8 An observer close to Barnard's star would see a pattern of radio emission with peaks and dips arising from all of the Earth's television transmitters. The pattern would repeat every time the Earth rotated.

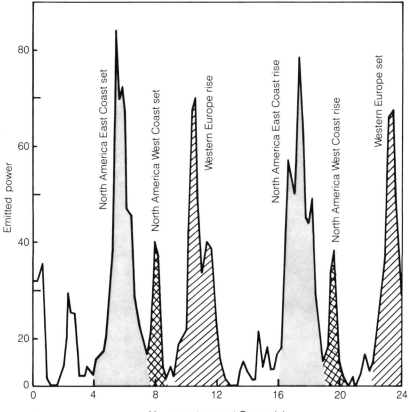

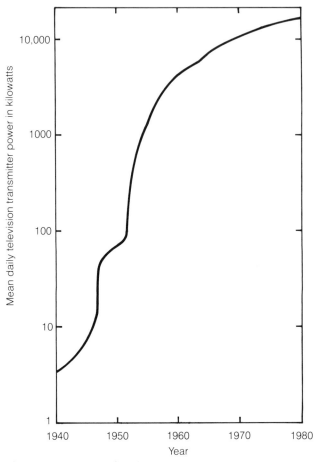

Fig. 20-9   The average power of radio and television transmission emitted from the Earth has grown steadily during the past 40 years.

bubble that is expanding around the Earth at the speed of light. Such stars have only had the chance to hear, and faintly at that, the great Amos 'n Andy and Jack Benny radio shows of the 1930s; most of them must wait years more for news of the Korean War and the Ed Sullivan show. Another 400 years must pass before our radio signals reach the million nearest stars, so these stars still have no news of an intelligent civilization on Earth. When we turn the picture around and think of other "young" civilizations, we see that only the 4000 nearest stars could have signaled their presence to *us* during the past 70 years. As we have seen (see page 353), this number of stars may not be large enough to include even one other civilization. To fulfill our goal of finding other civilizations and exchanging messages with them, we must hope that a fairly large fraction of civilizations last for at least several thousand years after they have developed interstellar communication ability (see page 356).

The Present State of Radio Searches for Other Civilizations

Until now, the human effort to search for other civilizations by radio has included only a few attempts at radio observatories, attempts that remain only a small part of the observatories' activities, which center, no doubt rightly, on the study of natural events that produce radio waves. Table 20-2 lists ten searches that have been made, or are continuing, in the United States, Canada, and the Soviet Union. These searches are hampered by the relatively small collecting areas of the telescopes used.

Calculations show that if we hope to find another civilization by over-hearing its radio broadcasts, we need a collecting area that is much larger even than that of the Arecibo antenna for a reasonable chance of success. In other words, to *eavesdrop* out to distances of several hundred parsecs, which would include a million or so stars, we need much larger antennas than we have now. Eavesdropping is much more difficult than trying to detect a signal from a known direction, at a known frequency, and with a known frequency bandpass. The Arecibo telescope can *now* communicate with a similar telescope *anywhere in our galaxy, if* these three key items are known. The Arecibo telescope represents the largest dish we have built so far, and its construction was relatively inexpensive because of a convenient topographic configuration (Figures 20-10 and 20-11). Bernard Oliver has suggested as an alternative that we should consider building many large antennas, as shown in Figure 20-12, and then analyze all the photons that arrive at these antennas *simultaneously.* This process will duplicate the results from a single, extremely large antenna, and at a lower (though still impressive) cost than that of a single dish. To build an array of a thousand 100-meter antennas, together with the appropriate receivers and electronics equipment and the master computer to run them, might cost 10 or 20 billion dollars. Is this too much to pay for a good chance of detecting another civilization? We, and other civilizations, must debate this issue before proceeding.

For now, let us note that we need not construct the entire array at once. Instead, we could begin with a dozen antennas, see whether we find any radio signals that show an apparently nonnatural origin, and then build another few dozen—if we want to. This carefully paced procedure would allow us to learn more about the frontiers of technology that govern the performance of a large array of radio telescopes and receivers. Naturally enough, if we found a signal that looked promising, though we could not determine for sure whether or not it was artificial, we would have an increased incentive to build more antennas to resolve this burning question.

To reach the full stage of a thousand antennas envisioned by Oliver, which is called the "Cyclops" after the one-eyed giant in Homer's *Odyssey,* might require a dozen years or more. To many scientists, Cyclops, rather than a trip to the moon or a colony of humans in space, represents

TABLE 20-2

SEARCHES FOR EXTRATERRESTRIAL RADIO SIGNALS

Year	Scientific Investigator	Antenna Diameter (m)	Frequency of Observation (MHz)	Frequency resolution (kHz)	Total frequency band (MHz)	Targets
1960	Frank Drake	26	1420	0.1	0.4	2 stars
1968	V. S. Troitskii	14	927	0.013	2.2	12 stars
1970 and onward	V. S. Troitskii		1875 1000 600			all directions
1972	Gerrit Verschuur	43 91	1420 1420	7.0 0.49	20.0 0.6	10 stars 3 stars
1972 and onward	B. Zuckerman and P. Palmer	91	1420	3.0		602 stars
1972 and onward	Stuart Bowyer and coworkers	26	variable	2.5	20.0	semirandom
1973 and onward	N. S. Kardashev					all directions
1973 and onward	F. Dixon and D. Cole	53	1420	20.0	0.38	$^1/_5$ of entire sky by areas
1974 and onward	A. Bridle and P. Feldman	46	22,200	30.0		500 stars
1975 and onward	Frank Drake and Carl Sagan	300	1420 1653 2380	1.0 1.0 1.0		several galaxies
1977	D. Black, J. Cuzzi, T. Clark, and J. Tarter	91	1665 1667	.005	1.32	200 stars
1977	M. Stull and F. Drake	300	1665	.0005	4	6 stars
1978	Paul Horowitz	300	1420	.000015	.001	180 stars

Fig. 20-10 The radio telescope of Arecibo, Puerto Rico, has a diameter of 300 meters, and could hold all the beer drunk on Earth in 1973.

the true flowering of human knowledge and inquisitiveness in this domain, for with Cyclops we would attempt to merge with the rest of the universe not by sending our bodies on a short and repetitive ride through our corner of the solar system, but by sharing in universal ideas that could arrive from other civilizations.

Fig. 20-11 Workmen who adjust the panels that form the reflecting surface of the Arecibo telescope must wear special shoes to avoid damaging the surface. The Arecibo Observatory is part of the National Astronomy and Ionosphere Center operated by Cornell University under contract with the National Science Foundation.

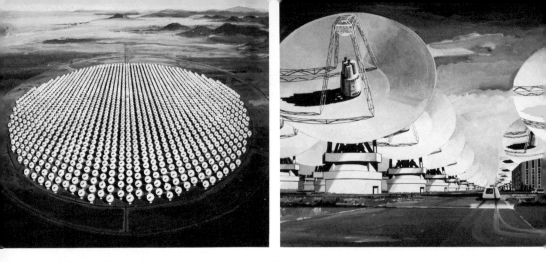

Fig. 20-12 To increase the effective collecting surface of a radio antenna, we can build many separate antennas and connect them together, so that we analyze the signals they receive simultaneously. The "eye" of Project Cyclops will consist of an array of 100-meter telescopes, all capable of pointing to the same place in the sky.

What Messages Could We Send or Receive?

In the last few pages, we have considered how we might detect another civilization by arranging to capture some of the stray radio photons that the civilization emits. Suppose that we did so, and that we could thus pinpoint the location, as well as the preferred radio frequencies, of a civilization whose existence seemed certain to us. What should we do about it? Should we send them a message? Or, conversely, should we expect other civilizations to send *us* a message, once they discover that *we* exist?

The answer is, maybe. Since we let stray radio photons leak into space all the time, why not add a bit more information? On the other hand, why should we or any other civilization encourage visitors, or even return messages? Human opinions vary widely on this subject, so we ought to bear in mind that more discussions will be needed to resolve the issue. Meanwhile, you may be surprised to learn that, acting entirely on their own, some astronomers have already sent a few messages to our unknown interstellar neighbors.

Figure 20-13 shows the most important of these messages, sent on November 16, 1974, from the Arecibo Observatory, using the great dish as a radio transmitter. This message, beamed in the direction of the great globular cluster in Hercules, contained 1679 bits of information. The bits appear in Figure 20-13 as 0's or 1's, but in the actual message each bit consisted of a photon pulse at one of two frequencies close to 2380 MHz. The total band of frequencies covered in the message transmission was 10 hertz, a few times greater than the minimum bandpass that we judged reasonable. Since the star cluster is 7700 parsecs from us, the message will reach it in about 25,000 years, ready for interpretation and a message back to us.

What could we expect another civilization to make of such a message? What would we think if such a message arrived? We would count on another civilization to recognize that it contains just 1679 bits. Then someone might ask, *why* 1679 bits? Mathematicians would notice that this number

Fig. 20-13 The message sent into space from the Arecibo Observatory consists of 1679 bits of information, basically "ons" and "offs" (see page 55).

0 0 0 0 0 0 1 0 1 0 1 0 1 0 0 0 0 0 0 0 0 0 0 0 0 1 0 1 0 0 0 0 0 1 0 1 0
0 0 0 0 0 0 1 0 0 1 0 0 0 1 0 0 0 1 0 0 0 1 0 0 1 0 1 1 0 0 1 0 1 0 1 0 1
0 1 0 1 0 1 0 1 0 1 0 0 1 0 0 1 0 0 0 0 0 0 0 0 0 0 0 0 0 0 0 0 0 0 0 0 0
0 0 0 0 0 0 0 0 0 0 0 0 0 0 0 0 1 1 0 0 0 0 0 0 0 0 0 0 0 0 0 0 0 0 0 0 0
1 1 0 1 0 0 0 0 0 0 0 0 0 0 0 0 0 0 0 0 0 0 0 1 1 0 1 0 0 0 0 0 0 0 0 0 0
0 0 0 0 0 0 0 0 1 0 1 0 1 0 0 0 0 0 0 0 0 0 0 0 0 0 0 0 0 0 0 1 1 1 1 1 0
0 0 0 0 0 0 0 0 0 0 0 0 0 0 0 0 0 0 0 0 0 0 0 0 0 0 0 0 0 0 0 1 1 0 0 0 0
1 1 1 0 0 0 1 1 0 0 0 0 1 1 0 0 0 1 0 0 0 0 0 0 0 0 0 0 0 0 0 1 1 0 0 1 0
0 0 0 1 1 0 1 0 0 0 1 1 0 0 0 1 1 0 0 0 0 1 1 0 1 0 1 1 1 1 1 0 1 1 1 1 1
0 1 1 1 1 1 0 1 1 1 1 1 0 0 0 0 0 0 0 0 0 0 0 0 0 0 0 0 0 0 0 0 0 0 0 0 0
0 1 0 0 0 0 0 0 0 0 0 0 0 0 0 0 0 0 0 1 0 0 0 0 0 0 0 0 0 0 0 0 0 0 0 0 0
0 0 0 0 0 0 0 0 0 0 0 1 0 0 0 0 0 0 0 0 0 0 0 0 0 0 0 0 0 1 1 1 1 1 1 0 0
0 0 0 0 0 0 0 0 0 0 0 1 1 1 1 1 0 0 0 0 0 0 0 0 0 0 0 0 0 0 0 0 0 0 0 0 0
0 0 1 1 0 0 0 0 1 1 0 0 0 0 1 1 1 0 0 0 1 1 0 0 0 1 0 0 0 0 0 0 0 1 0 0 0
0 0 0 0 0 0 1 0 0 0 0 1 1 0 1 0 0 0 0 1 1 0 0 0 1 1 1 0 0 1 1 0 1 0 1 1 1
1 1 0 1 1 1 1 1 0 1 1 1 1 1 0 1 1 1 1 1 0 0 0 0 0 0 0 0 0 0 0 0 0 0 0 0 0
0 0 0 0 0 0 0 0 0 1 0 0 0 0 0 0 1 1 0 0 0 0 0 0 0 0 0 1 0 0 0 0 0 0 0 0 0
0 0 1 1 0 0 0 0 0 0 0 0 0 0 0 0 0 0 0 1 0 0 0 0 0 1 1 0 0 0 0 0 0 0 0 0 0
1 1 1 1 1 1 0 0 0 0 0 1 1 0 0 0 0 0 0 1 1 1 1 1 0 0 0 0 0 0 0 0 0 0 1 1 0
0 0 0 0 0 0 0 0 0 0 0 0 1 0 0 0 0 0 0 0 0 1 0 0 0 0 0 0 0 0 1 0 0 0 0 0 1
0 0 0 0 0 0 1 1 0 0 0 0 0 0 0 1 0 0 0 0 0 0 0 1 1 0 0 0 0 1 1 0 0 0 0 0 0
1 0 0 0 0 0 0 0 0 0 0 1 1 0 0 0 1 0 0 0 0 1 1 0 0 0 0 0 0 0 0 0 0 0 0 0 0
0 1 1 0 0 1 1 0 0 0 0 0 0 0 0 0 0 0 0 0 1 1 0 0 0 1 0 0 0 0 1 1 0 0 0 0 0
0 0 0 0 1 1 0 0 0 0 1 1 0 0 0 0 0 0 1 0 0 0 0 0 0 0 1 0 0 0 0 0 0 1 0 0 0
0 0 0 0 0 1 0 0 0 0 0 1 0 0 0 0 0 0 0 1 1 0 0 0 0 0 0 0 0 1 0 0 0 1 0 0 0
0 0 0 0 0 1 1 0 0 0 0 0 0 0 0 1 0 0 0 1 0 0 0 0 0 0 0 0 1 0 0 0 0 0 0 0 0
1 0 0 0 0 0 1 0 0 0 0 0 0 0 1 0 0 0 0 0 0 0 1 0 0 0 0 0 0 0 1 0 0 0 0 0 0
0 0 0 0 0 0 1 1 0 0 0 0 0 0 0 0 0 1 1 0 0 0 0 0 0 0 0 1 1 0 0 0 0 0 0 0 0
0 1 0 0 0 1 1 1 0 1 0 1 1 0 0 0 0 0 0 0 0 0 0 0 1 0 0 0 0 0 0 0 1 0 0 0 0
0 0 0 0 0 0 0 0 0 0 1 0 0 0 0 0 1 1 1 1 1 0 0 0 0 0 0 0 0 0 0 0 0 1 0 0 0
0 1 0 1 1 1 0 1 0 0 1 0 1 1 0 1 1 0 0 0 0 0 0 1 0 0 1 1 1 0 0 1 0 0 1 1 1
1 1 1 1 0 1 1 1 0 0 0 0 1 1 1 0 0 0 0 0 1 1 0 1 1 1 0 0 0 0 0 0 0 0 0 1 0
1 0 0 0 0 0 1 1 1 0 1 1 0 0 1 0 0 0 0 0 0 1 0 1 0 0 0 0 0 1 1 1 1 1 1 0 0
1 0 0 0 0 0 0 1 0 1 0 0 0 0 0 1 1 0 0 0 0 0 0 1 0 0 0 0 0 1 1 0 1 1 0 0 0
0 0 0 0 0 0 0 0 0 0 0 0 0 0 0 0 0 0 0 0 0 0 0 0 0 0 0 0 0 0 0 0 1 1 1 0 0
0 0 0 1 0 0 0 0 0 0 0 0 0 0 0 0 0 0 1 1 1 0 1 0 1 0 0 0 1 0 1 0 1 0 1 0 1
0 1 0 0 1 1 1 0 0 0 0 0 0 0 0 0 1 0 1 0 1 0 1 0 0 0 0 0 0 0 0 0 0 0 0 0 0
0 0 1 0 1 0 0 0 0 0 0 0 0 0 0 0 0 0 0 1 1 1 1 1 0 0 0 0 0 0 0 0 0 0 0 0 0
0 0 0 1 1 1 1 1 1 1 1 1 0 0 0 0 0 0 0 0 0 0 0 0 1 1 1 0 0 0 0 0 0 0 1 1 1
0 0 0 0 0 0 0 0 0 1 1 0 0 0 0 0 0 0 0 0 0 0 1 1 0 0 0 0 0 0 0 1 1 0 1 0 0
0 0 0 0 0 0 0 1 0 1 1 0 0 0 0 0 1 1 0 0 1 1 0 0 0 0 0 0 0 1 1 0 0 1 1 0 0
0 0 1 0 0 0 1 0 1 0 0 0 0 0 1 0 1 0 0 0 1 0 0 0 0 1 0 0 0 1 0 0 1 0 0 0 1
0 0 1 0 0 0 1 0 0 0 0 0 0 0 0 1 0 0 0 1 0 1 0 0 0 1 0 0 0 0 0 0 0 0 0 0 0
0 1 0 0 0 0 1 0 0 0 0 1 0 0 0 0 0 0 0 0 0 0 0 0 1 0 0 0 0 0 0 0 0 0 1 0 0
0 0 0 0 0 0 0 0 0 0 0 0 1 0 0 1 0 1 0 0 0 0 0 0 0 0 0 0 0 1 1 1 1 0 0 1 1
1 1 1 0 1 0 0 1 1 1 1 0 0 0

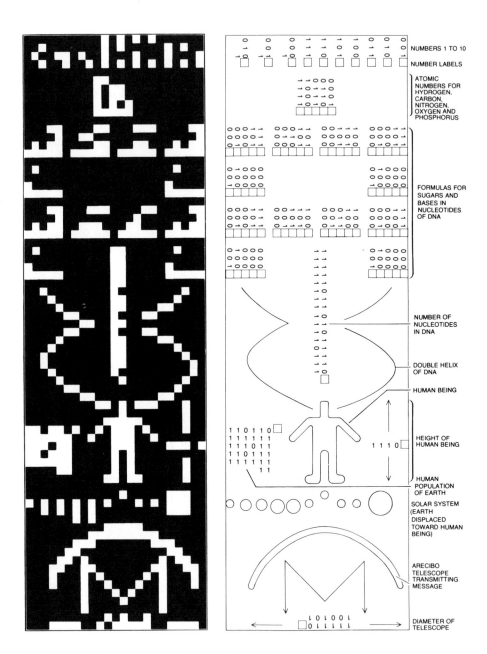

Fig. 20-14 If we arrange the 1679 bits into 23 columns of 73 bits each, and color the "ons" and "offs" differently, we find the pattern shown here.

equals the product of 23 and 73, and of no other numbers; 23 and 73 are prime numbers, not divisible by any others save themselves and unity no matter what counting system we use. This fact suggests, at least to us, that we might profit from arranging the bits of information in 73 columns of 23 bits each, or in 23 columns of 73 bits each. The first choice gives no discernible pattern, but the second produces the interesting array pictured in Figure 20-14, where we have replaced the zeroes by white squares and the ones by black squares to emphasize their contrast. The pattern seems clearly nonrandom, and some study (say, a few hours on the best computer a civilization has) should reveal what the astronomers were trying to say.

The top part of the message gives a lesson about the number system that the astronomers use: It shows the numbers 1 through 10 in binary notation, along with a "number marker" that tells when a symbol represents a number. Since the binary number system forms just about the simplest way to write numbers (although, significantly enough, most of the human race has never heard of it!), this part of the message should be recognizable as a starting point.

Next in the message comes the sequence of numbers 1, 6, 7, 8, and 15. Since this sequence seems odd on purely mathematical grounds, it must be trying to say something—namely, that we pay particular attention to the first, sixth, seventh, eighth, and fifteenth kinds of atoms, listed in order of the atoms' atomic number (number of protons). These atoms are, respectively, hydrogen, carbon, nitrogen, oxygen, and phosphorus. Any recipient of the message might be able to deduce that these five elements have key importance to us. (Notice that not everyone would assign phosphorus an equal rank with hydrogen, carbon, nitrogen, and oxygen; as we discussed in Chapter 8, the four most important elements seem far more important than the fifth, phosphorus.) Below these numbers we find twelve groups, each of five numbers. Each of these groups gives the chemical formula of molecules important to life, specifying the particular element (hydrogen, carbon, nitrogen, oxygen, and phosphorus) by using the same ordering of elements as the single group expressed earlier. These molecules include the four key bases, thymine, guanine, adenine, and cytosine, that we discussed on page 155, as well as the phosphate group (PO_4) and the sugar molecule deoxyribose.

Farther down the diagram, we encounter the chemical structure of DNA, *the* key molecule of life (see page 154). The double helix pattern appears, wound around the number of pairs of bases (about 4 billion) that exist in a single human chromosome, the basic carrier of genetic information. The twin helix ends at a crude picture of a human being, which, to the authors of the message, indicates the connection between DNA and the evolution of the intelligent beings who sent the message. To the human's right we see a line extending from head to foot, together with the number 14 (in binary notation), so the human must be 14 units tall. The only unit we share in common with the recipient of our message must be the *wavelength* of the photons that carry the news, so humans must be 14 wavelengths tall, or 14 × 12.6 cm = 176 cm, or 5 feet 10 inches. To the left of the human figure, the number 4 billion (actually a bit different, because binary notation does not round off in the same way that the decimal system does) gives the human population at the time that the message was sent.

Below the human, we find a sketch of the solar system, with the sun at the right and nine planets to its left. The third planet (Earth) stands out of line, which shows that the Earth is something special; its displacement toward the human figure establishes the connection that humans live on Earth.

Finally, below the solar system, we see a drawing of a telescope, whose function can be clearly seen from the fact that it "focuses" photons to a central point. The last line of information gives the size of the telescope, 2430 wavelengths or 306 meters.

Not bad for 1679 bits of information! We have conveyed (at least to ourselves) the idea that those who sent the message have a number system, consider certain elements to be of primary importance, judge certain molecules made from these elements to be also of key importance, attach great meaning to a certain spiral that springs from the central figure, whose size and numbers appear on either side, live on the third of nine planets around a star, and build telescopes 300 meters across. Even if the receiving party could not understand all of this message, certain facts should emerge clearly: We are here, primed with information and eager to talk about it. The intensity of photons upon arrival would make their point of origin appear as the brightest "star" in our galaxy at the message frequency during the time of reception. This would indicate a high technology.

This message, traveling at the speed of light, took only an *hour* to travel farther from the sun than Pioneer 10, which had been on its journey out of the solar system for more than *two years*.[4] By now, five years afterward, the message has reached past the distance to Alpha Centauri.[5] If a civilization that orbits some star in the globular star cluster M 13 should receive it, in about 25,000 years, and should then send a return signal, we would detect their signal on Earth about 50,000 years from now. Meanwhile, we may pause to wonder about the question of how many of these "Hello— are you there?" sorts of messages may be sailing through the galaxy now, and of the impact upon human civilization if another world's 1679 bits appeared in the daily newspaper. Would we consider it just another cross-word puzzle, or would most of us stop to look at that strange figure in the middle of the diagram?

---

[4]Furthermore, the cost of sending this message was about one ten thousandth the cost of Pioneer 10.

[5]Of course, the message is headed toward Hercules, not Centaurus.

Summary

Once we judge that photon messages are the best way for civilizations to find and to communicate with one another, we must ask certain important questions, and determine their correct answers, before we can hope to enter into communication with our closest neighbors. First of all, we must consider the most likely photon *frequencies* to be used for messages. The generally favored frequencies should lie in the radio domain of the spectrum, because most other spectral regions (save only the visible-light band) suffer from propagation difficulties, and radio photons are cheaper to produce than visible-light photons. Just *which* radio frequency will be used for deliberate searches remains a mystery, but it may be resolved by the fact that the 1420-MHz frequency of the photons emitted by hydrogen atoms represents the single most important frequency in the universe, known to every other civilization that studies the cosmos. Hence, many astronomers incline to the view that interstellar messages travel at frequencies close to 1420-MHz (or perhaps at an even multiple, or a simple fraction, of this basic frequency). If water is as important to another civilization as it is to ours, then the "waterhole" of frequencies between 1420 MHz (H) and 1612 MHz (OH) may be the frequency range to study.

We must also resolve the question of the *bandwidth* of frequencies over which the message signal spreads, if we hope to conduct our search for other civilizations' messages with high efficiency. Furthermore, we must decide on a total *range* in frequency over which we are willing to search—and hope that this includes the frequencies at which messages are actually traveling.

If we resemble other civilizations in our judgment and knowledge, then we stand a good chance of finding messages sent by the nearest civilizations if we search the most likely nearby stars, at the most likely frequencies, using the most likely frequency bandpass. Already human beings have sent several messages out into space with the deliberate, though hardly unanimous, attempt to announce our presence to any civilization that intercepts them. Such messages can be reassembled into a crude picture by any civilization as intelligent as ours. Even a few thousand bits of information suffice to tell another civilization the basic chemistry, location, population, and physical size and shape of the senders.

Even if other civilizations tend *not* to broadcast messages deliberately designed to reveal their presence, they may still, like ourselves, produce photons for their own local communications, some of which inevitably leak into space. Although the detection of such leaked photons presents a far more difficult task than the discovery and analysis of a signal designed for interplanetary communication, we could now construct an array of antennas capable of eavesdropping on a civilization like our own anywhere among the million nearest stars. Though we might not determine the actual content of the radio programs or whatever messages are being sent, the simple discovery of another civilization by eavesdropping techniques might have profound consequences for our own attitude toward life on Earth.

Questions

1. Why do we think that our galaxy may well contain a network of inter-communicating civilizations?

2. Which stars seem to be the best candidates to have planets on which civilizations have developed? Why?

3. Why do radio waves seem superior to photons of other frequencies for interstellar communication?

4. What is the "waterhole"? Why do some astronomers think that it may include interstellar radio transmissions?

5. Why does finding other civilizations by eavesdropping require more sensitive detection devices than communication with a civilization whose existence is known?

6. How could we hope to distinguish a planet that broadcasts much as we do from a natural source of radio emission, such as a pulsar?

7. Light waves and radio waves travel at 300,000 kilometers per second. How long did it take the radio message sent from Arecibo in 1974 to overtake the Pioneer spacecraft, which had then been traveling for two years at a speed of 10 kilometers per second? What does this tell us about the relative merits of spacecraft and radio messages?

8. Suppose that we received an apparent message from another civilization that consisted of "on" and "off" pulses that repeated after a total of 2117 "ons" and "offs." How could we begin to interpret this message? Is there more than one way to make such an interpretation?

Further Reading

Billingham, John, ed. 1973. *Project Cyclops: A design study of a system for detecting extraterrestrial intelligent life.* Moffett Field, Calif.: NASA/Ames Publication CR114445.

Morrison, Philip, Billingham, John, and Wolfe, John, eds. 1978. SETI: *The search for extraterrestrial intelligence.* Washington, D.C.: U.S. Government Printing Office.

Ridpath, Ian. 1978. *Worlds beyond: A report on the search for life in space.* New York: Harper & Row.

Sagan, Carl, ed. 1973. *Communication with extraterrestrial intelligence.* Cambridge, Mass.: The M.I.T. Press.

Wetherill, Chris, and Sullivan, Woodruff III. 1979. Eavesdropping on the Earth. *Mercury* 8:8 (March/April), 2.

Science Fiction

Gunn, James. 1972. *The listeners.* New York: Charles Scribner's Sons.

Hoyle, Fred, and Elliott, John. 1962. *A for Andromeda.* New York: Harper & Row.

CHAPTER 21

Extraterrestrial Visitors to Earth?

In the last three chapters, we have devoted our attention to the problem of how to find other civilizations and how to communicate with them. Human nature, with a certain economy of effort, immediately insists on the question: Why not let *them* find us? Why spend so much time and energy to send messages, or to listen for stray signals, if we could merely open our eyes and discover extraterrestrial visitors on Earth?

To be sure, if we could talk with extraterrestrial civilizations *here,* we would not need to go *there,* or to send messages across the enormous reaches of interstellar space. But we shall see that although some of the evidence for extraterrestrial visitors seems intriguing, we have no convincing evidence that we have been visited, or are being visited, by members of any civilization besides our own. The reports of such visits, whatever seeds of reality they may include, continue to signal an abiding belief that *we are special,* definitely worthy of an inspection tour, if not an extended holiday. No matter what we conclude about the evidence for extraterrestrial visitors, we can admire the confidence with which humans assume not only that "we are not alone," but also that someone else ought to do the hard work of traveling.[1]

When we ask what evidence does in fact exist of extraterrestrial sojourns on our planet, we can start with what would be the best evidence of all, an actual visitor, visible to crowds of people and ready for conversation. No such visitors have appeared on Earth recently, though a few *individuals* say they have been abducted by extraterrestrials. The next best

---

[1]Although one can find human beings who claim to have made journeys to Venus, Mars, and various other planets, in the opinion of the authors these claims are not worthy of detailed examination.

case for extraterrestrial visitation would come from some *device* of clearly nonterrestrial origin, such as an antigravity machine or an interstellar spacecraft. Again, no such artifact exists, despite some charlatans' claims to the contrary. Third best would be indisputable photographic and spectroscopic evidence that spacecraft capable of interstellar flight have passed close by the Earth's surface. Since astronomers and geologists now have taken thousands upon thousands of photographs to survey the sky and the Earth's surface, the absence of anything resembling an extraterrestrial spacecraft from any of these photographs argues against extraterrestrial visitors, at least those who might travel in a vehicle that can be photographed. A few of these survey photographs show interesting trails of light that are hard to explain as meteors, but this is a long way from photographic proof of the presence of another civilization's spacecraft.[2]

Fourth best in the list of possible evidence comes the set of reports from humans who claim to have seen extraterrestrial visitors or their spacecraft. These reports must always be subject to error, especially in times of stress, and the potential for error increases as the strangeness of the observation increases. If we wish to examine some of the evidence concerning human sightings of UFOs (unidentified flying objects), we must agree that we are putting aside the first, second, and third best possibilities for proving that extraterrestrial visitors are here, and must deal with eyewitness testimony with all its contradictions and confusion.

To sift through all the varied kinds of UFO reports; to interview known and potential witnesses; to attempt to evaluate all possible natural explanations (such as planets, clouds, and birds) and all possible human artifacts (sunlight reflected from weather balloons and airplanes, artificial satellites, unusual aircraft lights) represents an immense task, often unrewarding and depressing, as when UFO reports turn out to be the result of pranksters. Hence, it is not surprising that few scientists have devoted much effort to a study that they see mainly as the realm of sociologists and psychologists. Indeed, most scientists are poorly trained to deal with problems in which the observer plays a key role in the phenomenon. One of the best-known, most thorough investigators of UFO sightings is Philip Klass, a senior editor of the journal *Aviation Week & Space Technology.* In our discussion of UFOs, we have drawn heavily on the evidence that Klass has assembled to interpret some famous UFO reports.

Two Representative UFO Sightings

Consider, for example, the people who saw a UFO on July 24, 1948, while flying aboard an Eastern Airlines DC-3 near Montgomery, Alabama. This

---

[2]We shall discuss the "conspiracy theory" that such evidence exists, but has been concealed, on page 420.

sighting occurred a year after the first postwar UFO report, made by a private pilot named Kenneth Arnold who said he had seen mysterious flying disks near Mount Rainier in Washington. Arnold's 1947 sighting led to the coinage of the term "flying saucers," which has hung over UFO reports ever since. Six months after the initial burst of flying saucer sightings that followed Arnold's story, Air Force Captain Thomas Mantell crashed his jet airplane while chasing a giant UFO that was probably a new type of military balloon. Mantell's fatal accident, almost surely the result of his failure to use an oxygen mask while climbing toward the unidentified object, led to the suggestion that flying saucers had the ability to destroy those who pursued them, and this suggestion, much beloved by newpapers and magazines, must have been on the minds of many of the passengers and crew of the Eastern Airlines DC-3.

While this aircraft flew eastward in the early hours before dawn at an altitude of 5000 feet, the captain and copilot had a nearly full moon to light the night sky, except for a few broken clouds. Without warning, the crew spied what they thought was a giant jet aircraft coming from the east, which passed by at a distance estimated to be 700 feet, at a speed of 500 to 700 miles per hour. Both the captain and copilot said later that the object had two rows of windows that appeared to be lit from within; the captain said that "You could see right through the windows and out the other side." Both men guessed that the object was about 100 feet long, and 25 to 30 feet across (Figure 21-1). The object passed by for only 10 seconds, long

Fig. 21-1 These sketches, made by pilot Chiles and copilot Whitted, show the UFO that they saw while flying a DC-3 aircraft near Montgomery, Alabama on July 24, 1948. Notice that the two sketches show a significant difference of opinion.

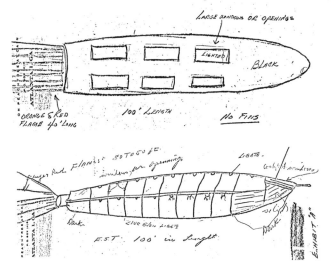

enough for a good look but not long enough to arouse other potential witnesses. One passenger aboard the plane reported seeing a similar streak
of light, while one other aircraft in the same vicinity saw a kind of jet or
rocket trail at the same time.

Both the pilot and copilot saw rows of windows on the UFO, and both
disagreed with the suggestion that they might have seen a meteor. Furthermore, the pilot said that the UFO had "pulled up with a tremendous
burst of flame out of its rear and zoomed up into the clouds." The Air
Force spent a considerable effort in investigating possible explanations of
this incident. One of their consultants, Dr. J. Allen Hynek, wondered
whether "the immediate trail of a bright meteor could produce the subjec-
tive impression of a ship with lighted windows." But the investigators felt
that the pilot and copilot could not have been so grossly deceived, or so
far wrong, as to see a meteor as a craft with double rows of windows.

Just 20 years later, however, on the night of March 3, 1968, three reliable
witnesses in Tennessee saw the same kind of phenomenon from the
ground: A bright light neared the trio, spouted an orange-colored flame from
behind, and passed overhead at an altitude that the three said was 1000
feet or less. In complete silence, the UFO floated by, "like a fat cigar,"
said one of the witnesses, and with at least ten large, square windows that
seemed lit from inside (Figure 21-2). The three observers talked about what
they had seen and agreed that the object could only be either a secret
military aircraft or an extraterrestrial spacecraft.

Now it happens that the same UFO was seen by six people in Indiana,
two hundred miles to the north, who also saw a cigar-shaped object with
many brightly lit windows (Figure 21-2). This object, the observers said,
was 150 to 200 feet long, and passed at treetop level without a sound. Other
witnesses in Ohio, entirely trustworthy, saw similar craft, but *three* of
them, also passing in silence. An apparently trustworthy observer reported

Fig. 21-2 Twenty years after the 1948 sighting, a Soviet rocket burning in the upper
atmosphere was seen as a UFO with a line of lighted windows by an observer in
Tennessee (top sketch) and another in Indiana (bottom sketch).

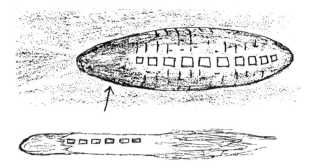

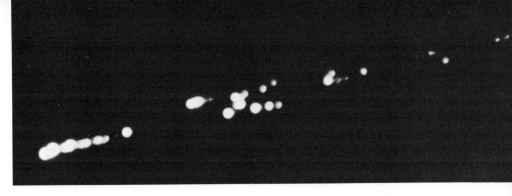

Fig. 21-3 These blobs of light are fragments of a large meteoroid breaking apart as it passes through the Earth's atmosphere.

that her dog whimpered in fright as she watched, and that she herself had an overpowering urge to sleep afterward.

On the night before these sightings, the Soviet Union had launched a set of rocket boosters to help put a spacecraft into orbit. On the following night, one of these rockets fell back to Earth, burning up in our atmosphere as an artificial meteor. American radars had kept track of this object from the time of its launching, and no doubt exists that the rocket booster was passing *several dozen miles* above the observers at the time they reported seeing one or more craft with windows at low altitudes above them.

We may conclude from the 1968 incident that ordinary, reliable people may not record what they see with good accuracy, because human minds supply details that "ought" to belong, in order that we may view the world in a way that seems logical. Our minds rely on a background of previous experience to interpret what we meet: If we see strange glowing objects moving together in the sky, we may tend to see "windows"; if we hear no noise from what looks like an airplane fuselage, it must be a spacecraft. The woman who fell asleep despite her attempts not to carried a flashlight every night to communicate in morse code with UFO crew members, if the need should arise. Thus, her drowsiness might have arisen from her joy at having finally seen an extraterrestrial spacecraft, while her dog might have reacted to her own tension, or might have been complaining about the cold (24° F), since the owner said the dog hated the cold. Her recounting of the entire experience might have been heavily biased by her preconceptions.

The 1948 incident, with its almost perfect match to the 1968 observations, thus seems likely to have been a sighting of a true meteor. This possibility gains weight from the fact that July 24 marks the peak of the well-known Delta Aquarid meteor showers. Once again, human brains could well have interpreted an unfamiliar sight in a manageable way, placing the meteor at a distance of several hundred feet rather than several dozen miles, and substituting illuminated windows for a series of flashes as the meteor broke up and burned in our atmosphere (Figure 21-3). *Without photographs, we can never be sure.* Even with photographs, arguments would

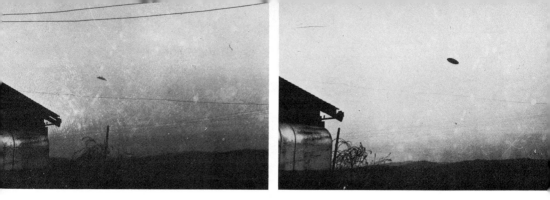

Fig. 21-4 These photographs, often cited as the most convincing evidence that UFOs are spacecraft, were taken near McMinnville, Oregon, in May, 1950 by the husband of a woman who had seen UFOs on several previous occasions, "but no one would believe me."

surely continue (Figure 21-4), but at least we would be discussing a piece of physical evidence rather than human recollections, with all the pitfalls of interpretation that enter any evaluation of someone else's experiences.

The Lubbock Lights

One hot night in August, 1951, three members of the faculty of Texas Tech University were looking at the sky from one of their backyards. All three, professors of science or engineering, were trying to count meteors when suddenly they saw 15 to 20 faint, yellowish white lights passing from north to south. An hour later, they saw another group of lights, moving in a semicircular formation. Near midnight, more than two hours after the first sighting, a third group passed overhead.

The three scientists estimated the altitude of the lights to be 50,000 feet, and their speed to be 5 miles per second. They telephoned the *Lubbock Evening Avalanche,* and their sighting received nationwide publicity as the "Lubbock Lights." But further investigation by the professors themselves solved the Lubbock mystery.

What were these mysterious lights that seemed to travel more rapidly than any airplane? They were birds! Migrating plover, reflecting the mercury-vapor light from the newly installed street lamps of Lubbock, had seemed to be not 1000 feet high (as they probably were), but 50,000 feet high, and therefore the estimated speed exceeded the true speed by a factor of 50. (Actually, the overestimate of speed must have involved a factor closer to 500 than to 50 times.) If three university instructors can make an error of this magnitude, we can hardly be surprised if most citizens can be equally confused. We simply cannot tell the altitude, and thus the speed, of an unknown object. Whenever you hear someone say, "It was a mile away," or, "It was a hundred yards away," of something seen *in the sky,* you are entitled to ask, "How do you know?" You will usually find that human intuition has provided the answer, rather than any scientific means of measurement.

414

Venus in Georgia

During the fall of 1967, police officers in more than 10 places in Georgia reported sighting UFOs. The first such case, typical of the remainder, began in the early hours of October 20, when an officer saw a "bright-red, football-shaped light" near the horizon, chased it with his associate in the patrol car for 8 miles, and lost sight of it only to find, as they returned to Milledgeville, that the UFO had again caught up with them! The UFO was so bright, the two officers said, that they could read the hands on their watches by its light. A third officer joined them at headquarters, where the three noted that the UFO had gained altitude. They watched for about half an hour as the UFO changed in color from bright red to orange to white, climbing in the sky until it resembled "a star."

The police report sparked a local interest in UFOs, and on the following dawn not one but two UFOs were seen in the east. On the fourth night, an aircraft aided the police officers in their chase and also saw the UFO, plus a second, fainter object above the eastern horizon; both objects seemed to "back off" and keep "moving higher and away from us" as the airplane chased them toward the east. More police officers in other Georgia towns reported observations similar to those made in Milledgeville; for a few weeks, central Georgia percolated with UFO sightings.

These observations recorded the early-morning rising of the planet Venus, which was then particulary bright, accompanied by a fainter planet, Jupiter. Though Venus has often been misidentified as a UFO, the number of places in Georgia that reported Venus in 1967 set a single-state record. Aside from the pervasive power of one report to trigger another, what is notable about the Venus sightings is, of course, that any true celestial object will seem to "back off" as you approach it, but will also seem to follow you wherever you go. The report that the officers could read the hands of their watch by the light from Venus shows that Venus appears remarkably bright, sometimes bright enough to cast a shadow, and the excitement of seeing an apparent UFO can easily exaggerate the brightness of an object.

Landing in Socorro

On April 24, 1964, a patrolman in Socorro, New Mexico, named Lonnie Zamora was following a speeding motorist when he heard a loud roar and saw a "flame in the sky" over a mesa less than a mile away. Zamora drove up a steep road to the mesa top, where he saw a "shiny-type object" with "two people in white overalls" nearby. Zamora parked his police car about a hundred feet from the object, got out, and heard a "very loud roar" as the object slowly rose from the mesa with a blast of flame underneath it. As Zamora ran back to his car, he knocked off his glasses, but he heard a brief, sharp whine, and saw the UFO fly off to the southeast.

Zamora drew sketches of the UFO as he recalled it, making it look something like a large egg with legs, and he found four indentations in the ground, plus some evidence of burned vegetation. The chief Air Force consultant on UFOs, J. Allen Hynek, interviewed Zamora and found him "basically sincere, honest, and reliable."

What can we make of Zamora's report? On one hand, Zamora said that he saw something highly unusual. On the other hand, the evidence to support Zamora's sightings—small indentations and a small amount of burned vegetation—could never be called convincing, if it had to stand alone. Furthermore, a man who lived with his wife a thousand feet south of the "landing site" was home and heard nothing, though his doors and windows were open, according to a report by Philip Klass.

The Socorro incident is important becuase it is usually included among the "most reliable" cases that offer evidence of extraterrestrial visitors to Earth. Although Zamora's report is intriguing, no court of law would consider the case proven. As is the case with most of the "best" UFO sightings, the evidence consists almost entirely of an eyewitness sighting, in this case by a single individual whose account conflicts with others.

Difficulties in Verifying the Spacecraft Hypothesis

With a fair degree of justice, we can summarize the best-known UFO reports by saying that they are no more nor less convincing than the Socorro incident. More spectacular reports exist, but the trustworthiness of those making the report remains open to question. A subject as loaded with emotion as the possibility of extraterrestrial visitors must inevitably draw its share of liars, charlatans, and hoaxes, and all of these have appeared in abundance in UFO reports. A good investigator can hope to uncover such frauds, and many hoaxes have indeed been exposed. There remain a few cases such as that in Socorro, when the trustworthiness of the report *as an accurate description of what the observer thinks he saw* seems reasonably good. The difficulty here, as we have said, is that the human mind invariably insists upon imposing its intuition on the scenes it records.

Even a group of people who observe the same event can influence one another, or can all be subject to much the same sort of interpretive interference. We saw this in the case of the reentering spacecraft, and it appears in many UFO reports submitted by two or more people, often close relatives.[3] The fact remains, quite simply, that eyewitness accounts can never by themselves prove that a given event did or did not occur. The problem

---

[3]An example of this effect comes from the famous sighting of UFOs (probably the planets Venus and Jupiter) in New Guinea in 1959, when an Anglican missionary and 37 others were present. Twenty-five of the 37 later signed the missionary's statement, but their relationship to the missionary (pupils, assistants, parishioners) made them far from independent in their interpretation and memory—let alone the fact that it remains unclear whether they knew exactly what they were attesting to.

of mental interpretation grows especially acute with strange objects in the sky, because people nowadays are unfamiliar with natural celestial objects—hence the numerous reports of Venus as a UFO.

Classification of UFO Reports

One thing that we can do is to classify the *reports* of unidentified flying objects. This entirely scientific procedure draws support from all sides of the extraterrestrial visitor argument. Moreover, in case after case of scientific inquiry, the classification of evidence by itself has helped to resolve a problem. In the late 1950s, Professor Hynek devised a sixfold classification for UFO sightings:

1. Nocturnal lights: bright lights seen at night
2. Daylight disks: usually oval or disklike
3. Radar-visual: those detected by radar
4. Close encounters of the first kind: visual sightings of an unidentified object
5. Close encounters of the second kind: visual sightings plus physical effects on animate and inanimate objects
6. Close encounters of the third kind: sightings of "occupants" in or around the UFO

With these categories, Hynek characterized the reports that reached him both during his service as Air Force consultant on UFOs, and later, when he had resigned from this post and founded the Center for UFO Studies. In his new position, Hynek has accused the Air Force of a "cover-up" similar to the Watergate affair.

Notice that Hynek's list omits "close encounters of the fourth kind," actual physical contact with occupants of UFOs. (The psychologist John Wasserman characterized these as meetings where telephone numbers are exchanged and there is dancing). In other words, aside from the claims of rather disturbed people, close encounters of the third kind are as near as the reports come to close contact. This suggests that either normal human imagination stops short of physical contact, or that extraterrestrial visitors shun such close encounters.

Hynek has derived from the three types of close encounters a general fact about UFO reports: Less unusual events are correlated with more reliable reports, and vice versa. In other words, some reports have a great degree of strangeness (for example, that a flying saucer landed in a cornfield), but a low degree of credibility (for example, that a single individual of limited eyesight saw this happen). Other reports have a low degree of strangeness (for example, that a strange light appeared to hover in the sky, then accelerate to great speed as it disappeared), but a great degree of credibility (for example, that five reliable witnesses all saw the same sequence of events). The elusive case remains the one with great strangeness and great credibility.

Consider the fact that the first three of Hynek's six categories—nocturnal lights, daylight disks, and radar-visual—are commonplace occurrences, distinguished chiefly by the degree of surprise that the observer reports. Airplanes, meteors, and planets account for the majority of nocturnal lights, with the extreme brilliance of Venus leading the pack. Daylight disks usually turn out to be blimps, weather balloons, or clouds. Occasional hoaxes, especially in the days after a UFO report, often involve balloons lit by candles or covered with aluminum foil. The later explanation of these hoaxes rarely receives as much media coverage as the original report. Finally, false radar echoes remain a common problem for every radar operator. By themselves, the radar spottings of objects in locations with no known aircraft do not provide much evidence for extraterrestrial spacecraft: Flocks of birds, swarms of insects, and what the technicians call "angels" or anomalous propagation of radar waves may be at work.

None of this discussion proves that all UFO reports have a completely natural explanation, or that all observers make mistakes. Instead, the many cases where UFO reports involve natural or humanmade objects indicate that the possibility of error will remain strong *whenever eyewitness testimony provides the sole basis for the report.*

Arguments for the Spacecraft Hypothesis

Without clear-cut physical evidence, the proponents of the extraterrestrial visitor hypothesis have a difficult job if they seek to *prove* that we are being visited by other civilizations. The proponents of this hypothesis usually then rely on the following line of argument:

1. The number of civilizations in our own galaxy, let alone those in other galaxies, may reach an enormous figure (see page 353).
2. Even if most of these presumed civilizations have no interest in or aptitude for spaceflight, *some* of them must have both the inclination and the ability.
3. Since a more advanced civilization would possess incredibly sophisticated technology, the civilizations that do decide to visit can send as many spacecraft as they like, and can perform whatever feats they choose with them.
4. Hence, the most likely explanation of mysterious objects with no natural cause that we can determine must be that extraterrestrial visitors are repeatedly visiting us.

Few astronomers would dispute points 1 and 2 of this argument. On point 3, as we have seen, a scientific analysis suggests that spaceflight always requires energy, though we must certainly allow for the possibility that another civilization *may* know some enormous secret we don't know. Therefore, we might be led to conclusion 4, that the sky is not the limit when we consider spaceflight by extraterrestrial civilizations. But in fact the scientific argument is that the sky *is* the limit, and it's an important limit. Look at the numbers: If a million civilizations exist in our galaxy, then only one in every 400,000 stars has a civilization nearby. That is, despite the great number of civilizations, the number of stars remains a tremendously greater number. Then if, in this example, one civilization in 100 decides to make a thorough exploration of its surroundings, there would be 10,000 such exploring civilizations in the Milky Way—one for every 40 million stars. The conclusion that follows is that we on Earth are unlikely to be visited by extraterrestrials (always assuming that they find space travel worth the effort) as part of a general survey, simply because the number of stars so far exceeds the number of exploring civilizations. To explain the many UFO sightings reported each year with the spacecraft hypothesis requires the assumption that *we are doing something special to draw attention*.

What could this special something be? Are our nuclear explosions seen as evidence of intelligence? Doubtful. Of dangerous stupidity? Not too dangerous; we can't yet go to other civilizations. Of pollution? Perhaps, but so far we pollute only our own solar system. We meet again the idea that humanity merits special attention, if we think that a thousand, or even a million, civilizations could suffice to explain the high number of UFO sightings on Earth.

If UFOs really are spacecraft from other, more advanced civilizations intent on gathering information about Earth, then we would expect them to be seen all over our planet as they attempt to examine our various land-scapes and life forms. In this case, sightings of extraterrestrial spacecraft and their inhabitants would be a worldwide experience shared by contemporary humans. On the other hand, if UFOs are a mental interpretation of natural phenomena, perhaps the unconscious attribution of mechanical properties by the inhabitants of highly industrialized countries, then we might expect that societies ignorant of the publicity given to UFO reports would know nothing about UFOs. A group of people unaware of the UFO sightings publicized in the United States, Europe, and the Soviet Union would not have a constant suggestion that objects in the sky are perhaps visitors from other planets. When these people saw unusual lights in the sky, they would probably assume that these lights arose from natural phenomena. To think otherwise, they would need another possibility in mind, and if they had never heard of UFOs, they might be content to ascribe their observations to terrestrial events. Such people would presumably have no experience of Hynek's UFO classes 2 through 6 (page 417).

Where could we find such a group of people among the hypercommunicative inhabitants of our planet? Approximately 900 million Chinese may be unaware of the agitation in the rest of the world that arises from excited reports of visits from, and abductions by, little men (why not women?) in flying saucers. We can see that a team of sociologists could perform fruitful research by comparing the Chinese UFO experience with that in the West. For now, we can report that a discussion with Chinese space scientists showed that none of them had heard of *any* reports of UFO sightings in China, although they themselves were aware of Western interest in this subject. We may draw a preliminary conclusion that different societies make quite different interpretations of the same phenomena. Or do extraterrestrial visitors avoid China because no one there cares about them?

Some Conclusions about UFOs

What, then, can we conclude from our discussion of UFOs? First, that many people with no familiarity with common celestial objects can be easily overimpressed with the planets Venus, Mars, and Jupiter. These planets can be exceptionally bright on certain occasions, when the Earth-planet distance nears a minimum; if they appear between cloud layers, their light can seem to cast a beam or a shadow. Furthermore, if an observer's mind sees an object in pursuit of the observer, then of course Venus, Mars, or Jupiter will keep pace, no matter how the observer tries to throw the planet off the track. Venus alone has caused more UFO reports, especially in summer—when people remain outdoors longer at night—than all other natural objects combined.

Second, we can be sure that human memories become steadily more confused with the passage of time. Because of the degradation of memories over long periods of time, an observer needs to record any sighting as quickly as possible, before the details of observation vanish. Even a *day* can play many tricks with human memories.

Our third conclusion from the study of UFO reports is that whatever the truth about UFOs may be, the fact that the United States (and other governments) may or may not have engaged in a "cover-up" does not prove that they have something to cover up. Recent history shows, instead, that even trivial facts may seem so dangerous to government officials that they deal with them in a time-honored way, by concealing them. Quite possibly, if extraterrestrial spacecraft *did* land in some out-of-the-way location, our military and political leaders would try to cover this up. But these leaders would also try to cover up evidence of incompetence, misjudgment, or just plain laziness on their part. All of these possibilities (and more) enter the explanation of why the government's opinion on UFOs does not carry much weight with most scientists.

Of course, this attitude may itself be a part of a cover-up, though it seems more likely that scientists are continuing to be simply skeptical. We cannot rule out the possibility that our military officers know all about UFOs—that, for example, they have already entered into communication with other civilizations—and find it useful to conceal this fact from us. This supposed conspiracy could include scientists too, so that this book might, for instance, have been written by extraterrestrial command (despite its sometimes skeptical tone). Indeed, it is worth wondering sometimes whether or not everyone you meet might not be from another planet, all collaborating to maintain a "normal" environment in which they can study you better. Or, to complete the picture, you too could be from another solar system, but have not yet grown up far enough to realize it.

Since this sort of approach leads us nowhere, let us return to the arena where UFO reports have left their greatest impact and importance: the human mind. These reports must, after all, be considered a purely human and terrestrial phenomenon until we judge that extraterrestrial visitors produce them. Thus, some investigators of UFO reports consider the UFO *experience*—the human involvement with UFO sightings and UFO reports— to have more importance than the attempt to discover the "reality" of the UFOs. If we try to generalize what various people say they saw in close encounters of the third kind, we find that most saw a spacecraft with inhabitants who looked a good deal like humans: two arms, two legs, a head, and so forth. Usually, the UFOs had bright lights either inside, or outside, or both. If the inhabitants had a sex, that sex is almost always male. These facts have a ready explanation if people see what they might like to see, or expect to see. If we accept a psychic explanation of UFOs, this match supports, rather than contradicts, the reality of the UFO experience.

As we look at past records of miracle-working "little people" and celestial visitors, we can see a continuity in human attitudes. This does not resolve the question of what UFO reports may imply, but we may reach the conclusion that the same phenomenon has occurred, in slightly different forms, for as long as humans have contemplated the vast and mysterious cosmos in which we live. Furthermore, the strand that unites the continuing flow of stories, tales, reports, or whatever we choose to call them consists of human attempts to connect our own experience with the universe at large, in a basically human-centered, trusting, and straightforward manner, by assuming that the inhabitants of unknown parts of the universe care about us, play with us, abduct us, or dance and sing for our benefit and entertainment.

Von Däniken: Charlatan of the Gods?

To take the theory of extraterrestrial visitors one step farther, we may consider the possibility that the record of what we consider to be human progress may consist of little more than outside intervention into a relatively dull and unimaginative human population. (If we take this theory one step too far, we *all* become visitors from other planets, as we discussed above). Approximately 50 million copies of Erich von Däniken's books *Chariots of the Gods?, Gold of the Gods,* and others, have been sold to readers who apparently like to learn that ancient astronauts visited the Earth (so von Däniken says), leaving behind many relics of their interactions with human beings.

The relics are most obvious to von Däniken when he examines "primitive" civilizations, those which he judges clearly incapable of large engineering feats. Thus, for example, von Däniken says of ancient Egypt:

> Great cities and numerous temples . . . pyramids of overwhelming size—these and many other wonderful things shot out of the ground, so to speak. Genuine miracles in a country that is suddenly capable of such achievements without recognizable prehistory.

To distort history as von Däniken does requires a sweeping ignorance, not to mention a bold charlatanism. In fact, Egypt has a long and quite recognizable prehistory, going back well before the time in which the "miracles" of which von Däniken writes were built.[4]

Although von Däniken on occasion rises to wholesale invention, as when he describes a trip that he never made into South American caves filled with ancient gold treasures, his most impressive feat consists of citing "amazing" coincidences which any reader could demonstrate to be false. Von Däniken obviously connects so well with the longing of many of us *to see ourselves as a part of the cosmos,* not adrift in a huge and uncaring universe, that he knows that most readers will not want to check up on him. If von Däniken had merely *suggested* that ancient astronauts built Egyptian, Maya, or Inca temples, he might seem to respect his audience. But let us look at a representative example of his confidence in his readers' gullibility.

To impress us with the impossibility that the "heathen" Egyptians built the great pyramid of Khufu, von Däniken asks:

> Is it really a coincidence that the height of the pyramid of Cheops [Khufu] multiplied by a thousand million—98,000,000 miles—corresponds approximately to the distance between the Earth and the sun?

Well, let's see. The pyramid's height equals 481 feet. If we multiply this by 10^9 (one thousand million), and then divide by the 5280 feet in a mile, we obtain 91,000,000 miles. In fact, this gives an even better result than von Däniken found because the true distance from the Earth to the sun is 93,000,000 miles. Is this a coincidence, or does this give an example of ancient beings with special knowledge?

One test might be to look at the heights of *other* buildings that were constructed before the distance from the Earth to the sun was well known. A nice example appears in the tower of the cathedral at Rouen, France, completed in the thirteenth century. This tower's height equals 485 feet, so if we perform the same multiplication by a thousand million, we find an answer of 92,000,000 miles, in better agreement with the Earth-sun distance than the great pyramid! Skeptics may suggest that the builders of Rouen cathedral *also* had special knowledge, but we think not.

---

[4]The same holds true for the Sumerians, whose legend of Oannes, the bringer of writing and many other inventions, has been cited as a possible example of extraterrestrial intervention.

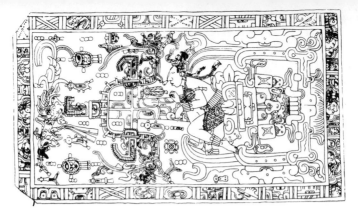

Fig. 21-5 This tomb lid from Palenque, Mexico shows a Mayan king, Pacal, together with important symbols of Mayan culture. Von Däniken sees Pacal as an astronaut riding a rocket, his foot on the pedals, his hands on the controls, and an oxygen mask on his face.

Once we start multiplying something by some really large numbers, we can find many interesting coincidences. For example, is it merely chance that the length of a common felt-tip pen, 5⅞ inches, *when multiplied by a million million* (10^{12}), gives the *exact* distance to the sun, 93,000,000 miles?[5] Are felt-tip pen manufacturers eager to show their extraterrestrial knowledge? And why do they produce a better product than the great pyramid? After all, if ancient astronauts really had built the pyramid, surely they could have added another 10 feet to its height, and thus furnished the exact value for the Earth-sun distance! Strange, too, that they didn't even try with the other two pyramids at Giza!

Another favorite ploy of von Däniken is the presentation of an ancient drawing or sculpture as portraying spacemen from another world, as in his claim that a Mayan illustration shows an astronaut with a backpack (Figure 21-5). In fact, these drawings show well-known religious motifs, far more evident when more of the drawing than von Däniken presents are included. Expressing wonder at the symbols carved on the Nazca plains of Peru, von Däniken suggests that these must have been landing strips for extraterrestrial spacecraft. These geometrical markings are clearly manmade, but it is hard to see why an advanced civilization would use such complicated figures to guide their landings. Far more likely is the explanation that these carvings expressed respect for the totemic figures of ancient cultures, figures that we also find on their pottery.

The key to von Däniken's approach remains his lack of respect for the intelligence of ''primitive'' cultures. Thus, for example, he claims that no primitive culture could have obtained a knowledge of human anatomy without X rays, when in fact far more direct ways to learn about the human skeleton and musculature have always been available.

---

[5]If you fail to perform this multiplication for yourself, you are entitled to ask yourself why you believe this book instead of von Däniken's.

Fig. 21-6 The enigmatic Sphinx at Giza, Egypt, is more than 4000 years old. According to von Däniken, this figure represents an extraterrestrial visitor to Earth.

Why do people take von Däniken seriously? Two basic factors provide the explanation. First, von Däniken gives a human dimension to the cosmos, connecting with our desires to find friends in the vast heavens around us and to believe that a benevolent providence looks after us. Second, many people have a lingering belief that ideas such as von Däniken's could not be printed unless they were true: The magic power of the printed word persists in a world that has seen books for only the past five centuries. In actual fact, whole forests have been sacrificed to various propaganda causes, and although this may be laudable in a free society, we should not fall prey to the intuitive belief that what appears in books has more validity than what you hear on the street, or what you can figure out for yourself!

Many authors bank on readers' intuitive beliefs that the world is a magic place, that if we can only look with different eyes, we shall see delights beyond mere amazement. The combination of the liberating joy in seeing the world a new way with the respect we give to books has made millionaires out of many a charlatan or out-of-his-depth thinker. But before we can feel entitled to judge these men (and notice that these supersalesmen have all been male!) to be wrong, we must consider the grounds on which we assess the truth or falsity of an assertion about the world. If we believe because it "sounds right," we deserve to be taken for a few dollars' worth of gullibility. If we believe because we swallow the "proof" that such authors provide, we receive low marks for critical assessment of what "proof" may be, as in the felt-tip pen example we mentioned earlier. But if we believe because we have considered both sides of the argument and have actually done some calculation ourselves, then we may feel entitled to some trust in ourselves, or at least to the joy of a good argument.

We should not, however, leave the subject of "ancient astronauts" without noticing its relevance to the problem of interstellar communication. When we puzzle over the pyramids, or try to read and to understand ancient

Egyptian inscriptions, we are "receiving messages" from a civilization that flourished over four thousand years ago (Figure 21-6). The fact that we cannot send messages in the other direction may be discouraging, but it does not dampen our interest in this culture. We would have the same experience in receiving messages from an advanced civilization on a star 4000 light years away. Once again, we would not have the chance for a pleasurable dialogue, but we shall be mightily intrigued by what those messages have to tell us.

Summary

The evidence concerning unidentified flying objects (UFOs) consists mainly of eyewitness accounts of strange objects in the sky whose presence, motions, colors, and appearance seem to rule out natural or human-made explanations. However, because even the best eyewitness testimony remains subject to the degradation of memory with time, and to the attempts of human minds to impose a familiar reality upon strange objects, this evidence remains unconvincing, from a scientific viewpoint, in providing proof of an extraterrestrial origin of UFOs.

Photographic evidence would be much better, but the photographs of UFOs that exist are either unreliable or unconvincing that anything extraterrestrial or even particularly strange has been photographed. Although photographs can be faked, they do at least provide evidence that can be studied with care, and, in contrast to human observations, the evidence does not change with time nor suffer from the filtering effects of human consciousness. Thus, if scientists ever become convinced that an extraterrestrial explanation of UFO reports has some merit, it will probably occur through photographs rather than through eyewitness accounts.

The statistics of galactic civilizations that we have considered show that even if a million civilizations exist in the Milky Way, and even if 1 percent of these civilizations are devoted to interstellar exploration on a grand scale, there would still be 40 million stars to explore for each civilization. This argues against the Earth's being frequently visited as part of a survey, so that we would have to assume that we are something special if we favor the extraterrestrial explanation of UFOs. The weight of the evidence favors the opposite viewpoint, that almost all UFO reports represent natural events, often misinterpreted but with no fakery on the part of the observer. The most impressive UFO sightings, from the point of view of proving that extraterrestrial visitors have reached the Earth, rely on far less verifiable evidence than the "typical" UFO sightings of meteors, weather balloons, or the planet Venus.

Public interest in UFOs as possible evidence of extraterrestrial visitors connects with a long human history of interest in the heavens and belief that the cosmos contains friendly and protective beings who watch over us. The success of such pseudoscientific efforts as those of Erich von Däniken shows the power of this feeling, since the "evidence" that von

Däniken cites in favor of his notion that extraterrestrial visitors built the pyramids and similar ancient monuments falls apart upon examination, or proves to be outright falsehood. The ancient monuments of Earth have much to tell us about communication across the centuries, but they deal with communication among humans, not with the interstellar communication we seek.

Questions

1. What are the chief factors that make eyewitness accounts of strange events unreliable?
2. Why are photographs superior to human observation for scientific analysis of a past event?
3. How can we explain the fact that observers who see a meteor many miles away may believe that it is moving silently at an altitude of only a few hundred feet?
4. Why is Venus the single object most commonly reported as a UFO?
5. Why does the theory that UFOs are mental projections from another world tell us little about UFOs?
6. Suppose that the average lifetime of a civilization in our galaxy equals 10,000 years, and that therefore about 10,000 civilizations exist in the Milky Way at any time. If one civilization in ten goes exploring, how many stars must each one explore to cover all 400 billion stars in the galaxy?
7. The circumference of the Earth is 40,000 kilometers, and the distance from the Earth to the sun is 150 million kilometers. How many copies of this book must be sold before this number *exactly* equals the ratio of the Earth-sun distance to the Earth's circumference?

Further Reading

Clarke, Arthur C. 1964. *Profiles of the future*. New York: Bantam Books.

Gardner, Martin. 1957. *Fads and fallacies in the name of science*. New York: Dover Books.

Hynek, J. Allen. 1973. *The UFO experience*. Chicago: Henry Regnery Company.

Klass, Philip. 1974. *UFO's explained*. New York: Vintage Books.

Playboy panel: UFO's. 1978. *Playboy,* January, 1978. [A pro and con discussion, with Philip Klass, J. Allen Hynek, Jacques Vallee, and four other participants.]

Sagan, Carl, and Page, Thornton, eds. 1972. *UFO's—A scientific debate*. Ithaca, N.Y.: Cornell University Press.

Vallee, Jacques. 1965. *Anatomy of a phenomenon: Unidentified objects in space—A scientific appraisal*. Chicago: Henry Regnery Company.

CHAPTER 22

Where Is Everybody?

Throughout our considerations of the search for other civilizations, we have emphasized the assumption that our own origin and development have followed a course that we believe to be representative of all civilizations in our galaxy. But if this assumption is valid, why has it taken so long for us to establish contact with other intelligence? If so many civilizations exist, the natural result of life on many planets, why haven't we been discovered by our galactic neighbors? In short, why are we still alone?

Within the framework of logic that we have been following, our answers to these questions are directly related to our expectation that contact between advanced civilizations occurs through radio waves, not through direct visits. And we can certainly agree that our civilization remains in infancy so far as radio communications are concerned. We have barely begun to search for radio signals from other possible civilizations; we have sent only a few messages on our own; and the leakage of radio waves from our planet has occurred for only the past 70 years or so. Thus, we who are alive today may simply have been "born too soon" to have attracted, or to have detected, another civilization.

Challenging this explanation of our apparent, even if temporary isolation, we encounter another possibility: Civilizations may be far less abundant than we have concluded. Our equation for estimating the number of civilizations may be correct, but the last term in the equation, the average lifetime of a civilization with communications ability and interest, may fall far below our estimates, thus reducing the number of civilizations in the Milky Way at any given time to a mere handful. Or we may have missed some important point about the way in which life has evolved on Earth, or in which civilizations can develop on a planet with life, that stamps our own civilization as unique, or nearly so. In other words, we may have unconsciously adopted a sort of religious belief in our averageness because we harbor some deep-seated hope of finding wisdom in the form of advanced civilizations throughout our galaxy and beyond.

Suppose, for example, that life on Earth is *not* an average example of life in the universe, because our planet has had one of the most rapid developments of intelligence from primordial soup. Perhaps the average time needed to pass from self-replicating molecules to eukaryotes (or their equivalent) is double the three billion years that characterized the Earth. Then we would have little hope of finding intelligent life on most planets. Recent studies of stellar evolution suggest that the oldest Population I stars like our sun may be only 6 or 7 billion years old. If such stars are essential to produce life-bearing planets (see Chapter 17), and if most planets need six billion years to produce eukaryotes from self-replicating molecules, then we may occupy a point in our galaxy's lifetime too early to find many advanced civilizations. In this event, only those few planets with unusually rapid development will be ready for communication now. Of course, with our present state of ignorance, we can assign an equal likelihood to the opposite scenario: If we are one of the *slowest* civilizations in development, and the average planet needs only 1.5 billion years to produce what took Earth 3 billion years, then the galaxy could be teeming with intelligence far beyond our own.

Humans have yet to find a single example of life outside the Earth. Because of this, efforts to apply the tools and knowledge of science to estimating the probability that life exists elsewhere—a field of endeavor often called *exobiology*—has been called a scientific discipline with no subject matter. We can reply that many fields of science began with just this handicap. For example, before astronomers could send equipment above our atmosphere, they had no way to detect gamma rays from celestial objects, since these high-energy photons could never penetrate to the ground. Gamma-ray astronomy was then a discipline with no subject matter, no observations, no way of checking theory against fact. But the theoretical framework that astronomers constructed before the era of satellites proved immensely useful when gamma-ray observations could at last be made. Perhaps the same will turn out to be true for the ''science'' of exobiology.

Meanwhile, we can point out that this discipline at the very least provides new and useful ways to think about life on Earth. When we consider the sweep of biological history on our planet, we cannot fail to be impressed by the speed with which great changes in life have occurred in recent eras. The history of life on Earth appears basically as a billion-year struggle (or longer!) of prokaryotes to develop into eukaryotes, of turning an oxygen-poor atmosphere into an oxygen-rich atmosphere. Once these events had taken place, the abundance of oxygen provided a tremendous source of energy, and eukaryotes were well set to develop more and more complex structures based on the use of this gas. The transition from prokaryotic to eukaryotic life may have been more difficult than the evolution of intelligent life from eukaryotes, or the evolution of long-lived civilizations from intelligent forms of life. We might therefore expect that life takes the longest for the first steps toward complex life, on planet after planet.

We cannot resolve these issues until we find other forms of life. Our experience does suggest in the strongest terms that since we do not consider the Earth unique among planets, or the sun unique among stars, whenever we have "world enough and time," life should develop and intelligence should evolve. But contact with another civilization is the only way we can make this conclusion certain.

Suppose that we are willing to conclude that many civilizations exist in our galaxy. Must we really abandon the idea that contact will occur by direct visits? The answer appears to be that we can expect interstellar spaceflight only among those civilizations that are immortal, or nearly so. Low-thrust spacecraft that take millions of years to travel from star to star have little appeal for us: We won't take a trip that long if the message we carry can cover the same distance in a few decades. We probably wouldn't go even if we knew how to freeze ourselves, or could otherwise induce a near-zero rate of living, because our human experiences leave us unprepared to accept such changes. Nor will our society elect to send automated probes that might generate a response in several million years, when next year's news seems quite far away. Only if we developed a much longer-term perspective on our existence would we be ready for million-year journeys, so we may speculate that only civilizations with great lifetimes are likely to make interstellar voyages. There may be no such civilizations anywhere in our galaxy, or there may be many. Can we decide which possibility lies closer to reality?

To a limited extent, we can. If we believe that the simple *possibility* of spaceflight inevitably leads to its use in exploration and colonization, then we must, on the evidence, be alone in the Milky Way. In other words, if civilizations as they develop follow the expansionist tendencies that our own civilization has shown, and spread through their neighboring planetary systems to the best of their ability, then the 10 billion-year age of the Milky Way has provided plenty of time for the Earth to have been colonized.[1] Yet as of today, our planet remains free from extraterrestrial settlers, unless we believe that humans are such immigrants. Does the absence of extraterrestrial colonization really show that no other civilizations exist in our galaxy?

---

[1]But notice that stars like our sun may be only half this old. If these stars are the sites of advanced civilizations, they may have had "only" a billion years or so to colonize other worlds.

Let us first examine the assumption that developing civilizations must strive to colonize their surroundings. This assumption rests on the fact that humans have indeed sought to inhabit as much of our planet as possible. We now confront the limitations of such a policy on a planet with a finite area (and all planets have finite areas): massive overpopulation.

The number of humans on Earth has been doubling about every 40 years for the past two centuries. In 1810, there were only a quarter of a billion humans; in 1850, half a billion; in 1890, 1 billion; in 1930, 2 billion; in 1970, 4 billion. We now increase in numbers every four years by the *total* population of the world in 1810. In fact, the rate of population growth has actually increased during the past 50 years over the rate of the previous century, so the present doubling time is only 35 years. Calculations made by some scientists show that if the rate of increase in population continues on its present course (that is, if the doubling time continues to decrease), then by the third decade of the next century, the number of people on Earth will become *infinite!* This calculation was made in 1960; 15 years later, in 1975, world census figures indicated that we were ahead of schedule! This faster-and-faster doubling of the population represents the fundamental problem of contemporary human society, and any developing civilization will have to solve it in order to survive.

If we ever learn to deal with this growth, and to recognize that a finite world demands a stable population, we may indeed acquire the long-term perspective that would promote interstellar spaceflight to a reasonable possibility. But the very change in attitude that induces a civilization's long-lived existence—the willingness to act upon the principle of finite resources—will remove the expansionist impulse that would promote the colonization of other worlds. We see this change on Earth as people realize that wars cannot solve our long-range problems in a positive way. Unfortunately, we seem far from a social attitude that will provide a successful replacement for the aggressive growth and frontier outlook that brought us to our present state.

We can also recognize that even in our current expansionist frame of mind, humans do not occupy all the space available on Earth. We could live in colonies in Antarctica, on the ocean bottoms, or in our deserts, all of which represent environments much more hospitable, accessible, and familiar than space colonies. But we don't. Our experience shows that our civilization flourishes best when it is least confined, and even palmy tropical islands are not considered ideal habitats by members of our questing, restless species. Similarly, humans do not take advantage of *every* technological opportunity that appears. We remain ambivalent about the use of supersonic aircraft, and despite great needs for energy, nuclear power has not captured the popular imagination. In astronomical exploration, we have seen that human success in landing on the moon has not produced a continuing program of space exploration by humans. Have we entered an era comparable to the pause that occurred after the discovery of the New

World, before humans began to colonize what they regarded as open territory? Or are we adopting an attitude that other, more pressing priorities must engage our attention and our resources?

If we think about the steps that would actually accompany a civilization's spread through the galaxy by colonization, we can see that repeated shifts in attitude would have to occur. First, such a civilization would have to develop expansionist tendencies; this much seems natural, if we use the Earth as a guide. Then the civilization must learn to deal with an ever-growing population and must achieve enough stability to build and to use interstellar spacecraft. Colonization can never relieve population pressure by itself, since the room made available by colonization tends to be quickly filled by the still-growing population. Thus the construction and use of interstellar spacecraft requires dedication and direction to that end on the part of the civilization that considers interstellar spaceflight. Once the descendants of the colonists from such a ''mature'' civilization have reached the planets of other solar systems, they must again revert to an expansionist attitude, in order to gather the resources from the asteroids or planets and build new ships for the next set of expeditions. And thus a galaxywide program of colonization requires alternating expansionist and ''mature'' periods. This may prove to be natural, but it may also turn out that *no* civilization will embrace enough cycles to spread through an entire galaxy. If this is so, then the absence of colonists from other civilizations here on Earth does not prove that no long-lived civilizations exist in the Milky Way.

This argument does not rule out the possibility that a ''mature'' civilization would send automated message probes throughout the galaxy, carrying messages or reconnaissance devices, perhaps even programmed to ''seed'' new life wherever possible. Alien probes could be watching us even now, enjoying our antics while cleverly guarding themselves against detection. But we must put this ''zoo'' hypothesis with the idea that there are new realms of physics so far outside our knowledge that advanced civilizations, who know these realms, can do simply anything. Both concepts, though intriguing, are empty of practical impact, since they cannot be tested in any effective way. We are led back to the search for other civilizations as the means to discover the general rules by which civilizations develop.

Because we still lack our first contact, we have had to discuss the motivations of hypothetical civilizations, including those far more sophisticated than ourselves, in a vacuum. Consider, for example, the difficulty of guessing how our *own* society will develop in the future. Famous novelists, such as George Orwell (in *1984*), Aldous Huxley (in *Brave New World*), and Herman Hesse (in *Magister Ludi*), have offered plausible visions of ''advanced'' societies in which the pursuit of scientific knowledge—let alone the exploration of space!—forms no part of human activity.

If we leave the question of motivation aside, our scientific knowledge does tell us that moving matter costs enormously more energy than sending radio waves, and this reality will be clear to every civilization. A preference for interstellar radio over interstellar spacecraft seems hard to avoid, except under the specific circumstances (immortality) that we have described. We can speculate endlessly on the possibilities for the existence of such civilizations and for their activities. But science rests on a framework of experiments: The only way to discover whether or not we are alone is a systematic search for evidence of extraterrestrial intelligence at work.

The Drake equation of Chapter 18 provides our summary and our guide. If we are an average civilization, except for being extremely young, then many other civilizations exist and our chances of making contact across the depths of space seem good. If our current lifetime with communications ability, some 70 years, is also average, then so few civilizations exist that we can probably never find any. (The third possibility, that our civilization's lifetime already exceeds the average, leads to conclusions of an extremely depressing nature!) If, as we suspect, civilizations in our galaxy number at least many hundred thousand, because their average lifetime spans at least a few hundred thousand years, then our civilization ranks with the youngest. Our abilities to guess what an advanced, more typical civilization may be doing could then be compared to the efforts of a six-year-old child of the Stone Age trying to speculate about civilization in the world today.

Still, amazingly enough, we have the means to advance our knowledge, to move from our awesomely small experience to a better understanding of what is happening in our galaxy. While direct exploration lies far beyond our present capabilities, our visible-light and radio telescopes have *already* brought us into contact with the distant reaches of the universe. The fact that radio astronomers keep discovering new molecules in the interstellar medium each year is indicative of the rapid increase in our knowledge. If we mean to be serious about finding other civilizations, our avenues are clear: Keep looking, keep listening, keep thinking, and our chances for eventually joining the galactic club will increase. We may even find new friends for humanity.

To succeed at last, we must embrace many failures. To provide an idea of how the search for other civilizations proceeds now, we can do no better than to read Frank Drake's account of one of his sessions at the giant radio telescope of Arecibo, searching for extraterrestrial intelligence[1]:

It is 1976. The night surf shimmers and breaks rhythmically on a dark Puerto Rican beach. In the neat, 30-room hotel above the beach, a sleepy night attendant (is he as old as Methuselah or does he only seem so?) for the third night in a row reluctantly awakens two Americans at the unthinkable hour of 4 A.M. He asks no questions, but looks on with hostility and suspicion. Any visitors at this resort who arise at four in the morning must be embarking on some evil

[1]From *Technology Review* 78:22 (June 1976).

deed. While the beautiful people sleep soundly, their expensive sunburns permitting, the two Americans drive off into the night. Chickens and toads scramble to escape the road in time. Carl Sagan is propped up in the front seat, eyes closed. He munches laboriously on scraps of dried-out garlic bread rescued from last night's dinner. It is all the breakfast there will be until the morning's observations are over.

An hour later the sky turns pink, and a Puerto Rican toad abandons its bell-like musical aria and crawls beneath an orchid plant for the day. The sun changes the wet misty quiet into a humid torpor. The brightness of the sunlight which warms the toad, however, will be equal only to that on Mars, for the toad is shielded by a canopy of 18 acres of perforated aluminum sheets—the reflector of the Earth's largest radio telescope. Five hundred feet over the canopy, murmuring machinery moves the 300-ton superstructure to follow a distant galaxy, the Great Nebula in Triangulum, across the sky. A nearby building houses radio receivers, far more sensitive than those of 1960 and capable of covering 3,024 frequency channels at once rather than the single channel covered in 1960. Scientists and telescope operators have meticulously arranged this vast ensemble of electronic equipment to fulfill the demanding requirements of a SETI [search for extraterrestrial intelligence] program. It has taken years for the talented people to construct the devices and computer programs that will deal with all the subtleties of the search such as the continuously changing effects of the earth's spin and its orbital sweep around the sun. Every 30 seconds, 100,000 transistors invisibly transmit the recorded information to the observatory's memory. The information captured in each of the 3,024 channels splashes a swath of twinkling green points across the face of an oscilloscope. It takes a hundredth of a second for the telescope to duplicate the two months of work done in 1960.

A split second of excitement. The astronomers, garlic bread and hunger now long forgotten, search the swath of points for a pattern which could not have been made by nature alone. When first their minds and later their hearts convince them that no such pattern is present, the command is given to move on, to look at another part of that distant galaxy—another billion stars—and to start the electronics searching and recording again. The hours fly by. One hundred billion stars will be searched before the task is completed and the hotel night attendant will no longer have to awaken at 4 A.M. No signals will be found this time. But there will come a time when that most thrilling of all cosmic treasures, news of life elsewhere, will be plucked from the vast vault of space.

Frank Drake, Carl Sagan, and several of their American colleagues have begun to search for life in other galaxies, as well as in our own Milky Way. Soviet astronomers have also begun to search, as have astronomers in Canada and in other countries. When enough effort has been made in these and similar searches, we shall have contact, and shall no longer be alone in the universe.

Subject Index